Porous Carbon Materials for Clean Energy

Porous carbons are widely used as electrode materials for supercapacitors owing to their high specific surface areas, abundant surface functionalities, well-controlled pore systems, and excellent conductivity and stability. New carbon materials with well-defined nanostructures and functionalization patterns have been developed to meet challenges of a growing global demand for energy-saving materials and sustainable materials to reduce negative environmental consequences. This book describes progress toward the conversion and efficient utilization of porous carbon and its derived precursor as electrode materials for clean energy.

- Explores the chemical structure, composition, properties, classification, and application of various porous carbon nanoparticles and nanostructured materials for clean energy uses.
- Proposes strategies for porous carbon production through featured examples.
- Covers a variety of materials, including those derived from biomass, graphene, aerogels, and carbon nanofibers.
- Discusses applications including electrocatalysts, batteries, hydrogen production, supercapacitors, and energy storage.
- Examines challenges and future opportunities.

This book will be of interest to materials and chemical engineers, scientists, researchers, and others active in advancing the development of renewable and clean energy technologies.

Jing Huang received his M.S. and Ph.D. degrees in organic chemistry in 2009 and 2012 from Southwest University, China, and served as a postdoctoral research fellow under the supervision of Prof. Chang Ming Li. He teaches at the College of Sericulture, Textile, and Biomass Sciences, Southwest University, China, and his research interests focus on nanomaterials derived from biomass for electrochemical applications such as energy storage and conversion.

Advances in Materials Science and Engineering
Series Editor
Sam Zhang

Semiconductor Nanocrystals and Metal Nanoparticles: Physical Properties and Device Applications
Tupei Chen and Yang Liu

Advances in Magnetic Materials: Processing, Properties, and Performance
Sam Zhang and Dongliang Zhao

Micro- and Macromechanical Properties of Materials
Yichun Zhou, Li Yang, and Yongli Huang

Nanobiomaterials: Development and Applications
Dong Kee Yi and Georgia C. Papaefthymiou

Biological and Biomedical Coatings Handbook: Applications
Sam Zhang

Hierarchical Micro/Nanostructured Materials: Fabrication, Properties, and Applications
Weiping Cai, Guotao Duan, and Yue Li

Biological and Biomedical Coatings Handbook, Two-Volume Set
Sam Zhang

Nanostructured and Advanced Materials for Fuel Cells
San Ping Jiang and Pei Kang Shen

Hydroxyapatite Coatings for Biomedical Applications
Sam Zhang

Carbon Nanomaterials: Modeling, Design, and Applications
Kun Zhou

Materials for Energy
Sam Zhang

Protective Thin Coatings and Functional Thin Films Technology, Two-Volume Set
Sam Zhang, Jyh-Ming Ting, Wan-Yu Wu

Fundamentals of Crystallography, Powder X-ray Diffraction, and Transmission Electron Microscopy for Materials Scientists
Dong ZhiLi

Materials for Devices
Sam Zhang

Materials in Advanced Manufacturing
Yinquan Yu and Sam Zhang

For more information about this series, please visit:
https://www.routledge.com/Advances-in-Materials-Science-and-Engineering/book-series/CRCADVMATSCIENG

Porous Carbon Materials for Clean Energy

Edited by Jing Huang

Designed cover image: © Jing Huang

First edition published 2025
by CRC Press
2385 NW Executive Center Drive, Suite 320, Boca Raton FL 33431

and by CRC Press
4 Park Square, Milton Park, Abingdon, Oxon, OX14 4RN

CRC Press is an imprint of Taylor & Francis Group, LLC

ISBN: 978-1-032-48173-9 (hbk)
ISBN: 978-1-032-48189-0 (pbk)
ISBN: 978-1-003-38783-1 (ebk)

DOI: 10.1201/9781003387831

Typeset in Times LT Std
by codeMantra

The electronic version of this book was funded to publish Open Access through Taylor & Francis' Pledge to Open, a collaborative funding open access books initiative. The full list of pledging institutions can be found on the Taylor & Francis Pledge to Open webpage.

Contents

Contributors

Paul K. Chu
Department of Physics
Department of Materials Science and Engineering, and Department of Biomedical Engineering
City University of Hong Kong
Kowloon, Hong Kong, P.R. China

Biao Gao
The State Key Laboratory of Refractories and Metallurgy
Institute of Advanced Materials and Nanotechnology
Wuhan University of Science and Technology
Wuhan, Hubei, P.R. China

Cheng Huang
School of New Energy and Materials
Southwest Petroleum University
Chengdu, P.R. China

Jing Huang
State Key Laboratory of Resource Insects
Key Laboratory of Sericultural Biology and Genetic Breeding
Ministry of Agriculture and Rural Affairs
College of Sericulture, Textile and Biomass Sciences
Southwest University
Chongqing, P.R. China

Tieqi Huang
College of Chemistry and Chemical Engineering
Central South University
Changsha, Hunan, P.R. China

Yongfa Huang
School of Light Industry and Engineering
State Key Laboratory of Pulp and Paper Engineering
South China University of Technology
Guangzhou, Guangdong, P.R. China

Kaifu Huo
Wuhan National Laboratory for Optoelectronics
Huazhong University of Science and Technology
Wuhan, Hubei, P.R. China

Gaojie Li
The State Key Laboratory of Refractories and Metallurgy
Institute of Advanced Materials and Nanotechnology
Wuhan University of Science and Technology
Wuhan, Hubei, P.R. China

Ruchun Li
Faculty of Chemistry and Chemical Engineering
Yunnan Normal University
Kunming, Yunnan, P.R. China

Ting Li
College of Chemistry
Xiangtan University
Xiangtan, Hunan Province, P.R. China

Yueming Li
State Key Laboratory of Metastable Materials Science and Technology
College of Materials Science and Engineering
Yanshan University
Qinhuangdao, P.R. China

Bei Liu
College of Chemistry
Xiangtan University
Xiangtan, Hunan Province, P.R. China

Wenxing Miao
Key Laboratory of Eco-functional Polymer Materials of the Ministry of Education
Key Laboratory of Polymer Materials of Gansu Province
College of Chemistry and Chemical Engineering
Northwest Normal University
Lanzhou, Gansu, P.R. China

Hui Peng
Key Laboratory of Eco-functional Polymer Materials of the Ministry of Education
Key Laboratory of Polymer Materials of Gansu Province
College of Chemistry and Chemical Engineering
Northwest Normal University
Lanzhou, Gansu, P.R. China

Xinwen Peng
School of Light Industry and Engineering
State Key Laboratory of Pulp and Paper Engineering
South China University of Technology
Guangzhou, Guangdong, P.R. China

Yan Song
Key Laboratory of Carbon Materials
Institute of Coal Chemistry
Chinese Academy of Sciences
Taiyuan, Shanxi, P.R. China

Xiaodong Tian
Key Laboratory of Carbon Materials
Institute of Coal Chemistry
Chinese Academy of Sciences
Taiyuan, Shanxi, P.R. China

Gang Wang
College of Materials and Engineering
Sichuan University
Chengdu, Sichuan Province, P.R. China

Jiale Xie
School of New Energy and Materials
Southwest Petroleum University
Chengdu, P.R. China

Pingping Yang
School of New Energy and Materials
Southwest Petroleum University
Chengdu, P.R. China

Tao Yang
Key Laboratory of Carbon Materials
Institute of Coal Chemistry
Chinese Academy of Sciences
Taiyuan, Shanxi, P.R. China

Yong Ye
College of Chemistry
Xiangtan University
Xiangtan, Hunan Province, P.R. China

Dingsheng Yuan
Department of Chemistry and Institute of Nanochemistry
Jinan University
Guangzhou, P.R. China

Linxin Zhong
School of Light Industry and Engineering
State Key Laboratory of Pulp and Paper Engineering
South China University of Technology
Guangzhou, Guangdong, P.R. China

Long Zou
College of Life Sciences
Jiangxi Normal University
Nanchang, Jiangxi, P.R. China

1 Porous Carbon Aerogel for Supercapacitors

Xinwen Peng, Yongfa Huang and Linxin Zhong

1 INTRODUCTION

So far, supercapacitors can be divided into double-layer capacitors, quasi-capacitors, and hybrid capacitors according to the energy storage mechanism of active substances with positive and negative electrodes. The positive and negative electrodes of the double-layer capacitor store energy by forming an electric double layer (EDL) between the electrode surface and the electrolyte. The electrode material is mainly porous carbon. The positive and negative electrodes of the quasi-capacitor store energy through the rapid redox reaction on the electrode surface. The electrode materials are mainly metal oxides and conductive polymers. In the hybrid capacitor, one electrode is made of double-layer energy storage material, and the other is made of quasi-capacitor energy storage material.

The electrode material is the core component of the supercapacitor and determines the charge storage capacity of the supercapacitor. The capacitance is proportional to the specific surface area of the material according to the classical double-layer capacitor equation. With further study of supercapacitors, the above proportional relationship is not genuine under certain conditions, especially for porous carbon materials. Among many porous carbon electrode materials, porous carbon aerogel has been favored by researchers due to its high electron conductivity, large specific surface area, abundant micro-mesoporous pores, and a wide range of raw materials. This chapter will focus on the following aspects: (1) the way to understand the charge storage mechanism of porous carbon materials: (2) understanding the relationship between porous carbon materials and capacitance; (3) the way to prepare porous carbon aerogel; and (4) the application of porous carbon aerogel electrode materials in supercapacitors.

2 POROUS CARBON DOUBLE-LAYER MODEL

2.1 Development of Classical Double-Layer Model

As shown in Figure 1.1, the concept of EDL was first proposed by von Helmholtz in the 19th century while studying the distribution of opposite charges at the interface of colloidal particles.[1] The Helmholtz bilayer model states that two layers of opposite charges form at the electrode/electrolyte interface and are separated by atomic distances. Because the Helmholtz double-layer model does not consider these factors,

DOI: 10.1201/9781003387831-1

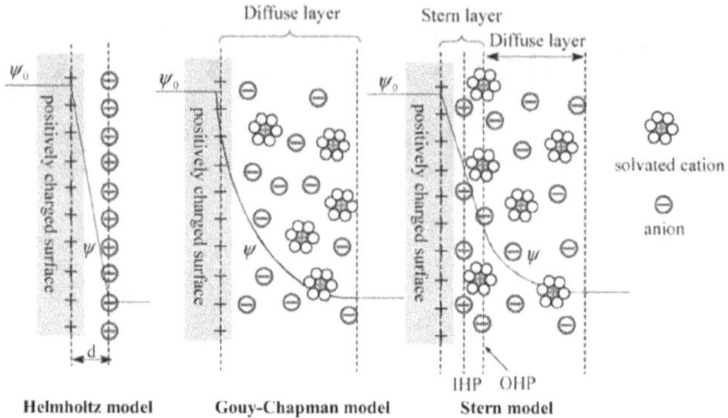

FIGURE 1.1 Models of the electrical double layer at a positively charged surface: the Helmholtz, Gouy–Chapman, and Stern models.

including diffusion and dipole motion of solvent molecules, this model was further modified by Gouy and Chapman, which is known as the diffusion layer.[2,3] However, in the Gouy–Chapman model, there exists an ion diffusion layer with decreasing potential between the electrode surface and the electrolyte body, which makes it unable to be used well in high-concentration electrolytes. Later, Stern combined the Helmholtz model and the Gouy–Chapman model, to identify the ion distribution in two regions – the inner region called the compact or Stern and diffuse layers (see Figure 1.1).[4] The electrode strongly adsorbed ions (usually hydrated) in the compact layer. In addition, the compact layer consists of specifically adsorbed ions (in most cases, they are anions, regardless of the charged nature of the electrode) and non-specific adsorbed counter ions. The inner Helmholtz plane (IHP) and the outer Helmholtz plane (OHP) were used to distinguish the two types of adsorbed ions. The Gouy–Chapman model defines the diffusion layer region.

In the Stern model, the electric double-layer capacitance (C_{DL}) is regarded as the sum of Stern layer capacitance (C_{H}) and diffusion capacitance (C_{D}). Therefore, C_{DL} can be expressed by the following formula (Eqn 1.1):

$$\frac{1}{\mathrm{C_{DL}}}=\frac{1}{\mathrm{C_H}}+\frac{1}{\mathrm{C_D}}\ \frac{1}{C_{DL}}=\frac{1}{C_H}+\frac{1}{C_D} \tag{1.1}$$

The electric field on the electrode, the type of electrolyte ions, the solvent used to dissolve the electrolyte ions, and the chemical affinity between the adsorbed ions and the electrode surface are the factors that determine the behavior of planar EDL. In practice, due to the high concentration of electrolyte, the double diffusion layer can be considered very thin, and the diffusion capacitance C_{D} can be ignored, resulting in $C_{\mathrm{DL}} \approx C_{\mathrm{H}}$.

EDL supercapacitors are similar to paddle capacitors. Its capacitance can be calculated using the classical Helmholtz double-layer model as follows (Eqn 1.2):

$$\mathrm{C}=\frac{\varepsilon_r\varepsilon_0}{\mathrm{d}}\mathrm{A}\ C=\frac{\varepsilon_r\varepsilon_0}{d}A \tag{1.2}$$

where C, ε_r, ε_0, d, and A respectively represent the dielectric constant of vacuum (F g^{-1}), the relative dielectric constant of electrolyte solution (no unit), the permittivity of a vacuum (no unit), the distance between electrolyte ions and electrode surface (m), and the accessible surface area of electrode materials ($m^2 g^{-1}$). In Eqn 1.2, the specific capacitance (C) and specific surface area (A) should show a proportional relationship. However, some experimental results show this simple linear relationship does not always hold. It is traditionally believed that the submicron pores of the electrode are not involved in the formation of EDL because the surface of the submicron pores is inaccessible to large solvated ions. However, Simon and Gogotsi reported[5] an abnormal increase in capacitance at carbon electrodes with pore sizes smaller than 1 nm. The most significant EDL capacitance was observed when the electrode aperture was close to the ion size.

2.2 EDCC AND EWCC MODEL

Huang and coworkers[6] reported a heuristic method based on the classical double-layer model. In this method, based on the assumption that the shape of the porous carbon pores is cylindrical and considering the influence of the curvature of the pore diameter, he proposes two electric double-layer models, namely electric two-cylinder capacitors (EDCC) and electric cylindrical internal wire capacitors (EWCC), as shown in Figure 1.2. The EDCC and EWCC model have their application scope, among which the EDCC describes the double layer of the mesoporous structure. The curvature of the large pore structure is minimal, so the parallel plate capacitor model can describe it. When the electrolyte interacts with the mesopore, the counter ions enter the pore and arrange along the inner wall to form the EDCC double-layer model. When interacting with micropores, solvated or desolvated counter ions are arranged along the central axis of the circular hole to form the EWCC double-layer model. For the EDCC and EWCC models, the capacitance estimates of the two models proposed are given in equations (Eqn 1.3) and (Eqn 1.4), respectively.

$$C = \frac{\varepsilon_r \varepsilon_0}{b In[b/(b-d)]} A \tag{1.3}$$

$$C = \frac{\varepsilon_r \varepsilon_0}{b In[b/a_0]} A \tag{1.4}$$

where b is the pore size, d is the distance close to the ion to the surface of the carbon electrode, and a_0 is the effective size of the counter ion (i.e. the degree of electron density around the ion).

The mathematical results obtained based on Eqns 1.3 and 1.4 fit well with the experimental data without considering the type of carbon material and electrolyte used. The proposed EWCC and EDCC models can be used to explain the abnormal increase of capacitance when the aperture is less than 1 nm and the tendency to increase slightly when the aperture is more than 2 nm.

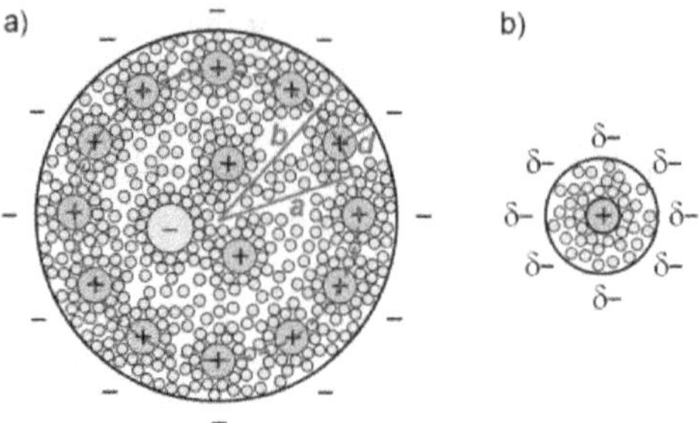

FIGURE 1.2 (a) EDCC double layers model (b) EWCC double layers model.

3 INFLUENCE OF POROUS CARBON PORE STRUCTURE ON CAPACITIVE PERFORMANCE

Based on the Stern model, the electric double-layer capacitance C_{DL} can be approximately equal to the tight layer C_H. According to Eqn 1.2, the specific capacitance should be directly proportional to the specific surface area of the electrode material. However, it is found that the specific capacitance of porous carbon does not follow the above relation when the specific surface area is more significant than 1,200 $m^2 g^{-1}$. For this phenomenon, Barbieri and coworkers believed that the possible reason was that the pore wall with a large specific surface area could not completely shield the charge effect of two adjacent holes compared with the pore wall with a small specific surface area.[7] The results of Barbieri and coworkers are similar to the situation in which the specific capacitance of a single layer of graphene is higher when the electrolyte is in contact with one side of the electrolyte than when the two sides are contacted. The above experimental results show that the double-layer capacitance cannot be improved by increasing the specific surface area of the pore carbon alone. In addition, due to the excessive porosity of the porous carbon material, its vibration density is significantly reduced, resulting in a decrease in the actual device filling, so the energy density and power density of the capacitor drop sharply.

In recent years, researchers have found that the main factor affecting the capacitance of porous carbon materials is their pore structure.[8] Moreover, the pore structure controls the specific surface area of porous carbon. Therefore, understanding the mechanism of the influence of pore structure on the porous carbon ratio capacitor is the key to the design of high-performance carbon electrode materials with pores, which can provide direction for improving the performance of double-layer capacitors.

3.1 Ion Screening Effect

Avraham and coworkers proposed that when the average pore size of activated carbon is compatible with the size of ions in an electrolyte, electrolyte ions are likely to be adsorbed by porous carbon and defined this selective adsorption based on size effect as ion decoration effect.[9]

Avraham and coworkers took activated carbon with different average pore sizes as research objects to carry out experiments through cyclic voltammetry tests and obtained the following results. When the average pore diameter (0.58 nm) is larger than the size of Mg^{2+} and SO_4^{2-} in the electrolyte, the cyclic voltammetry (CV) curves are a standard rectangle, indicating that ions in the electrolyte can enter the pores of activated carbon. When the pore diameter is 0.51 nm, the CV curves are almost a straight line, indicating that the diameter of the bath ions in the electrolyte exceeds the pore size, and Mg^{2+} and SO_4^{2-} are restricted from entering the pore structure of the activated carbon, resulting in almost no double-layer energy storage effect of the activated carbon. In order to further verify the above conclusions, Li_2SO_4 and $MgCl_2$ were selected as the electrolyte, and activated carbon with an average pore size of 0.51 nm was selected as the research object for the above experiments. The deformation of the right branch of activated carbon in the Li_2SO_4 electrolyte (related to the anion adsorption process) and the deformation of the left branch of activated carbon in the $MgCl_2$ electrolyte (related to the cation adsorption process) further confirmed that activated carbon could not absorb Mg^{2+} and SO_4^{2-} larger than its average pore size, and could only accept the smaller size of Li^+ and CI^-. In addition, Cachet-Vivier and coworkers have also demonstrated the existence of ion screening effects.[10]

In addition to the size-based ion selection, Noked and coworkers studied the ion adsorption properties of porous carbon with an average pore size smaller than 0.7 nm, which confirmed that the spatial geometry effect of ions is also one of the reasons for the ion screening effect.[11] For example, the slit pore structure can accommodate the planar ion SO_3^{2-}, but exclude the spherical ion Cl^-.

3.2 Ion Desolvation Effect

Due to the ion screening effect, porous carbon cannot absorb more solvated ions than its average pore size. To obtain high specific capacitance, the pore size of porous carbon in an aqueous solution system needs to be about 2 nm, and that of porous carbon in an organic electrolyte system needs to be about 5 nm. However, after studying the capacitive properties of template carbon and carbide-derived carbon (CDCs), Vix-Guterl and Gogotsi found that the ions in the electrolyte could be desolvated and then stored in micropores or ultra-micro pores.[12,13]

When studying porous carbon synthesized by different template methods, Vix-Guterl and coworkers found that the mass-specific capacitance of porous carbon was directly proportional to that of porous carbon micropores or ultrafine pores. The specific surface area of porous carbon has no greater effect on the relative capacitance than the proportion of micropores in the material. The above experimental phenomenon is contrary to the conventional perspective. In order to explain the above view, researchers combined experiments and calculations to find the effect of ion desolvation. Vix-Guterl and coworkers observed a positive correlation between the pore capacity of the micropores and the carbon mass ratio capacitance of the porous pores. Based on the experimental results, they hypothesized that the ions are removed from the solvated shells and then enter the micropores under the effect of electric field polarization.

To further confirm the existence of the ion desolvation effect, Gogotsi and coworkers studied the standard specific capacitance of different types of CDCs and compared it with a series of template carbons with different pore sizes in the literature and found that when the pore size was 1 nm larger in the electrolyte of 1.5 mol/L TEA-BF_4 acetonitrile solvent, the standard specific capacitance of porous carbon was positively correlated with the pore size. However, when the aperture is lower than 1 nm, the standard specific capacitance of CDCs prepared by any precursor system will increase significantly and reach the maximum when the aperture is 0.7 nm. The optimal pore size is equivalent to the theoretical diameter of the non-solvated cationic TEA^+ in the electrolyte (the calculated value is 0.74 nm). According to the experimental results, Gogotsi and coworkers concluded that ions could enter the micropore after desolvation, and the double-layer capacitance in the micropore is the central part of the entire porous carbon double-layer capacitance, which is consistent with the experimental results of Vix-gutter and coworkers. The above results were also found in the study of the capacitive properties of ordered mesoporous carbon. These experimental results have fully confirmed the existence of the ion desolvation effect.

In order to better display the relationship between mass-specific capacitance and micropore diameter and surface area, Ghosh and coworkers summarized many relevant research results in stream electrolytes. Experimental results show that the high point of the mass-specific capacitance is concentrated in the area where the small pores are well developed, and the specific surface area is high. When the pore size is larger than 1 nm, the specific capacitance of activated carbon begins to decrease. Especially when the pore size is larger than 2 nm, the downward trend is more significant. Even if the specific surface area has reached 2,000–2,500 $m^2 g^{-1}$, the material-specific capacitance is only 100–150 F g^{-1}, which is lower than the average 200 F g^{-1} of activated carbon in water-electrolyte. It is proved that the pore volume of small micropores significantly influences mass-specific capacitance more than the specific surface area.

3.3 Influence of Mesoporous Structure

Although the experiment has confirmed that the structure of small micropores is conducive to the energy storage of porous carbon, when the pore capacity of small micropores is too high, the conduction resistance of ions from the electrode surface to the depth of the porous carbon body will be increased, resulting in the reduction of specific capacitance during charging and discharging with high current density. In recent years, it has been found that mesoporous pores, which were previously thought to have no significant role in increasing the specific capacitance of porous carbon, play an essential role in the diffusion of ions from the electrode surface to the depth of the body. Therefore, a certain proportion of mesoporous structure is the basic condition for porous carbon to obtain high specific capacitance under high current density.

In addition to the influence on the mass-specific capacitance, Wu and coworkers found that a certain proportion of mesoporous structure was beneficial to improve the charge-discharge capacity of the material under a high current density.[14] The AC impedance test data showed that the slope of the ordered mesoporous carbon was significantly higher than that of the disordered activated carbon in the linear region of the ion transport process, which indicated that the ion transport impedance corresponding to the ordered mesoporous carbon was lower than that of the disordered activated carbon, proving that the mesoporous structure mainly affected the ion transport process in the electrode body, which in turn affected the performance of the whole electrode.

As for the mechanism of mesopores to improve the energy storage and high current density charging and discharging performance of activated carbon micropores, it is generally believed that an appropriate mesopore ratio ensures the rapid ion transport within the material, which indicates that mesopores are a vital bridge to form interconnected three-dimensional ion transport channels within the activated carbon. When activated carbon is only composed of micropores, the ion transport process will be limited by the narrow pore size channel, which is challenging to conduct to the inside of activated carbon, resulting in the energy storage performance of some micropores cannot be well played. When there is a certain proportion of mesoporous pores, multiple micropores can be connected through mesoporous pores, effectively improving the efficiency of ion transport, conducive to the diffusion of electrolyte ions from the electrode surface to the material body, and promote the micropores to maximize the role of charge storage. When the three-dimensional ion transport channel is formed through the mesoporous structure, the ion transport resistance of activated carbon will be significantly reduced, and the charging and discharging ability of high current density will be substantially improved.

3.4 Effect of Saturation

When studying the effect of pore size distribution on mass-specific capacitance, it is found that activated carbon has a saturation effect. When the charging voltage exceeds a particular value, activated carbon cannot continue to store charge. In the CV curves, the saturation effect is usually manifested as a current-voltage curve that is not strictly rectangular, but there is significant current decay in the high-potential region. Furthermore, in the charge-discharge curve, the saturation effect usually manifests as a significant deviation from linearity in the discharge part of the voltage-time curve.

Mysyk and coworkers synthesized two kinds of activated carbon with different pore size distributions by other preparation methods. They studied their CV curves and charge-discharge performance in 1.5 mol L^{-1} TEA-BF_4 acetonitrile solvent electrolytes.[15] The pore size distribution of PC-activated carbon is narrow, mainly lower than 0.68 nm. In contrast, the pore size distribution of VC-activated carbon is vast, with many pores around 1 nm. When the charging voltage exceeds 1.5 V, PC appears to have apparent current attenuation, while VC maintains a strict short shape in the voltage region of 0–3 V. The discharge part of the charge and discharge curve of PC-activated carbon also deviates from linearity. The above phenomenon is the typical characteristic of the saturation effect. The above experimental results show that the saturation effect is closely related to the micropore size distribution. In this regard, they proposed the pore saturation effect. This effect means that when the voltage reaches a specific value, the pore structure of the activated carbon is in a saturated state and cannot continue to store ions, resulting in a sharp decrease in capacitive current when the voltage continues to increase.

To exclude the influence of solvation on the study of the saturation effect, Mysyk and coworkers designed ionic liquids composed of four alkyl chains of different lengths: $[(C_2H_5)\ N_4]^+[TfN]^-$, $[(C_3H_7)_4N]^+[Tf_2N]^-$, $[(C_4H_9)_4N]^+[Tf_2N]^-$, $[(C_5H_{11})N]^+[Tf_2N]^-$, and tested the CV curves of VC in these four ionic liquids, because there is no solvation effect in ionic liquids. Therefore, the actual size of ions should only be considered when discussing their interaction with the pore structure of activated carbon. The results demonstrate that with an increase in alkyl chain length, there is a tendency for VC to manifest as a capacitor current attenuation phenomenon. Furthermore, the attenuation degree appears to be proportional to the alkyl chain length. This current attenuation phenomenon is because as the size of cations gradually increases, the proportion of microholes matching the diameter of cations in VC becomes smaller and smaller, resulting in the negative electrode storing cations being saturated before the positive electrode at low voltage, thus limiting the capacitance of capacitors.

The above analysis confirms that the root cause of the porous carbon saturation effect is that the charge storage capacity of positive and negative electrodes is not equal in the working voltage range. Furthermore, it should be emphasized that the reason for the unequal charge storage capacity of positive and negative electrodes is not the difference in pore size distribution but the difference in the effective specific surface area of the positive and negative electrodes to absorb ions. The porous carbon saturation effect also indicates that the positive and negative electrodes in the real "symmetric" double-layer capacitor are not only fabricated from the same material, but also have comparable charge storage capacity. Therefore, to design a real "symmetrical" double-layer capacitor and avoid the saturation effect, the effective charge adsorption area of the positive and negative electrodes must be equal.

3.5 Ion Deformation, Ion Embedding/Ejection

Hahn and coworkers studied the electrode volume change phenomenon of porous carbon and graphite in the acetonitrile electrolyte of TEA-BF_4 during the charge-discharge process. They concluded that electrolyte ions might also have the embedding/removal effect for porous carbon in electrochemical capacitors.[16] When the polarization potential is less than 1.5 V (vs. reference electrode), the thickness of porous carbon sources is significantly lower than that with a polarization potential of 2 V (vs. reference electrode). Moreover, at positive polarization (corresponding to adsorption of negative ions) and with a maximum polarization potential of 2 V (vs. reference electrode), the degree of thickness of porous carbon electrodes increases with the number of cycles. The changes in the thickness of the porous carbon electrode confirm that there may be an ion embedding/ejection process during the charge-discharge process. Moreover, the volume change caused by ion embedding/ejection is the main factor affecting the cycle life of the activated carbon electrode.

4 FABRICATION TECHNIQUES OF CARBON AEROGELS

Carbon aerogels with high specific surface area and controlled micro-nano pore features have captured great attention in energy storage and conversation. The formation of carbon aerogels needs several intermediate processes, including hydrogel formation, drying, and carbonization. Therefore, this chapter aims to present the different methods of the above processes and to reveal how the combination of these methods affects the properties of the final carbon aerogel materials.

4.1 Hydrogels formation

Currently, there are mainly several methods of hydrogel formation: sol-gel processing, hydrothermal synthesis, molecular and colloidal approaches, templated synthesis, and chemical vapor deposition.

4.1.1 Sol-gel processing

Sol-gel method is simple and convenient, and the pore size is adjustable, which is an effective method for preparing porous materials. The specific operation method is that the precursor compounds containing high chemical active components are mixed, hydrolyzed, and condensed under the condition of a liquid phase to form a sol-gel system. When cellulose is used as raw material, acid hydrolysis, enzyme hydrolysis, mechanical treatment, and other methods are usually used to obtain cellulose micron or nanometer unit dispersion. The molecules in the sol are crosslinked through chemical bonds, hydrogen bonds, Van der Waals force, and other interactions to obtain a solid gel with a network structure. In addition, by ultrafast sol-gel synthesis, Pauzauskie and coworkers prepared graphene aerogels with high electrical conductivity, chemical stability, and low cost.

4.1.2 Hydrothermal synthesis

Hydrothermal synthesis refers to chemical reactions in a sealed environment under specific temperature and pressure conditions. For instance, Lee and coworkers synthesized three-dimensional reduced graphene oxide/TiO_2 aerogel by a facile one-step hydrothermal treatment. The obtained aerogel exhibited high adsorption and photodegradation ability.

4.1.3 Molecular and colloidal approaches

The method mainly includes molecular approach-chemical crosslinking, molecular approach-physical interactions, molecular approach-radiation crosslinking, and colloidal process. Among them, the molecular way refers to forming a network through the bottom-up route via driving the aggregation of molecular precursors by chemical, physical, and radiation crosslinking. At the same time, the colloidal method uses colloidal objects in dispersion to form aggregates by changing solvent conditions. For example, Zhang and coworkers synthesized poly rotaxanes aerogels with high specific surface area (232 $m^2 g^{-1}$), high strength (74.7 MPa), and temperature-responsiveness by introducing poly(*N*-isopropyl acrylamide) to the molecular necklaces followed by chemical crosslinking with a rigid crosslinker.

4.1.4 Templated synthesis

The template method can be divided into the hard template method and the soft template method. The hard template method adds the carbon precursor to the prepared template for polymerization and carbonization and erodes the template. Unlike the hard template method, the soft template method is a method of self-assembly to synthesize porous materials by reacting the precursor with the surfactant through polymerization and carbonization. In the case of aerogels, a network can be created on the existing structure, or the formed aerogels can be used for further functionalization. For instance, Yu and coworkers prepared a new kind of RF-GO-metal aerogel with tunable densities and mechanical properties through a low-temperature process using GO sheets as template skeletons and metal ions (Co^{2+}, Ni^{2+} or Ca^{2+}) as catalysts and linkers. Further, they synthesized metal-organic framework (MOF) aerogels by self-assembling uniform and monodisperse MOF nanofibers. The prepared aerogels were used as carbon precursors further to obtain well-dispersed hollow porous carbon nanofibers by calcination.

4.1.5 Chemical vapor deposition

Chemical vapor deposition refers to chemical gas or steam reacting on the substrate surface to synthesize coatings or nanomaterials. This method introduces two or more gaseous raw materials into a reaction chamber. Then they react with each other to form a new material, which is deposited on the surface of a wafer. Aerogels are ideal substrates for chemical vapor deposition processing because of their high porosity and specific surface area. For example, Abdullah and coworkers synthesized CNT aerogel from waste engine oil by applying the floating catalyst chemical vapor deposition method, and the obtained aerogel possessed a mesopore distribution of $142.9\,m^2g^{-1}$ as its specific surface area, 39.079 nm as average pore diameter, and $1.3964\,cm^3g^{-1}$ as total pore volume.

4.2 Drying methods

The drying process is one of the most significant steps in the construction of aerogel. In order to obtain aerogels with good structural properties, it is crucial to dry hydrogels by various methods so they will not shrink or collapse. Several drying methods are often used to construct aerogel, including ambient pressure drying, freeze-drying, and supercritical drying.

4.2.1 Ambient pressure drying

Ambient pressure drying is a viable technique for producing large-scale aerogels. The drying process usually consists of washing the wet gel with a low surface tension solvent, silylation with organosilanes, and evaporation under ambient pressure. However, aerogel usually has difficulty maintaining its complete structure and shows a large shrinkage during ambient pressure drying. Therefore, preparing flexible aerogels with transparency and superinsulating by ambient pressure drying is still a major challenge. Malfait and coworkers reported large monolithic aerogel plates with high thermal conductivity and larger surface area using ambient pressure drying methods.

4.2.2 Freeze-drying

Freeze-drying is the most common and safe way to dry aerogels, which has the advantages of simple operation and complete skeleton structure. In this process, the aqueous material is frozen below the freezing point, converting water to ice and steam under a higher vacuum, resulting in a loose, porous aerogel with good structural properties. Especially the porous structure of aerogels can be improved by adjusting the ice crystal growth process of the gel solution. The freezing rate and precursor concentration are important factors affecting the porous structure of aerogel. Among them, a low freezing rate will produce large gel pores, while a high precursor concentration will hinder the growth of ice crystals and thus affect the porous structure of the gel. For example, Tuo and coworkers prepared all para-aramid aerogel from PANF hydrogel by a modified freeze-drying method, and the obtained low-density PANF aerogels have high specific compressive strengths and low thermal conductivities.

4.2.3 Supercritical drying

Compared with other drying methods, supercritical drying can maintain the textural structure of the aerogels to the maximum extent, and the resulting products have higher porosity and larger specific surface area. The technology is to increase the pressure and temperature to make the solvent exceed the CO_2 critical point and become a supercritical fluid, thus weakening the surface tension of the gas-liquid interface and maintaining the crosslinked structure of the gel. Supercritical drying includes two kinds: one is high-temperature supercritical drying with methanol and other organic solvents as supercritical liquid, and the other is low-temperature supercritical drying with CO_2 as a supercritical liquid. The aerogel obtained by the former will not cause structural cracking due to water absorption in the air and has stronger stability, while the aerogel obtained by the latter has hydrophilic OH groups on its surface and higher microporosity in its structure. For example, Bras and coworkers prepared aerogels of cellulose nanofibrils and TEMPO-oxidized CNFs by an optimized supercritical drying.

4.3 Carbonization

For carbonization of aerogels, the aerogels, after drying, are calcined at high temperatures under an inert atmosphere or vacuum conditions to remove the oxygen and hydrogen-containing functional groups and convert them into corresponding carbon aerogels. In the carbonization process, the carbonization temperature, heating rate, carbonization time, and other conditions should be strictly controlled. At present, carbonization mainly includes pyrolysis and hydrothermal carbonization.

4.3.1 Pyrolysis

Pyrolysis is one of the most common methods of obtaining carbon materials, which is a typical heating process (400–1,300°C) in an inert atmosphere. It can be divided into four stages: free water removal (25–200°C), pyrolysis stage (200–500°C), amorphous carbon formation (500–900°C), and construction of the new structure (higher than 1,000°C). By controlling the temperature and heating rate, the properties of carbon aerogel can be adjusted. For example, the specific surface area of the obtained aerogels decreases with the increase of heating rate, while the conductivity increases with the temperature rise.

4.3.2 Hydrothermal carbonization

As a relatively low-energy consumption strategy, the hydrothermal method is to mix aerogels and water in a particular proportion and put them into the reactor at a low temperature, reaction time, and pressure to obtain solid product carbon aerogels. Under this condition, the required temperature and pressure are low, and the conditions are relatively mild. The products obtained by hydrothermal carbonization treatment have many inherent advantages, such as uniform size, regular morphology, good physical and chemical stability, and wealthy oxygen-containing functional groups on the surface.

5 CARBON AEROGEL USED IN SUPERCAPACITORS

It is well known that the electrochemical performance of a supercapacitor depends largely on the conductivity of electrode material, specific surface area, electrolyte diffusion, and the wettability of the electrode itself. Among all kinds of electrode materials for supercapacitors, carbon-based materials have the advantages of vast sources, low cost, outstanding conductivity, large specific surface area, excellent chemical/electrochemical stability, and adjustable pore structure. Therefore, carbon-based materials fit the requirements of electrode materials for supercapacitors well. For these reasons, most supercapacitors reported are constructed with carbon-based electrodes. In fact, many kinds of carbon materials, such as activated carbon, graphite, graphene, carbon nanotubes, carbon fiber, carbon foam, and carbon aerogel, are commonly used as electrode materials in the reported supercapacitors. And among all kinds of carbon materials, carbon aerogel has unique advantages such as super high specific surface area, reasonable absorption rate, high porosity, pore structure classification, high-temperature resistance, and low density, so much so that supercapacitors based on carbon aerogel electrodes have received wide attention and research. In addition, it is worth noting that some previous reports have shown that the specific surface area and pore size of carbon materials are the two most important parameters affecting the energy storage capacity and power transfer for supercapacitors using carbon-based electrodes, while the ultrahigh specific surface area and highly developed and easily regulated pore structure happen to be the advantages of carbon aerogel. This is why many reports have been about carbon aerogel electrode-based supercapacitors in recent years. Hence, this chapter introduces the application of different kinds of carbon aerogel materials in supercapacitors and the corresponding mechanism research.

5.1 FUNDAMENTALS OF SUPERCAPACITORS

The configuration of supercapacitors is similar to that of batteries. In the typical form, two electrodes (positive and negative electrodes) are separated by a diaphragm (separator) to prevent short circuits. Conventional electrochemical capacitors store electrical energy in EDL between the electrolyte and the electron conductor (electrodes). The charge storage depends mainly on redox reactions for pseudocapacitors or hybrid supercapacitors. Specifically, the positive electrode (cathode) is related to reductive chemical reactions that gain electrons from the external circuit, while the negative electrode (anode) is associated with oxidative chemical reactions which release electrons into the external circuit. Moreover, the separator acts as a physical barrier between the positive and negative electrodes to prevent electrical short circuits and as a carrier of electrolytes. The primary forms of separators are porous membranes, porous inert materials filled with electrolytes, and flexible hydrogel. Notably, the separator must endure the ion and inert supercapacitor environment. In addition, two current collectors (such as copper and nickel foil) are usually attached to the electrode surface to direct the current from the electrode. The electrodes, separator, and current collectors are immersed in an electrolyte, enabling ionic current flow between electrodes while preventing electronic current from discharging the capacitor, as illustrated in Figure 1.3.[17]

The energy storage principle of supercapacitors is the same as that of conventional capacitors, which is achieved through the charging and discharging process at the electrode-electrolyte interface. As a result, supercapacitors can release and store energy quickly. Compared to conventional capacitors, supercapacitors' electrodes have a much larger effective surface area, which allows their capacitance to be significantly increased by about 10,000 times compared to traditional capacitors. At the same time, supercapacitors maintain low equivalent series resistance and high working ratio power. There are two main mechanisms of energy storage in supercapacitors: (1) EDLC is related to the adsorption of coulombic charges near the electrode-electrolyte boundary (EDLC mechanism for supercapacitors is the same as that for conventional dielectric capacitors, while the energy storage mechanism for dielectric capacitors can be referred to the classical double electric layer model above); (2) pseudocapacitors are generated by surface redox reactions related to their respective potentials.

Pseudocapacitors using conductive polymers and transition metal oxides/hydroxides/sulfides/nitride generally offer higher specific capacitance and energy density than EDLCs, due to the energy boost in the pseudo-electrode encapsulated by the fast reversible redox reactions of Faraday properties. Pseudocapacitors are non-electrostatic and result from electrochemical charge transfer accompanying a finite active substance. In electrochemistry, the term pseudocapacitance refers to a

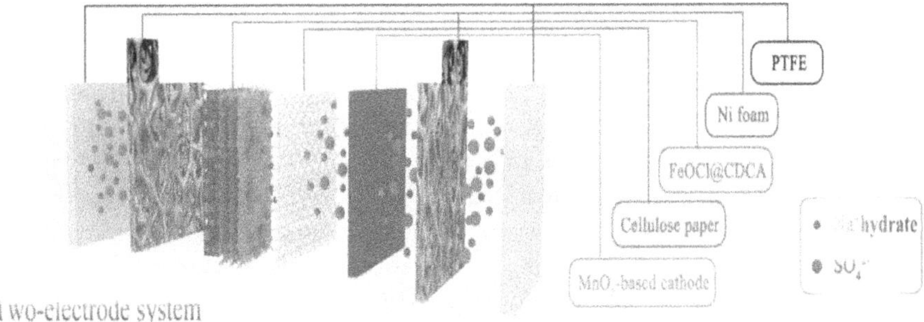

FIGURE 1.3 A schematic diagram of an asymmetric supercapacitor (positive electrode: MnO_2; negative electrode: FeOCl@CDCA; separator: cellulose paper; current collector: Ni foam; packaging material: PTFE plate).

capacitive material with electrochemical properties whose stored charge is linear depending on the width of the potential window. In a pseudocapacitor, the capacitance is related to the amount of charge received and the potential change. Reactions such as electroadsorption or redox reactions of electroactive materials play a crucial role in generating Faraday current in pseudocapacitive electrodes. Electroadsorption occurs due to the chemical adsorption of electron-giving anions such as Cl^- and B^-. In the case of redox reactions, the static separation of charges is replaced by the exchange of charges. For supercapacitors, two charge storage mechanisms (EDLC and pseudocapacitors) usually coexist. The pseudocapacitive storage mechanism significantly contributes to the specific capacitance, while the other contributes little. Pseudocapacitive dominant electrode materials include metal oxides (such as ruthenium oxides, manganese oxides, nickel oxides, cobalt oxides, copper oxides, and vanadium nitride), conductive polymers (such as poly-pyro [PPy] and polyaniline [PANI]), carbon-based heteroatoms (such as N, P, or S-doped carbon), and porous carbon nanoparticles that electroadsorb hydrogen. Pseudocapacitive materials have high specific capacitance and energy density, but they have low conductivity, poor rate capacity, and poor structural stability due to repeated ion insertion.

5.2 Pure Carbon and Activated Carbon Aerogels for Supercapacitors

First, it should be noted that due to the different physical structures of different carbon aerogels, there are significant differences in specific surface area, pore size distribution, flexibility, mechanical strength, etc., which play an important role in determining their application form. For instance, fragile carbon aerogels are usually used in powder form and mixed with conductive agents (such as acetylene black) and adhesive agents (such as PVDF). The mixture is coated on a nickel foam surface and pressed into a nickel foam-supported electrode. In contrast, the flexible carbon aerogel can be directly used as the substrate of the independent electrode, which can avoid the introduction of adhesive, thus improving the load of electrode material and avoiding the parasitic reaction caused by the binder.

Yang and coworkers used NaOH/urea aqueous solution to dissolve bamboo cellulose fibers and prepared porous carbon aerogel.[18] After pyrolysis and further activation of KOH (to increase the specific surface area), the activated carbon aerogel was mixed with carbon black and PVDF at a weight ratio of 90:5:5 and fixed between two nickel foams at a pressure of 10 MPa. This carbon aerogel electrode exhibited a high specific capacitance of 381 F g^{-1} in a 6 M KOH electrolyte, an increase of 150% compared to the unactivated sample. Reasonable design of micropores and mesopores can help it achieve a 90% retention rate even at the high scanning rate of 200 mV s^{-1}. Low equivalent series resistance and small charge transfer resistance in carbon aerogel also indicate excitation capacitance behavior. Zu and coworkers dissolved microcrystalline cellulose in NaOH solution and proceeded with gelation, regeneration, and supercritical CO_2 drying to prepare a carbon aerogel with a high specific surface area.[19] The resulting carbon aerogel was further treated by a CO_2 activation process. Therefore, the final carbon aerogel product not only has an extremely high specific surface area (1,873 $m^2 g^{-1}$) and interconnected three-dimensional pore structure but also has a large number of nanopores, which undoubtedly provides a large number of fast channels for electrolyte ions and electron migration. The specific capacitance of the obtained activated carbon aerogel is 302 and 205 F g^{-1} when the current density is 0.5 and 20 A g^{-1}, respectively, indicating that it has outstanding rate performance. It is found that the activation treatment above helps to improve the specific surface area and thus the energy storage capacity. For pure carbon materials, only the EDLC mechanism can store energy. However, incomplete pyrolytic carbon aerogels retain a small number of oxygen-containing groups, so Faraday reactions that occur on their surfaces, such as the reaction between the carbonyl group (C=O) and the hydroxyl group (C–OF) (i.e., $C{=}O + H^+ + e^- \leftrightarrow C{-}OH$), may produce the pseudocapacitance.

The successful application of these pure carbon and activated carbon aerogels confirms the feasibility and great potential of carbon aerogel materials in supercapacitors. At the same time, it has been proved that it is entirely feasible to optimize the electrochemical performance of supercapacitors by regulating the pore structure and composition of carbon aerogel materials, and its structure-activity relationship and action mechanism are worthy of further study.

5.3 Heteroatoms-Doped Carbon Aerogels for Supercapacitors

To further improve the energy storage capacity of carbon aerogel in supercapacitors, some effective methods, such as heteroatom doping and integration with pseudocapacitive materials, have been well developed. Due to the porous structure and strong adsorption capacity of carbon aerogels, dye adsorption methods to achieve heteroatom doping have become a focus of attention. For example, some researchers used an alkaline peroxide mechanical pulp (APMP) fiber aerogel as the sorbent and adopted the aerogel saturated with Rhodamine B (RhB) dye to obtain N-doped carbon aerogels for supercapacitors.[20] The RhB molecule adsorbed on the APMP aerogel was used as the nitrogen source. The composite was carbonized to obtain N-doped carbon aerogel. When the current density is 1 A g^{-1}, the capacitance of the N-doped APMP aerogel was 19.4% higher than that of the original APMP aerogel. In addition, Liao and coworkers obtained a porous N-doped carbon aerogel by synthesizing Schiff-base porous organic polymer aerogel and then pyrolyzing it.[21] Supercapacitors based on this N-doped carbon aerogels displayed high specific capacitance (300 F g^{-1} at 0.5 A g^{-1}), fast rate (charge to 221 F g^{-1} within only 17 s), and high stability (retained >98% capacity after 5,000 cycles). Asymmetric supercapacitors assembled with these N-doped carbon aerogels showed high energy and power density, with maximum values of 30.5 Wh kg^{-1} and 7,088 W kg^{-1}, respectively.

In addition to a single type of heteroatom doping, various heteroatom co-doping strategies can significantly improve the electrochemical performance of supercapacitors using carbon aerogel electrodes. For example, some researchers have dissolved α-lipoic acid with disulfide bonds and carboxyl groups in a NaOH/urea solution and used this solvent to dissolve cellulose.[22] Then gelation and carbonization were performed to obtain N, S co-doped hierarchical porous carbon aerogel. Due to the fabricated carbon material has a proper structure and uniform heteroatom doping, its capacitance can reach 329 F g^{-1} at 0.5 A g^{-1}, 1,647.5 mF cm^{-2} at 2.5 mA cm^{-2}, and the acceptance rate property is 215 F g^{-1} at 10 A g^{-1} and 1,075 mF cm^{-2} at 50 mA cm^{-2}, respectively. An asymmetric supercapacitor (SSC) assembled from the carbon aerogel produces a high energy density of 10.3 Wh kg^{-1} at a power density of 130 W kg^{-1}. Together, these reports demonstrate the role of heteroatomic doping in improving the energy storage capacity of carbon electrode materials.

5.4 Metal-containing Carbon Aerogel Composites for Supercapacitors

In contrast to heteroatom doping, integrating carbon aerogel with pseudocapacitive materials (e.g., metallic compounds) is usually more powerful. For example, a highly conductive and porous cellulose-derived carbon aerogel substrate was introduced in combination with an orthogonal FeOCl with a self-stacking layered structure.[23] Due to the presence of carbon aerogel, the mechanical stability and charge storage dynamics of FeOCl are significantly enhanced. When applied to supercapacitors, this composite material provides an ultrahigh area ratio capacitance of 1,618 mF cm^{-2} (647 F g^{-1}) at 2 mA cm^{-2} and excellent cycle stability, with a capacitance loss of less than 10% after 10,000 cycles. Asymmetric supercapacitors prepared using this composite material (as an anode) and MnO_2 cathode showed a competitive energy/power density (288 mWh cm^{-2} at 1.8 mW cm^{-2}) and excellent rate capacity and cyclic stability. For example, Liu and coworkers used a novel doping method to prepare three-dimensional interconnected Mn_2O_3/carbon aerogel supercapacitor electrodes using MnO_2 coordinated by *N,N*-dimethylformamide (DMF).[24] The coordinated MnO_2 (DMF/$MnO_{2)}$ plays a key role in the sol-gel process of resorcinol and formaldehyde. Carbon aerogel composite material has a high specific surface area of 859 $m^2 g^{-1}$ and a good pore size distribution of 10–15 nm. When 5% DMF/MnO_2 was added to the precursor solution, the resulting carbon aerogel composite had a specific capacitance of up to 100 F g^{-1} at a current density of 10.0 A g^{-1} and retained 97% of the initial capacitance for 1,000 cycles at a current density of 5.0 A g^{-1}. The doped carbon aerogel shows high coulomb efficiency (up to 99.8%) and excellent rate capacity.

6 CONCLUSION AND PROSPECTS

The energy storage mechanism of porous carbon materials does not fully conform to the classical double-layer theory. To deeply understand the energy storage mechanism of porous carbon double layers, researchers have proposed various mechanism models. The EDCC and EWCC models based on surface curvature factors can better match the experimental data. However, the assumption of cylindrical shape and the neglect of electrolyte factor in these two models make them have some limitations.

To improve the energy density of porous carbon double-layer capacitors, a lot of research has been carried out on the energy storage mechanism of porous carbon in the last ten years to obtain a method to improve the porous carbon double-layer capacitor. When the pore size is equivalent to the ion diameter after desolvation (or deformation), the porous carbon specific capacitance is significantly increased, indicating that the pore diameter of porous carbon should be about 0.7 nm to obtain a high specific capacitance. However, it is necessary to have a proper proportion of mesoporous pores in porous carbon to improve the ion transport capacity in the material body and the charge storage capacity of micropores.

Porous carbon aerogel with a high specific surface area and controllable micro-mesoporous is an ideal supercapacitor material. The specific capacitance of porous carbon aerogel can be improved by adjusting the hydrogel formation, drying, and carbonization during the preparation of porous carbon aerogel. The charge storage capacity can also be enhanced by heteroatom doping to form abundant active sites.

In the future, further in situ characterization of supercapacitors with different electrolytes and carbon materials (including electrochemical quartz crystal microbalance, nuclear magnetic resonance spectroscopy, infrared spectroscopy, small angle X-ray scattering (SAXS), and small angle neutron scattering (SANS)) will help to elucidate the factors controlling the charging mechanism of supercapacitors fully. Beyond experiments, advanced theoretical methods that can adequately consider the electronic structure of ion and carbon electrodes can provide new insights. Suppose one can sufficiently understand the factors that affect the charge storage mechanism. In that case, it should be possible to develop high-performance supercapacitors by choosing the right electrolytic-electrode combination. In addition, the control or design of porous structures is a great challenge worthy of extensive exploration. 3D printing of aerogel is an exciting area because models can determine the structure of aerogel. Therefore, in the future, the combination of 3D printed carbon aerogel preparation and further study of its evolutionary transformation mechanism will help to obtain porous carbon aerogel with an ideal pore structure.

REFERENCES

1. H. Helmholtz, Ueber einige Gesetze der Vertheilung elektrischer Ströme in körperlichen Leitern, mit Anwendung auf die thierisch-elektrischen Versuche (Schluss.), *Ann. Phys.*, 1853, **165**, 353–377.
2. D.L. Chapman, LI. A contribution to the theory of electrocapillarity, *Lond. Edinb. Dubl. Philos. Mag. J. Sci.*, 1913, **25**, 475–481.
3. M. Gouy, Sur la constitution de la charge électrique à la surface d'un électrolyte, *J. Phys. Theor. Appl.*, 1910, **9**, 457–468.
4. L.L. Zhang and X.S. Zhao, Carbon-based materials as supercapacitor electrodes, *Chem. Soc. Rev.*, 2009, **38**, 2520–2531.
5. P. Simon and Y. Gogotsi, Materials for electrochemical capacitors, *Nat. Mater.*, 2008, **7**, 845–854.
6. J. Huang, B.G. Sumpter and V. Meunier, A universal model for nanoporous carbon supercapacitors applicable to diverse pore regimes, carbon materials, and electrolytes, *Chem. Eur. J.*, 2008, **14**, 6614–6626.
7. O. Barbieri, M. Hahn, A. Herzog and R. Kötz, Capacitance limits of high surface area activated carbons for double layer capacitors, *Carbon*, 2005, **43**, 1303–1310.
8. S. Dutta, A. Bhaumik and K.C.W. Wu, Hierarchically porous carbon derived from polymers and biomass: Effect of interconnected pores on energy applications, *Energy Environ. Sci.*, 2014, **7**, 3574–3592.
9. E. Avraham, B. Yaniv, A. Soffer and D. Aurbach, Developing ion electroadsorption stereoselectivity, by pore size adjustment with chemical vapor deposition onto active carbon fiber electrodes. Case of Ca^{2+}/Na^{+} separation in water capacitive desalination, *J. Phys. Chem. C*, 2008, **112**, 7385–7389.
10. C. Cachet-Vivier, V. Vivier, C.S. Cha, J.Y. Nedelec and L.T. Yu, Electrochemistry of powder material studied by means of the cavity microelectrode (CME), *Electrochim. Acta*, 2001, **47**, 181–189.

11. M. Noked, E. Avraham, A. Soffer and D. Aurbach, The rate-determining step of electroadsorption processes into nanoporous carbon electrodes related to water desalination, *J. Phys. Chem. C,* 2009, **113**, 21319–21327.
12. C. Vix-Guterl, E. Frackowiak, K. Jurewicz, M. Friebe, J. Parmentier and F. Béguin, Electrochemical energy storage in ordered porous carbon materials, *Carbon*, 2005, **43**, 1293–1302.
13. J. Chmiola, G. Yushin, Y. Gogotsi, C. Portet, P. Simon and P.L. Taberna, Anomalous increase in carbon capacitance at pore sizes less than 1 nanometer, *Science*, 2006, **313**, 1760–1763.
14. M. Wu, P. Ai, M. Tan, B. Jiang, Y. Li, J. Zheng, W. Wu, Z. Li, Q. Zhang and X. He, Synthesis of starch-derived mesoporous carbon for electric double layer capacitor, *Chem. Eng. J.*, 2014, **245**, 166–172.
15. R. Mysyk, E. Raymundo-Piñero, J. Pernak and F. Béguin, Confinement of symmetric tetraalkylammonium ions in nanoporous carbon electrodes of electric double-layer capacitors, *J. Phys. Chem. C,* 2009, **113**, 13443–13449.
16. M. Hahn, O. Barbieri, F.P. Campana, R. Kötz and R. Gallay, Carbon based double layer capacitors with aprotic electrolyte solutions: The possible role of intercalation/insertion processes, *Appl. Phys. A*, 2006, **82**, 633–638.
17. C. Zhang, H. Wang, Y. Gao and C. Wan, Cellulose-derived carbon aerogels: A novel porous platform for supercapacitor electrodes, *Mater. Des.*, 2022, **219**, 110778.
18. X. Yang, B. Fei, J. Ma, X. Liu, S. Yang, G. Tian and Z. Jiang, Porous nanoplatelets wrapped carbon aerogels by pyrolysis of regenerated bamboo cellulose aerogels as supercapacitor electrodes, *Carbohydr. Polym.,* 2018, **180**, 385–392.
19. G. Zu, J. Shen, L. Zou, F. Wang, X. Wang, Y. Zhang and X. Yao, Nanocellulose-derived highly porous carbon aerogels for supercapacitors, *Carbon*, 2016, **99**, 203–211.
20. L. E, W. Li, J. Sun, Z. Wu and S. Liu, N-doped carbon aerogels obtained from APMP fiber aerogels saturated with rhodamine dye and their application as supercapacitor electrodes, *Appl. Sci.,* 2019, **9**, 618.
21. H. Li, J. Li, A. Thomas and Y. Liao, Ultra-high surface area nitrogen-doped carbon aerogels derived from a Schiff-base porous organic polymer aerogel for CO_2 storage and supercapacitors, *Adv. Funct. Mater.*, 2019, **29**, 1904785.
22. C. Dang, Z. Huang, Y. Chen, S. Zhou, X. Feng, G. Chen, F. Dai and H. Qi, Direct dissolution of cellulose in NaOH/urea/α-lipoic acid aqueous solution to fabricate all biomass-based nitrogen, sulfur dual-doped hierarchical porous carbon aerogels for supercapacitors, *ACS Appl. Mater. Interfaces,* 2020, **12**, 21528–21538.
23. C. Wan, Y. Jiao, W. Bao, H. Gao, Y. Wu and J. Li, Self-stacked multilayer FeOCl supported on a cellulose-derived carbon aerogel: A new and high-performance anode material for supercapacitors, *J. Mater. Chem. A*, 2019, **7**, 9556–9564.
24. Y. Xu, S. Wang, B. Ren, J. Zhao, L. Zhang, X. Dong and Z. Liu, Manganese oxide doping carbon aerogels prepared with MnO_2 coordinated by N, N - dimethylmethanamide for supercapacitors, *J. Colloid Interface Sci.,* 2019, **537**, 486–495.

2 Porous Carbon-Based Materials for Supercapacitors

Hui Peng and Wenxing Miao

1 INTRODUCTION

Due to the over-exploitation of fossil fuels and the gradual deterioration of the environment, finding environmentally friendly clean energy sources that can replace fossil fuels (coal, oil, and natural gas) has become an urgent task for researchers.[1] Currently, clean energy sources such as wind, tidal, and solar have been utilized on a large scale, but how to store clean energy is a key issue. Scientists are again focusing their attention on the research of energy storage devices.[2] Electrochemical energy storage has a broad application prospect in the current clean energy storage field. Among them, supercapacitors (SCs) and batteries are two classic energy storage devices with complementary advantages. Although the energy density of SCs is much lower than that of batteries (only one-tenth to one-fifth of batteries), the power density of SCs is hundreds or even thousands of times than that of batteries.[3] Differently, the energy storage of SCs originates from the reversible physical adsorption at the interface between electrodes and electrolytes (electric double-layer capacitors, EDLC), or the highly reversible chemical adsorption/desorption or redox reactions by underpotential deposition on the electrode surface (pseudocapacitor).[4] Since the energy storage of SCs happens only through the rapid and reversible adsorption of ions on the surface of electrode materials, it does not involve similar energy storage processes as redox inside the electrode material of a battery; it also has excellent power density and ultra-long cycle life with fast charging and discharging behaviors.[5]

Porous carbon-based materials, such as carbon nanotubes (CNTs), graphene, activated carbon (AC), etc., are considered to be promising electrode materials for SCs due to their elevated specific surface area, high electrical conductivity, adjustable pore structure, extra pseudocapacitance, abundant resources, and environmental friendliness.[6–8] Unfortunately, the improvement of the overall electrochemical energy storage properties, especially the energy density of SCs from carbon-based materials is still very limited. Scientists have tried various methods to increase the energy density of SCs, mainly including the following strategies: (1) By increasing the degree of graphitization, the conductivity of electrode material is increased to accelerate the charge transfer ability of carbon materials. (2) Use heteroatom doping to provide additional pseudocapacitance, and (3) increase the defect structure and pore density of carbon materials to improve energy storage active sites (Figure 2.1).[9–12]

DOI: 10.1201/9781003387831-2

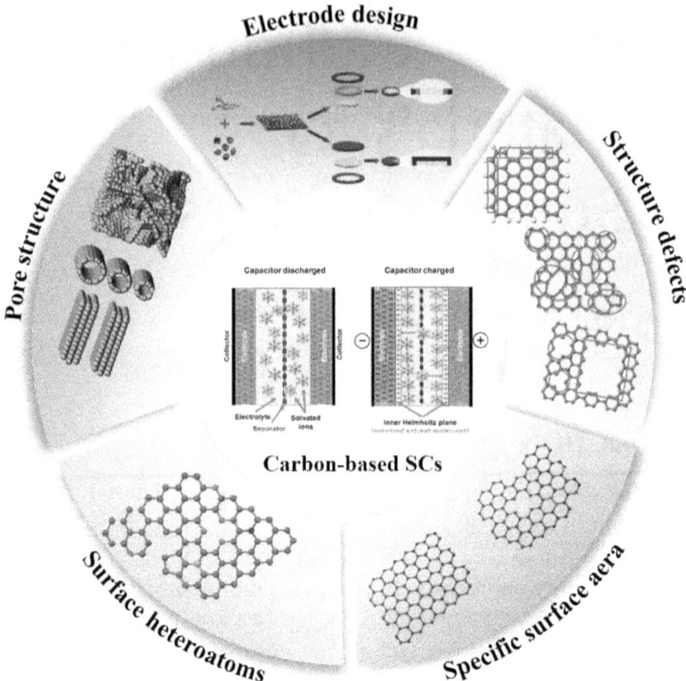

FIGURE 2.1 Illustration of strategies to improve the performance of carbon-based SCs.

In addition to the direct improvement of electrode materials, the electrolyte also plays an important role in the SCs. The energy density of electrochemical energy storage devices is directly proportional to the voltage and the capacity, and the energy density of SCs can be indirectly enhanced by improving the composition and type of the electrolyte to increase the working voltage of SCs.[13] Therefore, the development of an electrolyte with a wide operating voltage window, high electrochemical stability, and high ionic concentration is also the goal pursued by researchers. Currently, the main electrolytes commonly used in SCs are aqueous electrolytes, organic electrolytes, ionic liquid-type electrolytes and quasi-solid electrolytes.[14,15] Generally speaking, the relatively high operating voltage can be obtained in non-aqueous electrolytes (including organic and ionic liquid electrolytes), which results in a higher energy density than that in aqueous electrolytes. In contrast, higher power density can be achieved when the SCs are operated in an aqueous electrolyte compared to operating in a non-aqueous electrolyte,[15] which is due to the fact that the ionic conductivity of aqueous electrolytes is much higher than that of organic electrolytes. Thus, developing carbon materials with high specific capacitance and matching aqueous electrolytes with high ion diffusivity has become an ideal combination for SCs systems.[16]

This chapter first introduces the theoretical background and working mechanism of carbon-based SCs, and then discusses in detail the various components of carbon-based SCs, including electrolytes, porous carbon-based electrode materials, and several key factors that affect the performance of carbon materials. We summarize the research work on the application of porous carbon-based materials to SCs in recent years and discuss the effects of morphology, pore structure, heteroatom doping, and electrical conductivity on the performance of SCs. Finally, we elaborate on the current bottlenecks faced by carbon-based SCs and discuss their future perspectives.

2 THEORETICAL BACKGROUND FOR SUPERCAPACITORS

As illustrated in Figure 2.2a, the device structure of SCs is similar in design and manufacture to that of a battery, consisting of two working electrodes composed of active materials deposited on the current collector, an electrolyte, and a separator that electrically isolates the two electrodes.[17] In SCs, ions move rapidly through the liquid electrolyte to the electrode surface, while electrons are rapidly transported through the highly conductive carbon electrode to the external circuit.[18, 19] Therefore, SCs can provide excellent power density and ultra-long cycle life compared to fuel cells and batteries.

2.1 THE ENERGY STORAGE MECHANISM

Theoretically, there are two energy storage mechanisms in SCs systems, namely EDLCs and pseudocapacitors (Figure 2.2b). EDLCs are based on the reversible physical adsorption of electrolyte ions at the interface between the electrode material and the electrolyte to accumulate charge, while pseudocapacitors are based on the rapid reversible quasi-redox reactions between the electrode material and the electrolyte to absorb or discharge charge to achieve charge storage.[20]

2.1.1 The energy storage mechanism of EDLCs

The EDLCs use the reversible physical adsorption of electrolyte ions at the interface between the positive and negative electrodes and the electrolyte to accumulate charge for energy storage. The concept of the electric double layer (EDL) was first described and modeled by von Helmholtz in the 19th century when he investigated the distribution of opposite charges at the interface of colloidal particles.[21] The Helmholtz model describes the separation of charges at the electrode/electrolyte interface when the electrode with surface area S (m^2) is polarized. In this way, ions of opposite signs diffuse through the electrolyte, forming a condensed layer a few nanometers thick on a plane parallel to the electrode surface to ensure a neutral charge. This form of charge accumulation is called an EDL. Then, as the distance d (m) between the ions and the electrode increases, the potential near the electrode decreases.[21, 22]

This simplified Helmholtz double-layer (DL) can be regarded as an electrical capacitor of capacitance C_H defined by Eqn 2.1:

$$C_H = \varepsilon_0 \, \varepsilon_r \cdot \frac{S}{d} \tag{2.1}$$

where ε_0 is the vacuum permittivity ($\varepsilon_0 = 8.854 \times 10^{-12}$ F m^{-1}), ε_r is the relative permittivity of the dielectric electrolyte, and d is the effective thickness of the DL, often approximated as the Debye length.[22, 23] Taking into account the very large specific surface area of porous carbon electrodes (up to 3,000 m^2g^{-1}) and the Debye length in the range of <1 nm, the resulting capacitance of DL will be much higher than that of plate capacitors. However, the Helmholtz model does not consider factors such as the diffusion of ions in solution, the dipole moment of the solvent, and the interaction between electrodes.

In 1913, Gouy-Chapman came up with a better model (Figure 2.2c). This theoretical model assumes that the opposite counter ion is not strictly attached to the surface, but tends to diffuse into the liquid phase. The thickness of the resulting double layer is influenced by the kinetic energy of the counter ions. The model is named the diffusive DL model.[23, 24] However, the Gouy-Chapman model does not apply in the case of high charge density bilayers. In 1924, Stern suggested a model combining the Helmholtz and Gouy-Chapman models by accounting for the hydrodynamic motion of the ionic species in the diffuse layer and the accumulation of ions close to the electrode surface (Figure 2.2d). These two layers are equivalent to two capacitors in series, C_H (Helmholtz layer) and C_D (diffuse layer), and the total capacitance of the electrode (C_{DL}) is given by Eqn 2.2:

$$\frac{1}{C_{DL}} = \frac{1}{C_H} + \frac{1}{C_D} \tag{2.2}$$

The capacitance estimation for an EDL-type supercapacitor is generally assumed to follow that of a parallel-plate capacitor as Eqn 2.3:

$$C = (\varepsilon_0 \varepsilon_r / d) \times A \tag{2.3}$$

where ε_r is the electrolyte dielectric constant, ε_0 is the permittivity of a vacuum, A is the specific surface area of the electrode accessible to the electrolyte ions, and d is the effective thickness of the EDL (the Debye length). As can be seen from the above equation, the specific capacitance of SCs is very much influenced by the contact area between the electrode material and the electrolyte. Therefore, the preparation of electrode materials with large specific surface area and well-developed pores is a reliable way to improve the capacitance of a double layer.

2.1.2 The energy storage mechanism of pseudocapacitor

Pseudocapacitors, also called Faraday capacitors, differ from EDLCs in that they are capacitance generated by a highly reversible chemical absorption/desorption or redox reaction on the two-dimensional or quasi-two-dimensional space of the electrode surface, where the electroactive material is deposited under potential.[20, 25, 26] This type of energy storage does not produce a continuous current as in the conventional Faraday reaction process, but is an indirect form of energy storage and similar to that of a battery, so it is called pseudocapacitor. The most common active materials in pseudocapacitors are metal compounds (such as ruthenium oxide, manganese oxide, vanadium nitride, and so on) and conductive polymers (such as polyaniline and polypyrrole). Although the specific capacitance of pseudocapacitance can be higher than EDLCs, it suffers from the disadvantages of low power density (due to poor conductivity) and lack of stability during cycling.

2.2 The performance of supercapacitors

The performance of SCs is mainly evaluated on the basis of the following criteria: (1) high specific capacitance, (2) high power density, (3) high energy density, (4) fast charge/discharge processes within seconds, (5) excellent cyclability, (6) safe operation, and (7) low cost.[21, 26] The total capacitance of the supercapacitor device (C_T) is therefore calculated according to Eqn 2.4:

$$1/C_T = 1/C_{\text{positive}} + 1/C_{\text{negative}} \tag{2.4}$$

Energy density and power density are two important parameters for assessing the electrochemical performance of SCs. The maximum energy density (E, Wh kg^{-1}) and power density (P, W kg^{-1}) of a supercapacitor can be calculated by Eqns 2.5–2.6:

$$E = \frac{1}{2} C_T V^2 \tag{2.5}$$

$$P = \frac{V^2}{4R_s} \tag{2.6}$$

where C_T is the total capacitance of the supercapacitor device (F g^{-1}), V is the device voltage (V), and R_s is the equivalent series resistance (ESR, Ω).[21, 27–30] ESR mainly consists of the intrinsic resistance of the active material, the contact resistance between the electroactive material and the current collector, the diffusion resistance of the ions in the electrode material and separator, and the resistance of the electrolyte ions. In order to obtain SCs with excellent electrochemical performance, simultaneously achieving high specific capacitance, wide operating voltage, and the lowest possible equivalent series resistance are key factors.

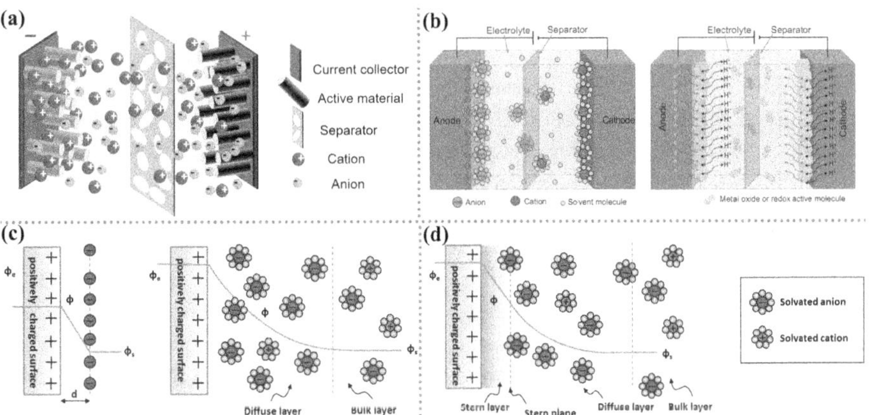

FIGURE 2.2 (a) The schematic illustration of the structure of a typical supercapacitor.[17] (b) Schematic representation of EDLCs and pseudocapacitor.[23] (c) Helmholtz and Gouy-Chapman, and (d) Stern model of the electrical DL formed at a positively charged electrode in an aqueous electrolyte.[22]

3 ELECTROLYTES

Electrolytes used in SCs can be divided into aqueous electrolytes, organic electrolytes, ionic liquid electrolytes, etc. This chapter will discuss in detail the specific classification and selection principles of electrolytes used in SCs. The different types of electrolytes used in carbon-based SCs are shown in Figure 2.3.[32]

3.1 AQUEOUS ELECTROLYTES

Aqueous electrolytes usually refer to the electrolyte solution composed of water, inorganic salts (such as KOH, NaOH, H_2SO_4, K_2SO_4, Li_2SO_4, Na_2SO_4, and NaCl) and additives, which have the advantages of smaller ionic radius, higher ionic concentration and lower internal resistance. For example, the conductivity of 1 M H_2SO_4 at 25°C is 0.8 S cm^{-1}, which is much higher than other aqueous electrolytes and organic electrolytes, so it can effectively reduce the resistance of the device and achieve higher power output. More importantly, during the device assembly process, aqueous electrolytes do not have high requirements in the assembly environment and can be assembled in the air completely. The only drawback is that aqueous electrolytes are severely limited by the water splitting (1.23 V), which limits their operating voltage window and energy density, making them difficult to be used on a large scale. The voltage window of aqueous electrolytes shifts with pH, it is mainly limited by the precipitation of oxygen and hydrogen.

Theoretically, the differences in the electrochemical energy storage behaviors for the SCs in various electrolytes are probably attributed to: (1) the ionic radius of the electrolytes, (2) the radius of the ionic hydration sphere of the electrolytes, (3) the conductivity of the ions, and (4) the mobility of the ions. Research results show that only changing the types of cation species (H^+, Li^+, Na^+, and K^+) in various electrolytes has significant differences in the specific capacitance of SCs.[32] Furthermore, long-term cycling stability is also strongly dependent on the cationic species.[33] For example, the specific capacitance of polypyrrole/graphene electrode in various aqueous electrolytes including HCl, KCl, NaCl, and LiCl with the same concentration decreases sequentially. The significant difference in specific capacitance is due to cation mobility, hydrated cation radius, conductivity, and their effect on charge/ion exchange and diffusion. Moreover, the specific capacitance of electrodes in HCl, LiCl, NaCl, and KCl electrolytes were 92.0%, 73.9%, 58.7%, and 37.9% after 10,000 cycles, respectively, which was due to the significant composition damage caused by the insertion/de-insertion of large size cation ($H^+ < Li^+ < Na^+ < K^+$) during the charging-discharging cycles. The hydrated cationic radius was in the order: H^+-$H_2O^{\delta-} < K^+$-$H_2O^{\delta-} < Na^+$-$H_2O^{\delta-} < Li^+$-$H_2O^{\delta-}$. The Li^+ ion had the largest hydration sphere radius because of the Li^+-$H_2O^{\delta-}$ strong interactions caused by the large surface charge density. The large ionic mobility and small hydrated sphere of H^+ were due to the jumping transference mode between water molecules by hydrogen bonds. The H^+ ion also had the highest molar ionic conductivity as compared to K^+, Na^+, and Li^+. The higher conductivity and ionic mobility help in fast charge transfer, and the smaller hydration sphere radius also offers more ion adsorption on the interface of electrolyte/electrode to further facilitate the Faraday reaction. Therefore, due to the smallest hydrated

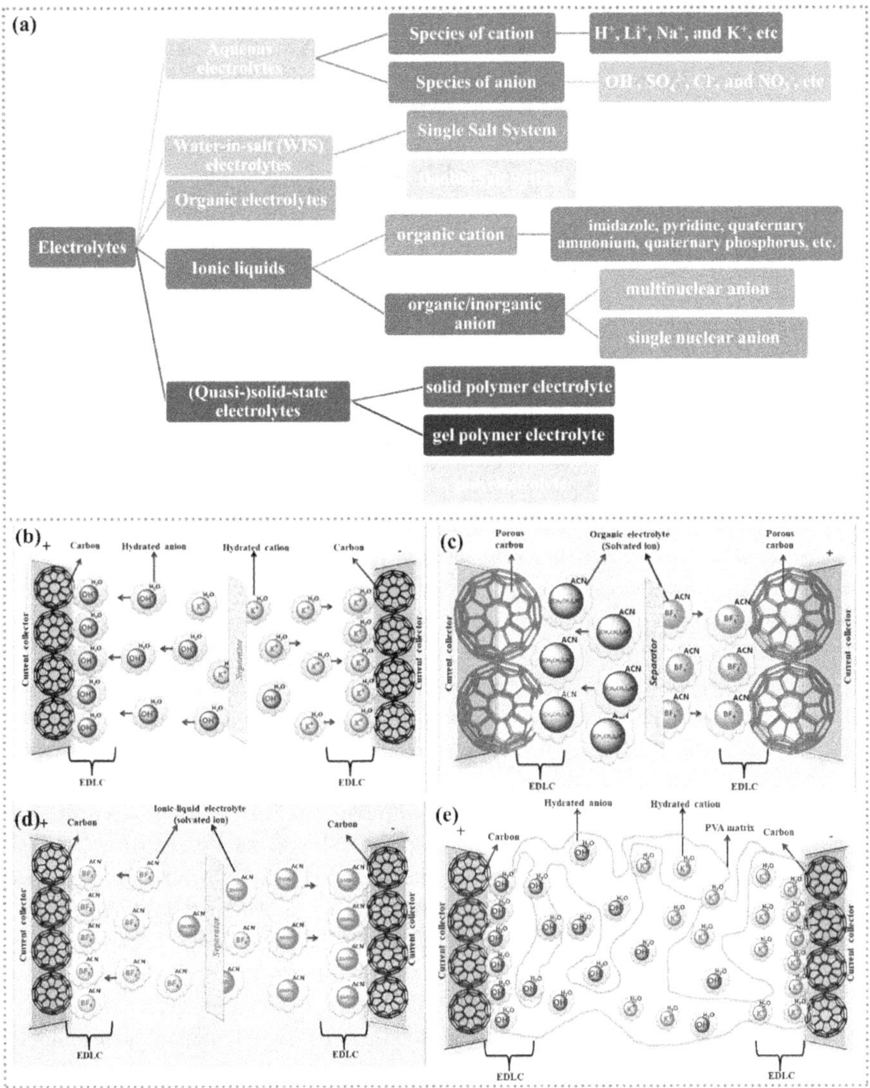

FIGURE 2.3 (a) Classification of electrolytes for SCs. Diagram of the reaction mechanism of (b) aqueous electrolyte, (c) organic electrolyte, (d) IL electrolyte, and (e) gel electrolyte between carbon materials of EDLCs.[32]

ions and highest conductivity and ionic mobility, the supercapacitor with HCl as an electrolyte shows the largest specific capacitance.

The influence of anion species (OH^-, SO_4^{2-}, Cl^-, and NO_3^-) on the electrochemi cal performance of SCs has been also studied.[32] Since the mobility of hydrated ions in the inner pores increases in the order of $NO_3^- < Cl^- < SO_4^{2-} < OH^-$, the different electrolytes (KOH, KCl, KNO_3, and K_2SO_4) with different conductivities and ion mobilities have significant differences in specific capacitance. The high conductivity and ionic mobility of OH^- can cause much good capacitive behavior. The radius of the hydration sphere of OH^- was slightly smaller than the radii of the hydration sphere of Cl^- and NO_3^- ions and the most probable apertures are equal to about 2.1 times the size of the radii of the hydration spheres. The radius of the hydration sphere of SO_4^{2-} ion was the bigger, and led to a decrease in the quantity of ions entering into pores, thus lower EDL is formed. Furthermore, the conductivity and ionic mobility of SO_4^{2-} ion were also low as compared to other anions. Compared with the capacitance of the electrode materials using the electrolytes containing KOH, NaOH, LiOH, $LiNO_3$, and Na_2SO_4 with the same concentration, the maximum specific capacitance is obtained when using the KOH electrolyte, followed by NaOH, LiOH and $LiNO_3$ and Na_2SO_4 is the smallest.

3.2 Water-in-Salt (WIS) Electrolytes

Recently, a series of highly concentration aqueous electrolytes, called "water-in-salt (WIS)" electrolytes, have been reported and can expand the working voltage of SCs to ≈3.0 V due to the formation of an electrode-electrolyte interphase. The most obvious change is that the electrochemical stability operating voltage window of the WIS electrolytes is widened and the potentials for both HER and OER were pushed beyond their thermodynamic potentials. In a WIS electrolyte, the electrochemical activity of water is remarkably reduced due to the formation of a strong coordination between the high-concentration metal-ion solvated sheath and water molecules. Owing to the impressive physicochemical properties of moisture resistance, nonflammability, and large electrochemical stability window, WIS electrolytes also show great opportunities for aqueous SCs applications.

The lithium bis(trifluoro methane sulfonyl)imide (LiTFSI) was first used in WIS electrolytes due to its high solubility and chemical thermal stability.[2] Currently, there is a handle of other WIS electrolytes that have been developed with similar properties. Specifically, the electrochemical stability windows of 30 M $ZnCl_2$, 17 M $NaClO_4$, 22 M $LiNO_3$, and 12 M $NaNO_3$ could reach 2.30, 2.80, 2.55, and 2.56 V, respectively.[34–37] Among them, the $NaClO_4$ and $NaNO_3$ are inexpensive and have a bright future for commercialization. Theoretically, the lower the free water content, the wider the electrochemical stability window of the WIS electrolytes. The amount of free water in WIS electrolytes is mainly controlled by the ratio of water-to-cation for the common double salt WIS system. According to the cation solvation rule, the free water will decrease in WIS electrolytes with the reduction of the water-to-cation ratio. To lower the water-to-cation ratio and further reduce the free water, a double salt system has been explored by adding another salt into the binary system. With this idea in mind, a new super-concentrated WIS electrolyte composed of 21 M LiTFSI and 7 M lithium trifluoro methane sulfonate (LiOTf) was proposed, which delivered a wider electrochemical stability window of 3.1 V, higher than that of 21 M LiTFSI (3 V).[38] Meanwhile, the ionic conductivity of the water-in-bisalt (WIBS) electrolyte has not changed significantly compared with 21 M LiTFSI. A research has reported that the addition of inert tetraethylammonium (TEA^+) cation to the WIS electrolytes, named inert-cation-assisted WIS, has made the concentration of the electrolyte reach 31 M (9 M NaOTF + 22 M TEAOTF), resulting in a wide electrochemical window of 3.3 V.[39]

Subsequently, Wang et al. found that the concentration of LiTFSI in water would reach 42 M when an asymmetric ammonium salt ($Me_3EtN{\cdot}TFSI$) was added to the LiTFSI electrolyte, and a 63 M WIBS electrolyte (42 M LiTFSI + 21 M $Me_3EtN{\cdot}TFSI$) was prepared with the water-to-cation ratio of 1.13.[40] Molecular dynamics simulations indicated that the amount of free water in 63 M WIBS was quarter or even half of that in 21 M WIS electrolytes and 42 M WIBS, respectively, bringing about an electrochemical stability window of 3.25 V. For the salts with low solubility used as active charge carriers, they could not directly formulate WIS electrolytes. At the same time, WIBS electrolytes could be achieved with a low water-to-cation ratio through mixing with other salts with high solubility.

3.3 Organic electrolytes

Organic electrolytes for SCs usually consist of organic solvents and conducting salts dissolved in them. Because of their high voltage window (2.6–2.9 V), SC devices based on organic electrolytes are currently dominating the commercial market. However, the use of organic electrolytes will have some inherent shortcomings that are difficult to avoid, such as high cost, low electrical conductivity compared to aqueous electrolytes resulting in reduced power density, low dielectric constant resulting in smaller capacitance, complex purification process, and safety issues due to the flammability and toxicity of organic solvents. At present, quaternary ammonium salts (R_4N^+), such as tetramethylammonium (Me_4N^+), tetraethylammonium (Et_4N^+), tetrabutylammonium (Bu_4N^+), trimethyl-ethyl ammonium (Me_3EtN^+), etc. are the main cations of organic electrolytes used in EDLCs. In addition, lithium-ion salts and quaternary phosphorus salts have also been studied. The main anions are ClO_4^-, BF_4^-, PF_6^-, AsP_6^-, etc. The most commonly used organic solvents are propylene carbonate (PC), γ- butyrolactone (GBL), *N,N*-dimethylformamide (DMF), vinyl carbonate (EC), sulfoxide, 3-methylsulfoxide (3-MeSL), acetonitrile (AN) and other nitrile derivatives and other aprotic solvents.

Organic electrolytes have also been used for pseudocapacitors with pseudocapacitive materials, such as metal oxides and conductive polymers.[41–44] Most of the organic electrolytes used for pseudocapacitors contain Li ions due to their small bare ion size and easy ion intercalation/de-intercalation. The most used salts for organic electrolytes were $LiPF_6$ and $LiClO_4$ as reported in the literature.[45,46] Organic electrolytes in asymmetric supercapacitors have also drowned significant attention due to further enhancement in the energy density. A number of organic electrolytes used in asymmetric supercapacitors such as carbon//TiO_2 (1 M $LiPF_6$/EC-DMC), graphite//AC (1.5 M $TEMABF_4$/PC), carbon//V_2O_5 (1 M LiTFSI/ACN), have been reported.[47] These asymmetric supercapacitors with organic electrolytes can deliver higher energy density due to a much wider voltage window (3–4 V) than aqueous-based asymmetric supercapacitors.

3.4 Ionic liquids

The molten salts composed of ions (organic cation and organic/inorganic anion) with melting points below 100°C are called ionic liquids (ILs). ILs are liquid salts that can be used alone as electrolytes for SCs or as supporting electrolytes for solvent-based electrolytes. As support electrolytes, ILs have the same functions as other types of electrolytes, with cations and anions mainly used for charge transfer. When ILs are used as electrolytes, two important factors need to be considered, one is the electrochemical stability window, and the other is the conductivity. The common ILs as the electrolyte of SCs and battery mainly includes the following categories: alkyl-substituted imidazole cations, alkyl-substituted quaternary ammonium salts cations, alkyl-substituted pyrrolidine cations, and other types.[48] Among them, imidazoles are the most studied cationic in ILs. The main anions are bis (trifluoromethyl sulfonyl) imine anions ($TFSI^-$), BF_4^-, and PF_6^-, which are stable in water. In particular, the ILs composed of $TFSI^-$ or BF_4 anion have the advantages of low viscosity, low melting point, and high conductivity, so they are widely used in SCs. For instance, the supercapacitor fabricated based on wine lees-based porous carbon framework electrodes and [Emim]BF_4 electrolyte delivers an ultra-high output-voltage of up to 4 V, a high energy density of 54 Wh kg^{-1} at a high power density of 401 W kg^{-1}, and a superior capacitance retention of 90% after 100,000 cycles.[49]

The fatal weakness of ILs used as electrolytes is that the viscosity is too high, but mixing with an appropriate amount of organic solvents can reduce the viscosity and improve the ionic conductivity without damaging its high boiling point, low vapor pressure, wide electrochemical window, etc. Thus, the mixed solvent system is suitable for high-power types of SCs. On the other hand, the solidification of ILs by gelation, mixing, and in situ polymerization of unsaturated functional groups is another important solution to solve the problem of high viscosity of ILs. At the same time, the relatively high conductivity of ionic liquid polymers also helps to overcome the disadvantages of the low conductivity of traditional all-solid electrolytes.

3.5 (Quasi-)Solid-State Electrolytes

Solid-state electrolytes are divided into three types: solid polymer electrolytes, gel polymer electrolytes, and polyelectrolytes. The gel polymer electrolytes are also called quasi-solid-state electrolytes, because the electrolyte form is neither liquid nor completely solid, but a soft gel-like form like jelly. The gel polymer electrolytes also play the role of electrode separator by simplifying the manufacturing process of supercapacitor ion-conducting media, free of liquid leakage, and packaging. Generally, gel polymer electrolytes have the highest ionic conductivity among the above three types of solid-state electrolytes due to the presence of a liquid phase. Due to its high ionic conductivity, gel polymer electrolytes are currently dominating solid electrolyte-based SCs devices, the focus here will be on gel electrolytes.

The gel polymer electrolytes are composed of a polymer matrix host (e.g., PEO, PEG, and PVA) and an aqueous electrolyte (e.g., KOH, H_2SO_4, and Na_2SO_4) or a conducting salt dissolved in a solvent. Copolymers with various constituent units have also been developed as polymer gel electrolyte hosts for SCs. According to the composition of gel electrolytes, it can be mainly divided into three categories: hydrogel, organogel, and ionic liquid gel. Regarding hydrogels, it is a three-dimensional polymer network formed by chemical or physical cross-linking of hydrophilic polymer chains (such as starch, cellulose, chitosan, poly-L-lysine, polyacrylic acid, polymethacrylic acid, polyacrylamide, etc.) using water as the dispersion medium. Some organic solvents such as EC, DMF, PC, and their mixtures (PC-EC and PC-EC-DMC) have also been used as plasticizers in organogel polymer electrolytes.[50–53] Organic reagents generally have high permittivity but tend to have high viscosity, thus hindering the electron mobility of ions in the medium. A strategy to address high viscosities is to blend these solvents with other solvents of lower viscosity. Besides, the specific capacitance of SCs assembled with organogel electrolytes is usually lower than that with hydrogel electrolytes, mainly because their solvated ions are larger, limiting their access to the internal pores of the electrode. Ionic liquid gel electrolyte is mainly composed of IL and polymer body. Since ILs usually have excellent fluidity and freeze resistance, some SCs assembled with ionic gel electrolytes have high specific capacitance, energy density, and cycle life, and can even operate at low temperatures. Flexible supercapacitor based on the dual-network PVA/Agar-EMIMBF$_4$-Li$_2$SO$_4$ gel electrolyte with a small amount of IL can deliver high capacities under a wide operating temperature range of –30°C to 80°C.[54] Presently, the variety of host polymers studied for organogel and ionic liquid gel electrolytes include PEO, PMMA, PVA, poly(ethylene glycol) diacrylate, and PVDF-HFP.[54–58]

4 POROUS CARBON-BASED ELECTRODE MATERIALS

4.1 Carbon nanotubes

Since the discovery of carbon nanotubes (CNTs) in 1991, they have become excellent electrode materials for SCs due to their light weight, inherent elasticity, high specific surface area, and high electrical conductivity.[59] CNTs are mainly obtained through the catalytic decomposition of hydrocarbons, and they can be divided into multi-walled carbon nanotubes (MWCNT) and single-walled carbon nanotubes (SWCNT) according to their carbon wall structure. Both types of carbon nanotubes have high electrical conductivity and a large external specific surface. In recent years, much research has been focused on the development of high density, nano-ordered, and oriented carbon nanotube arrays to increase their specific capacity by precisely modulating the distance between the tubes.[60]

Due to the inherently small specific surface area of CNTs ($<400\,m^2 g^{-1}$), their capacitance values tend to be low, typically 20–80 F g^{-1}. This is mainly due to the less developed microporous volume of CNTs, which mainly utilizes the capacitance of the boundary bilayer on the outer surface for energy storage. In order to increase the specific capacity of CNTs, they can be activated or chemically modified to improve their structure and function.[61] In addition, CNTs can also be combined with other materials to form composite materials, which can rely on the structural characteristics of CNTs to provide higher contact surface area and conductivity to improve overall performance. For example, Hu et al. used a simple and mild (room temperature) method to composite CNTs on nickel foam (Figure 2.4a).[62] The CNT provides a strong chemical bond to increase the bridge between NiCoO$_2$ and nickel foam to achieve improved energy storage and stability. The introduction of CNTs frameworks as self-supporting or for electrodeposition has also attracted much attention in recent years. Jiang et al. achieved continuous and uniform growth of polyaniline (PANI) on carbon nanotube (CNT) frame by electrodepositing method (Figure 2.4b).[63] The resultant free-standing can thus exhibit a high specific capacitance of 578.57 F g^{-1}. This CNT-based nanostructured electrode is very promising for microelectronic device applications.

4.2 Graphene

Graphene is a two-dimensional planar single molecular layer crystalline material consisting of a tightly arranged hexagonal lattice of sp^2 hybridized carbon. Although it contains only carbon, arranging the atoms in different ways modulates its properties over a wide range.[64] For example, graphene has a long cycle life and excellent electrochemical properties and has great potential in energy storage devices such as SCs and batteries.[65] The high theoretical specific surface area (2,630 $m^2 g^{-1}$) and effective specific surface area of graphene are much higher than that of carbon black (<900 $m^2 g^{-1}$) and carbon nanotubes (<1,300 $m^2 g^{-1}$).

The advent of graphene has undoubtedly changed the status of SCs technology through its extraordinary electrochemical properties and other unique properties, and the development of graphene composites has become the mainstay of current research on supercapacitor electrode materials. Graphene is a discrete monolayer of graphite with a huge accessible outer surface, which is particularly advantageous in the effective formation of double electric layers. In recent years, a great deal of work has focused on the development of methods for the preparation of monolayer or few-layer graphene, especially for the macro preparation of structurally stable graphene, in order to systematically investigate in depth, the physicochemical properties and applications of this novel material. Yang et al.[66] reported that graphene flakes mixed with a certain content of KOH were first stirred and sonicated at room temperature, followed by drying the solvent in an oven at 120°C (Figure 2.4c). Finally, activated graphene flakes with a high specific surface area (2,662 $m^2 g^{-1}$) and graded pore structure were prepared by heat treatment of the complexes at 850°C in a nitrogen atmosphere. The activated graphene has a high specific capacity of 238 F g^{-1} at a current density of 1 A g^{-1} and retains 99% of its initial capacity after 20,000 cycles at 5 A g^{-1}. Mohamed et al.[67] directly deposited functionalized graphene frameworks using electrostatic interactions between graphene oxide (GO) and a cationic surfactant. The surfactant improved the adsorption of GO sheets on the electrode surface, allowing the integration of individual graphene sheets into 3D structures with large electrochemically active surface areas. The functionalized graphene framework has a high specific capacitance (320 F g^{-1}) and area capacitance (≈400 mF cm^{-2} for a two-electrode cell with a single-sided coating) without the use of adhesives or conductive additives.

4.3 AC

AC is the earliest and most widely used electrode material in commercial applications of SCs due to its large specific surface area, good electrical properties, and moderate cost. AC is mainly obtained by heat treatment of carbon-rich organic

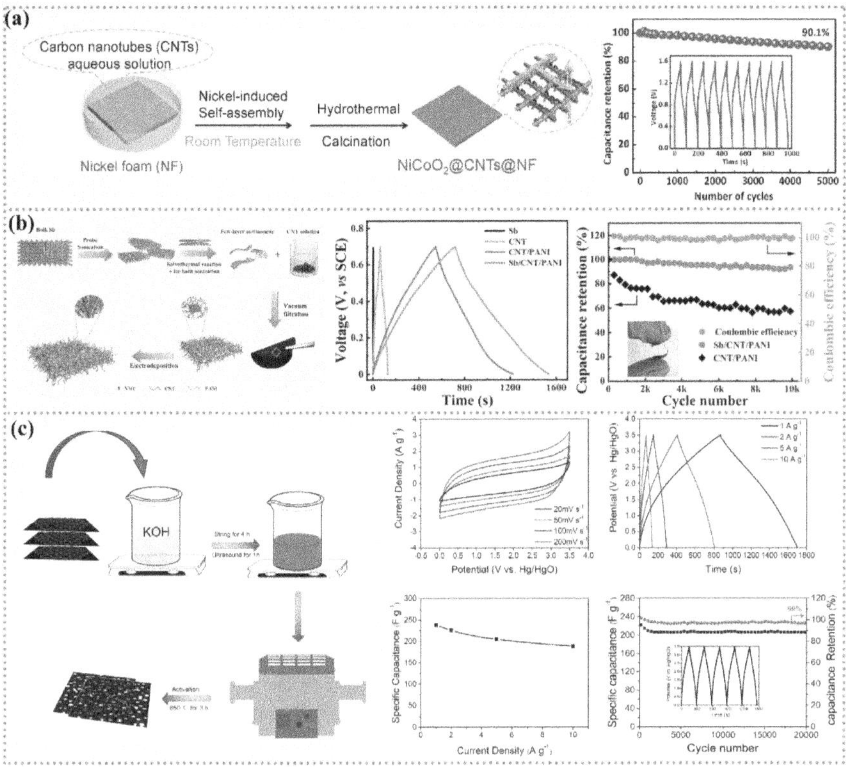

FIGURE 2.4 (a) Synthesis of $NiCoO_2$@CNTs@NF hybrids.[62] (b) Scheme presentation and electrochemical properties of Sb/CNT/PANI electrode.[63] (c) Scheme presentation and electrochemical properties of activated graphene flakes.[66]

precursors in an inert atmosphere and by physical or chemical activation to increase their surface area. Physical activation is mainly a process of locally controlled pore formation by CO_2 or water vapors on the precursor. Chemical activation routes using cheap organic/inorganic salts (KOH, $ZnCl_2$, phytic acid, etc.) as activating agents are also widely used.[68, 69] With the development of research, the surface area of AC has now been increased to 3,000 $m^2 g^{-1}$.[70, 71] There is no direct linear relationship between the specific surface area and the specific capacity of AC.[72] For example, the theoretical capacitance of an EDLC is 15–25 mF cm^{-2} for AC with a very high specific surface area (3,000 $m^2 g^{-1}$), while the actual capacitance of the AC is less than 10 mF cm^{-2}.[73] This is because the performance of EDLC is determined by many factors. Specifically, on the one hand, the performance of the EDLCs depends on the effective specific surface area. On the other hand, the pore size distribution, pore structure and shape, and electrical conductivity also have a great influence on the performance of the EDLCs.[74, 75] Generally speaking, the specific capacitance of AC is higher in aqueous electrolytes (about 100–300 F g^{-1}) than in organic electrolytes (<150 F g^{-1}), mainly because the effective sizes of electrolyte ions in organic solutions are larger than the aqueous electrolyte ions, which poorly penetrate the micropores of AC.

4.3.1 Biomass-derived carbons

In recent years, numerous researchers have prepared biomass-derived carbon materials using different treatment methods with different biomass resources as carbon sources for energy storage because of their widespread availability, renewable nature, and low cost. What's more, their inherently uniform and precise biological structure can be used as a template for fabricating electrode materials with well-defined geometries. Meanwhile, the basic elements of biomass are carbon, nitrogen, sulfur, and phosphorus, which can achieve in situ heteroatom doping. Peng et al.[11] reported a simultaneous activation, carbonization, and nitrogen-doped method to prepare nitrogen-doped porous carbon nanosheets (N-PMNC) by using an easily gained and green plant-waste (pomelo mesocarps) as carbon precursor (Figure 2.5a). The N-PMNC shows an interconnected sheet morphology, high specific surface area ($974.6\,m^2 g^{-1}$), and high nitrogen doping content (9.12 wt%). The N-PMNC-based supercapacitors exhibit excellent electrochemical performance with a high specific capacitance of 245 F g^{-1} at 0.5 A g^{-1} and excellent rate capability (72% capacitance retention even at 20 A g^{-1}). Xu et al.[76] developed a unique three-dimensional pepper leaf-derived nitrogen-doped porous carbon skeleton (ZLPC) with a hollow nanostructure. The method allows the combination of carbonization, activation, and heteroatom doping for improved synthetic efficiency compared to conventional template-carbonization or two-step carbonization strategies (Figure 2.5b,c). The symmetric supercapacitors based on ZLPC electrode materials have a high specific energy of 18.68 Wh kg^{-1} at a specific power of 225 W kg^{-1} operated in the voltage of 1.8 V in aqueous electrolyte and excellent cycle stability (92% after 20,000 charge cycle). So far, a series of porous carbon materials have been prepared by various methods using biomass as raw materials, which is an important field in the manufacturing of electrode materials for SCs. Therefore, researchers should explore more convenient, environmentally friendly, and scalable synthesis strategies using low-cost biomass feedstocks, and use mild chemicals for activation during preparation to obtain suitable supercapacitor electrode materials.

4.3.2 Template carbons

Template carbon is a kind of carbon material with special morphology and size which is synthesized by using the template as the main configuration. Templating methods enable the design and control of carbon precursor structures at the nanoscale and preserve this microstructure through a special carbonization process. As a result, the template method is the most effective way to precisely regulate the pore structure of carbon materials. Depending on the type of template used, template methods can be divided into hard template methods and soft template methods.

The process of preparing mesoporous carbon materials by the hard template method mainly involves the infusion of carbon precursors, carbonization, and removal of the template. The widely used hard templates include microporous molecular sieves, layered inorganic materials, mesoporous silicon oxide, porous alumina films, and colloidal crystalline materials.[77] Carbon materials with specific morphologies such as carbon nanosheets and spherical carbon can be obtained using hard templates (e.g., NaCl, SiO_2 and metal oxides [MgO, etc.]).[78] For example, Chen et al.[79] used mesoporous silicon (SBA-15) as a hard template to synthesize ordered parallel

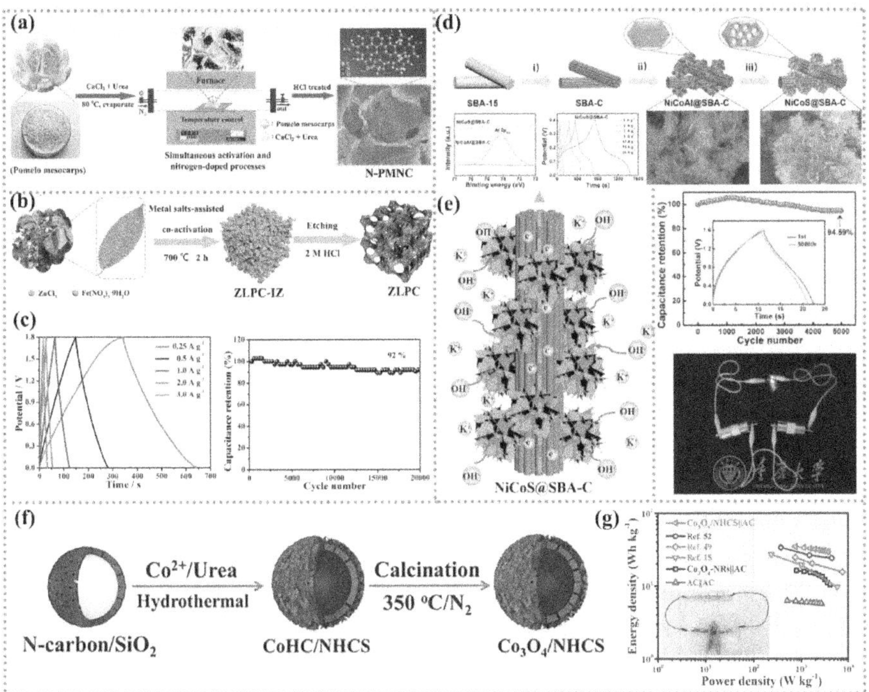

FIGURE 2.5 (a) Synthesis of N-PMNC electrode.[11] (b–c) Synthesis and electrochemical properties of ZLPC electrode.[76] (d–e) Synthesis and electrochemical properties of NiCoS@SBA-C electrode.[79] (f) Synthesis of Co_3O_4/NHCS electrode.[80]

mesoporous carbon (SBA-C), and further induced the growth of porous nickel cobalt sulfide (NiCoS) nanoflowers to form NiCoS@SBA-C composites (Figure 2.5d,e). The NiCoS@SBA-C electrode exhibited an excellent capacitance of 1,757 F g^{-1} at a current density of 1 A g^{-1} and good cycling stability. Liu et al. synthesized Co_3O_4/nitrogen-doped carbon hollow spheres (Co_3O_4/NHCSs) by hydrothermal method and subsequent calcination treatment using SiO_2 spheres as sacrificial templates (Figure 2.5f).[80] The prepared Co_3O_4/NHCS composites as electrode active materials exhibited higher specific capacitance of 581 F g^{-1} at a current density of 1 A g^{-1} than Co_3O_4 itself and achieved a capacity retention of 91.6% at 20 A g^{-1}. In addition, various synthetic strategies to fabricate highly wrinkled nitrogen-doped graphene or carbon nanosheets have been reported, and the wrinkled sheet morphology increases the effective surface area to provide excellent capacitive performance. Peng et al.[10] developed a facile one-step activation method combined with nitrogen doping using the combination of $Ca(OH)_2$ and NH_4Cl as hard templates and a nitrogen source to prepare novel highly crumpled nitrogen-doped graphene-like carbon nanosheets (N-CNSs). The N-CNSs have a rich wrinkled structure and an ultra-high pore volume (3.19 $cm^3 g^{-1}$) and are used as high-performance electrode materials for SCs.

The preparation of porous carbon by soft template method is usually based on suitable soluble zwitterions (such as surfactants) as the precursor, and the ordered polymer/structure-oriented composite is obtained by self-assembly and structure-oriented, and then the carbon material with an ordered and complete skeleton structure is obtained by removing the soft template and high-temperature carbonization steps. In the process of preparing carbon materials by the soft template method, the polymer skeleton shrinks severely during high-temperature treatment, resulting in a small specific surface area and pore size of the obtained carbon materials. Therefore, a combination of soft and hard templates (organic-inorganic-organic three-phase self-assembly) is used to introduce a rigid hard template into the polymer system to reduce the shrinkage of the skeleton and increase the pore size and specific surface area of the carbon materials.[81]

4.4 Other Carbon Structures

The design of carbon materials used in the field of SCs should have a high ion-accessible specific surface area and an appropriate pore structure.[82] Metal-organic framework (MOF)-derived carbon materials and carbide-derived carbon (CDC) are considered to be one of the promising materials in the future.

MOF-derived carbon is a porous carbon material obtained by high-temperature carbonization and removal of metal ions using MOF materials as precursors.[83] Liang et al.[84] proposed a facile strategy for simultaneous activation and catalytic carbonization of rare-earth metal lanthanide organic framework (La-MOF) precursors to prepare a three-dimensional (3D) mesoporous interconnected carbon nanosheet (LMCN) with large surface area and high degree of graphitization (Figure 2.6a). The symmetric supercapacitor assembled based on LMCN electrodes exhibited a high energy density of 20.7 Wh kg^{-1} at a power density of 228 W kg^{-1} with a wide operating voltage of 1.8 V, as well as a capacity retention of 90.7% after 10,000 cycles. Sun et al.[85] created two new PANI encapsulating bimetallic organic frameworks (ZnCe-MOF and CuCe-MOF). As electrode materials for SCs, the specific capacitance of CuCe-MOF@PANI-1 and ZnCe-MOF@PANI-1 electrodes can reach 724.4 F g^{-1} and 713.2 F g^{-1} at 1 A g^{-1} (Figure 2.6b). Although electrode materials composed of MOF or MOF-based composites show good performance, their specific structural changes and precise effects during electrochemical processes remain ambiguous. Understanding their mechanisms is important for the construction of inexpensive and easily fabricated high-performance electrode materials.

CDC is a class of carbon materials produced by selectively etching metals from metal carbides using chlorine at elevated temperatures in a process. CDC has all the performance characteristics of an ideal bilayer electrode material, including large specific surface area, concentrated and precisely tunable pore size distribution, high electron conductivity, and high density, and has been called "a new breakthrough in supercapacitors."[86] The CDC powders were usually synthesized from titanium carbide using selective etching of TiC by chlorine at temperature 800–1,000°C. The most important feature of the CDC preparation process is that the etching process does not cause any change in the shape and size of the carbide itself and allows atomic-level control of the carbon material. Rose et al.[87] prepared ordered mesoporous OM-CDC materials with a high ratio of micropores by nanocasting of ordered mesoporous silica templates. It was found that the specific capacitance of the OM-CDC materials exceeds 200 F g^{-1} in the aqueous electrolyte and 185 F g^{-1} in the ionic liquid. The ordered mesoporous channels in the produced OM-CDC materials

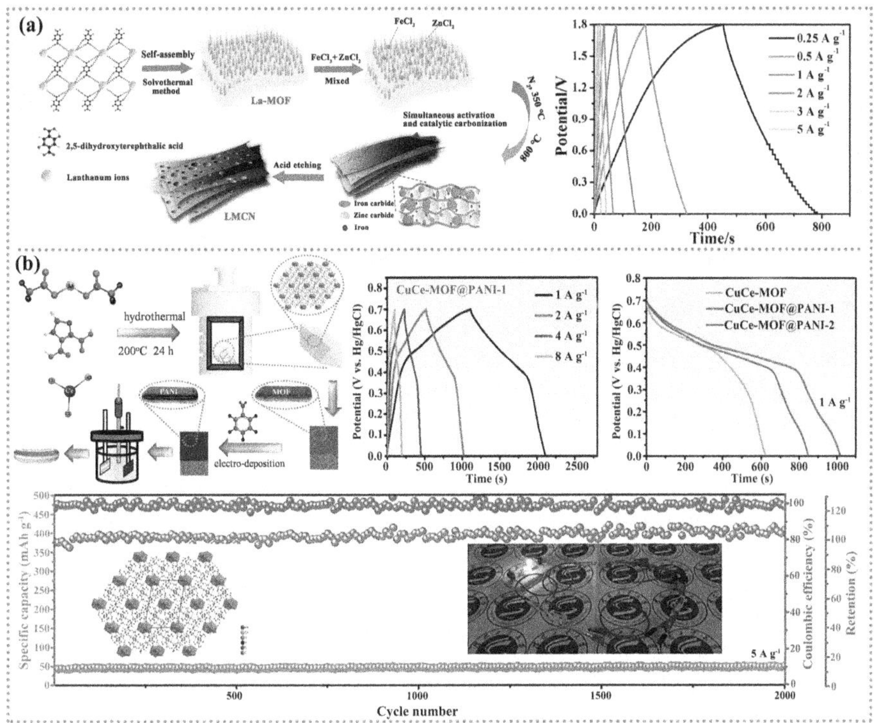

FIGURE 2.6 (a) Synthesis and electrochemical properties of LMCN electrode.[84] (b) Synthesis and electrochemical properties of ZnCe-MOF and CuCe-MOF.[85]

serve as ion-highways and allow for very fast ionic transport into the bulk of the OM-CDC particles. Despite numerous substantial advances in the field of CDC in SCs, their comprehensive performance is still limited.

4.5 Carbon-Based Composites

The fabrication of carbon-based composites by using carbon as a support not only increases the effective utilization of the electroactive materials but also improves the electrical conductivity and mechanical strength of the composite materials. Several studies have reported improved electrochemical properties of the carbon-based composites using various types of carbons, such as graphene, AC, CNT, and templated carbon, as a support. SCs based on carbon-based composite or hybrid electrodes exhibit excellent performance in terms of energy density and stability. For example, V_2O_5 and 3D graphene composite synthesized by hydrothermal method showed a high specific capacitance of 612 F g^{-1} at a current density of 1 A g^{-1} with good cyclic efficiency of 89.6% after 10,000 cycles, which is higher than that of pure V_2O_5 and pure graphene.[88] The MnO_2/CNTs composite was prepared by Li et al. through a modified one-pot reaction process, in which CNTs were coated by cross-linked MnO_2 flakes uniformly.[89] The MnO_2/CNTs composite exhibits high specific capacitance of 201 F g^{-1} and rate capability (the specific capacitance at 20 A g^{-1} is 70% of that at 1 A g^{-1}). Liu et al. reported a composite of polyaniline nanofiber/large mesoporous carbon (PANI-F/LMC) composite prepared by an in situ chemical oxidative polymerization of aniline monomer with nano-$CaCO_3$ templated mesoporous carbon as a matrix.[90] The PANI-F/LMC composite has a high capacitance of 473 F g^{-1} at a current load of 0.1 A g^{-1} with good rate performance and cycling stability.

5 FACTORS INFLUENCING THE PERFORMANCE OF CARBON MATERIALS

5.1 Morphology

Carbon materials with various structures have been developed, such as 0D nanospheres or hollow structures, 1D nanofibers or nanotubes, 2D nanosheets, and 3D porous frameworks. Zhou et al.[91] demonstrated an original strategy to design a high-performance 0D nominally non-porous carbon microspheres (CMs) co-doped with fluorine and nitrogen, achieved by a low-temperature solvent thermal path (Figure 2.7a,b). The co-doping of nitrogen and fluorine greatly improves the specific capacitance of the CM electrodes. The CM-NF electrode has a significant specific capacitance of 270 F g^{-1} at a current density of 0.2 A g^{-1}, corresponding to an ultra-high capacity of 521 F cm^{-3} in 1 M H_2SO_4. In addition, there is no loss of specific capacitance after 10,000 cycles in H_2SO_4 and 20,000 cycles in KOH during long-term charging and discharging at a current density of 5 A g^{-1}. Peng et al.[92] prepared the nitrogen-doped holey carbon nanosheets (N-HCNs) via one-step activation and self-generated template-assisted carbonization route using polyimide (PI) nanoflowers as carbon precursor and ethylenediaminetetraacetic acid disodium zinc salt hydrate (EDTA-Na_2Zn) as activating agent and self-generated template. The N-HCNs exhibit satisfactory specific capacitance of 205 F g^{-1} at 0.5 A g^{-1}. More importantly, a symmetric aqueous supercapacitor assembled based on N-HCNs electrodes carries a wide operating voltage of 2.0 V, and possesses a high energy density of 17.92 Wh kg^{-1} at a power density of 500 W kg^{-1}. Moreover, a facile integrated oxidation polymerization and catalytic carbonization method was introduced to prepare three-dimensional porous nitrogen-doped carbon networks (3D N-CNWs) with high nitrogen content (about 8.4 wt %) directly from poly(p-phenylenediamine).[93] In the synthesis process, the $FeCl_3$ serves not only as an oxidant for oxidative polymerization of p-phenylenediamine monomers but also as the carbonization catalyst to promote porous carbon network formation (Figure 2.7c,d). The maximum specific capacitance of 3D N-CNWs can be reached up to 304 F g^{-1} at 0.5 A g^{-1}, and retains the high values of 226 F g^{-1} even at a high current density of 20 A g^{-1}. It can be seen that carbon materials with different morphologies can obtain excellent electrochemical properties. Compared with the specific surface area and pore structure of carbon materials, the impact of morphology on their properties is not significant.

5.2 Pore Structure

The pore structure (porosity and size distribution) of carbon materials is an important factor affecting the energy storage performance of functional carbon materials

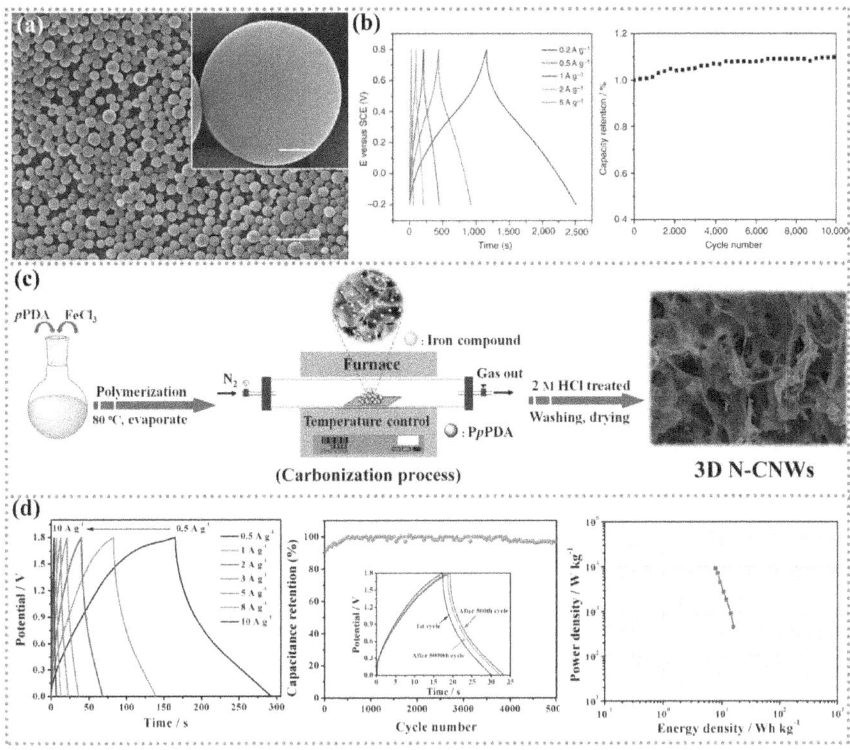

FIGURE 2.7 (a,b) SEM images and electrochemical properties of the CM-NFs.[91] (c) Scheme presentation of 3D N-CNWs. (d) Electrochemical properties of 3D N-CNWs.[93]

in SCs. According to the pore size, the pores are classified as microporous (<2 nm), mesoporous (2–50 nm), and macroporous (>50 nm). Different electrolytes require different pore sizes, and the capacitive behavior is theoretically optimal only when the pore size matches the electrolyte ions. For porous carbon materials, micropores are usually considered to play a dominant role in capacitive behavior, mesoporous structures mainly provide interconnected ion transfer channels for the electrolyte, and macropores serve as storage areas for the electrolyte to buffer rapid changes in ion concentration.[94]

Porous carbon materials with different pore sizes can be usually prepared by adjusting experimental parameters such as activator dosage, carbonization temperature, activation time, and activator type. Longer activation time or higher temperature leads to larger mean pore size. The AC for EDLCs can be reached to 100–120 F g^{-1} or 150–300 F g^{-1} in organic electrolytes and in aqueous electrolytes, respectively. The investigation on pore-dependent capacitance properties has revealed that the micropores adjacent to the open mouths of pores are effective in charge storage, and larger pores that are interconnected are important for smooth electrolyte transportation. A study carried out with CDCs in an ionic liquid electrolyte ([EMI^+,$TFSI^-$]), in which both ions have a maximum size of about 0.7 nm, showed the maximum capacitance for samples with the 0.7 nm pore size.[95] In most cases, macroporous carbon materials inevitably contain micro/mesopores. Spherical polymers of submicron size can be used as standard soft templates for macroporosity. By adjusting the diameter of the spherical particles, the resulting carbon macroporous size is adjusted accordingly. For example, Yamada et al.[96] have synthesized ordered porous carbons containing meso/macro/micropores with large specific surface area. They found that the ordered porous carbon having well-controlled hierarchical porous structures would be favorable for designing high-performance EDLCs electrodes.

5.3 Heteroatom Doping

Tuning the physicochemical and electronic properties of porous carbon materials by heteroatom doping is a promising approach to meet the requirements of high-performance SCs. The introduction of heteroatoms (N, S, O, P, B, and so on) into the carbon skeleton has a great impact on electrochemical behavior, mainly because heteroatom doping can modify the carbon-based structure and improve the electronic conductivity and wettability of carbon-based electrodes, thus becoming another way to improve its electrochemical performance.[97, 98] As different heteroatoms have different sizes and electro-negativities, their effects on tuning the properties of carbon materials may vary. Doped heteroatoms, such as O, N, S, F, and B would change the surface chemical heterogeneity of carbon materials, thereby inducing additional pseudocapacitance through reversible redox reactions.[99] Oxygen doping in carbon matrix can effectively change the surface wettability and provide more active sites for charge storage. Oxygen doping by means of oxygen functional groups may be obtained through the pyrolysis of oxygen-containing precursors such as most biomass.[100] For nitrogen doping, it has been proven that the conjugation of the lone-pair electrons of N and the graphitic p-bonds of carbon materials can distort the carbon structure to create defects and available active sites. Specially, pyridinic nitrogen and pyrrolic nitrogen are considered representing the pseudocapacitance effect, while quaternary nitrogen and pyridine-N-oxide showed enhancing effects on the capacitance due to their positive charge and thus an improved electron transfer.[101] Other heteroatoms, such as S, P, B, and F have also been extensively studied to regulate the surface properties and improve the electrochemical properties of carbon materials.[98]

In addition, dual and multiple heteroatoms doping in comparison with sole-heteroatom doping has further optimized the nano/microstructure of carbon electrodes. Dual or more heteroatoms co-doped carbon materials are more effective in most situations since they can deliver excellent capacitance performances than single atom doped-materials due to synergy among different heteroatoms.[98] Despite the great progress in heteroatom doping research of porous carbon-based nanotubes, their comprehensive properties are still limited, especially in terms of the high energy density required for practical applications. The ability to develop a simple, viable, and efficient route to heteroatom-doped carbon materials that simultaneously achieve high specific power and energy would render carbon-based SCs potentially highly competitive.

5.4 Electrical conductivity

Retaining fast electron and ion transport of electrode materials at high mass loading holds significant importance to SCs. The key to obtain excellent power performance is to increase the electron and ion transportation rates in electrode materials. An ideal electrode material for SCs needs the following characteristics: (1) high electron conductivity, (2) short ion diffusion distance, and (3) high ion-accessible surface areas. To address such issues, tremendous efforts have been focused on improving the specific capacitance and energy density without sacrificing high power density.

It has been shown that the gradual improvement of electrochemical performance of carbon-based SCs at high current density after heat treatment up to 900°C is mainly the result of the simultaneous increase in electrical conductivity.[102] Research showed that a transition between an insulating and conducting state typically occurs when carbon materials are heated to around 600–700°C.[103] Removal of strongly electron-withdrawing heteroatoms from the carbon surface facilitates electron delocalization and significantly enhances conductivity. Sánchez-González et al. also demonstrated that heat treatment induces significant changes in the electrical conductivity of carbon materials, characterized by an important jump between 550°C (4.6×10^{-6} S m^{-1}) and 700°C (7.3 S m^{-1}).[102] Li et al. synthesized predominant few-layer graphene (FLG) sheets with high electrical conductivity (~3.2×10^{4} S m^{-1}) by a multi-step intercalation and reduction method, which is comparable to that of pristine graphite.[104] The FLG-based electrode can reach a high specific capacitance of 180 F g^{-1} in 1 M Na_2SO_4 aqueous electrolyte.

6 SUMMARY AND FUTURE PERSPECTIVES

As a new type of electrochemical energy storage device, supercapacitors (SCs) stand out among many energy storage devices due to their advantages such as low cost, high power density, fast charge and discharge, and excellent long cycle performance. Unfortunately, although carbon-based SCs exhibit excellent power density, they still face greater challenges in energy density compared with energy storage devices such as batteries. This chapter presents a comprehensive review of the research progress related to carbon-based SCs in terms of theoretical background, electrolyte, modification, and application of carbon-based electrode materials. For electrolytes, non-aqueous electrolytes (including organic and ionic liquid electrolytes) can obtain relatively high operating voltages, resulting in higher energy densities than those obtained; in quasi-solid electrolytes, gel electrolytes have higher ionic conductivity due to the presence of a liquid phase, which has attracted extensive attention from researchers. For carbon-based electrode materials, common carbon materials such as AC, biomass-derived porous carbon, graphene, etc. are widely used as the substrate materials for SCs. Researchers usually expand the extra pseudocapacitance of the carbon materials by doping with heteroatoms, modifying the pore size, and controlling the morphology of the materials, so as to increase the energy density of carbon-based SCs.

Overall, the unique energy storage mechanism and excellent electrochemical performance of SCs make them a promising energy storage device. Based on the inherently lower energy density of SCs, the future development of carbon-based SCs can considered in the following: (1) Develop new organic solvents with high ionic conductivity, especially combining the respective advantages of ILs and organic solvents to optimize different composite solvents to obtain excellent electrochemical performance. (2) Develop and select new electrolyte salts, and then combine them with optimized organic solvents to form organic electrolytes with excellent performance. (3) For ionic liquid electrolytes, ILs and gel electrolytes can be combined to prepare new gel electrolytes to solve the problem of high viscosity of ILs and improve the problems of poor conductivity and temperature tolerance of gel electrolytes. (4) For carbon-based electrodes, researchers should focus on the optimization of graphitized structure and porous structure of carbon materials to balance the relationship between electronic conductivity and ionic conductivity. In addition, researchers can also try to construct some novel carbon nanostructures, such as polyatomic doping and hollow core-shell structures, to increase the volume energy density of carbon-based materials. (5) Combining the respective advantages of metal-ion batteries and SCs to construct metal-ion capacitors is a strategy to simultaneously achieve high energy density and high power density in an integrated device. In summary, carbon-based SCs have excellent power density, fast charging, and other good properties, but they still face many challenges in the commercialization process.

REFERENCES

1. J. Liu, Z. Huang, M. Fan, J. Yang, J. Xiao and Y. Wang, Future energy infrastructure, energy platform and energy storage, *Nano Energy*, 2022, **104**, 10.
2. L. Trahey, F. Brushett, N. Balsara, G. Ceder, L. Cheng, Y. Chiang, N. Hahn, B. Ingram, S. Minteer, J. Moore, K. Mueller, L. Nazar, K. Persson, D. Siegel, K. Xu, K. Zavadil, V. Srinivasan and G. Crabtree, Energy storage emerging: A perspective from the Joint Center for Energy Storage Research, *Proc. Natl Acad. Sci. U. S. A.*, 2020, **117**, 12550–12557.
3. X. Chen, R. Paul and L. Dai, Carbon-based supercapacitors for efficient energy storage, *Natl Sci. Rev.*, 2017, **4**, 453–489.
4. R. Reece, C. Lekakou and P. Smith, A high-performance structural supercapacitor, *ACS Appl. Mater. Interfaces*, 2020, **12**, 25683–25692.
5. Y. Wang, X. Wu, Y. Han and T. Li, Flexible supercapacitor: Overview and outlooks, *J. Energy Storage*, 2021, **42**, 14.
6. G. Pour, L. Aval and M. Mirzaee, CNTs supercapacitor based on the PVDF/PVA gel electrolytes, *Recent Pat. Nanotechnol.*, 2020, **14**, 163–170.
7. M. Celiktas and F. Alptekin, Conversion of model biomass to carbon-based material with high conductivity by using carbonization, *Energy*, 2019, **188**, 11.
8. H. Wang and J. Wen, Biomass porous carbon-based composite for high performance supercapacitor, *Mater. Res. Express*, 2020, **7**, 11.
9. T. Wang, X. Zang, X. Wang, X. Gu, Q. Shao and N. Cao, Recent advances in fluorine-doped/fluorinated carbon-based materials for supercapacitors, *Energy Storage Mater.*, 2020, **30**, 367–384.
10. H. Peng, G. Ma, K. Sun, J. Mu and Z. Lei, One-step preparation of ultrathin nitrogen-doped carbon nanosheets with ultrahigh pore volume for high-performance supercapacitors, *J. Mater. Chem. A*, 2014, **2**, 17297–17301.
11. H. Peng, G. Ma, K. Sun, Z. Zhang, Q. Yang and Z. Lei, Nitrogen-doped interconnected carbon nanosheets from pomelo mesocarps for high performance supercapacitors, *Electrochim. Acta*, 2016, **190**, 862–871.

12. G. Ma, Y. Wu, K. Sun, H. Peng, H. Wang and Z. Lei, High performance nitrogen-doped carbon for supercapacitor obtained by carbonizing eco-friendly and cheap polyaspartic acid, *Mater. Lett.*, 2014, **132**, 41–44.
13. S. Samantaray, D. Mohanty, I.M. Hung, M. Moniruzzaman and S.K. Satpathy, Unleashing recent electrolyte materials for next-generation supercapacitor applications: A comprehensive review, *J. Energy Storage*, 2023, **72**, 27.
14. J. Shaikh, N. Shaikh, R. Kharade, S. Beknalkar, J. Patil, M. Suryawanshi, P. Kanjanaboos, C. Hong, J. Kim and P. Patil, Symmetric supercapacitor: Sulphurized graphene and ionic liquid, *J. Colloid Interface Sci.*, 2018, **527**, 40–48.
15. L. Yu and G. Chen, Ionic liquid-based electrolytes for supercapacitor and supercapattery, *Front. Chem.*, 2019, **7**, 15.
16. X. He, D. Wu, Y. Shang, H. Shen, S. Xi, X. Wang, W. Li and Q. Wang, Regenerated hydrogel electrolyte towards an all-gel supercapacitor, *Sci. China-Mater.*, 2022, **65**, 115–123.
17. T. Wang, H. Chen, F. Yu, X. Zhao and H. Wang, Boosting the cycling stability of transition metal compounds-based supercapacitors, *Energy Storage Mater.*, 2019, **16**, 545–573.
18. N. Choi, Z. Chen, S. Freunberger, X. Ji, Y. Sun, K. Amine, G. Yushin, L. Nazar, J. Cho and P. Bruce, Challenges facing lithium batteries and electrical double-layer capacitors, *Angew Chem. Int. Ed. Engl.*, 2012, **51**, 9994–10024.
19. T. Mathis, N. Kurra, X. Wang, D. Pinto, P. Simon and Y. Gogotsi, Energy storage data reporting in perspective-guidelines for interpreting the performance of electrochemical energy storage systems, *Adv. Energy Mater.*, 2019, **9**, 1902007.
20. J. Huang, K. Yuan and Y. Chen, Wide voltage aqueous asymmetric supercapacitors: Advances, strategies, and challenges, *Adv. Funct. Mater.*, 2021, **32**, 2108107.
21. L. Zhang and X. Zhao, Carbon-based materials as supercapacitor electrodes, *Chem. Soc. Rev.*, 2009, **38**, 2520–2531.
22. F. Beguin, V. Presser, A. Balducci and E. Frackowiak, Carbons and electrolytes for advanced supercapacitors, *Adv. Mater.*, 2014, **26**, 2219–2251.
23. X. Chen, R. Paul and L. Dai, Carbon-based supercapacitors for efficient energy storage, *Natl Sci. Rev.*, 2017, **4**, 453–489.
24. F. Permatasari, M. Irham, S. Bisri and F. Iskandar, Carbon-based quantum dots for supercapacitors: Recent advances and future challenges, *Nanomaterials*, 2021, **11**, 91.
25. Q. Dou and H. Park, Perspective on high-energy carbon-based supercapacitors, *Energy Environ. Mater.*, 2020, **3**, 286–305.
26. J. Wang, X. Zhang, Z. Li, Y. Ma and L. Ma, Recent progress of biomass-derived carbon materials for supercapacitors, *J. Power Sources*, 2020, **451**, 227794.
27. C. Liu, Y. Liu, T. Yi and C. Hu, Carbon materials for high-voltage supercapacitors, *Carbon*, 2019, **145**, 529–548.
28. C. Zhu, Y. Ye, X. Guo and F. Cheng, Design and synthesis of carbon-based nanomaterials for electrochemical energy storage, *New Carbon Mater.*, 2022, **37**, 59–92.
29. L. Miao, Z. Song, D. Zhu, L. Li, L. Gan and M. Liu, Recent advances in carbon-based supercapacitors, *Mater. Adv.*, 2020, **1**, 945–966.
30. S. Saini, P. Chand and A. Joshi, Biomass derived carbon for supercapacitor applications, *J. Energy Storage*, 2021, **39**, 102646.
31. M. Islam, S. Afroj, K. Novoselov and N. Karim, Smart electronic textile-based wearable supercapacitors, *Adv. Sci.*, 2022, **9**, e2203856.
32. B. Pal, S. Yang, S. Ramesh, V. Thangadurai and R. Jose, Electrolyte selection for supercapacitive devices: A critical review, *Nanosc. Adv.*, 2019, **1**, 3807–3835.
33. H. Wu, X. Wang, L. Jiang, C. Wu, Q. Zhao, X. Liu, B.A. Hu and L. Yi, The effects of electrolyte on the supercapacitive performance of activated calcium carbide-derived carbon, *J. Power Sources*, 2013, **226**, 202–209.

34. J. Zhu, Y. Xu, J. Wang, J. Lin, X. Sun and S. Mao, The effect of various electrolyte cations on electrochemical performance of polypyrrole/RGO based supercapacitors, *Phys. Chem. Chem. Phys.*, 2015, **17**, 28666–28673.
35. J. Holoubek, H. Jiang, D. Leonard, Y. Qi, G.C. Bustamante and X. Ji, Amorphous titanic acid electrode: Its electrochemical storage of ammonium in a new water-in-salt electrolyte, *Chem. Commun.*, 2018, **54**, 9805–9808.
36. J. Guo, Y. Ma, K. Zhao, Y. Wang, B. Yang, J. Cui and X. Yan, High-performance and ultra-stable aqueous supercapacitors based on a green and low-cost water-in-salt electrolyte, *ChemElectroChem*, 2019, **6**, 5433–5438.
37. J. Zheng, G. Tan, P. Shan, T. Liu, J. Hu, Y. Feng, L. Yang, M. Zhang, Z. Chen, Y. Lin, J. Lu, J. Neuefeind, Y. Ren, K. Amine, L. Wang, K. Xu and F. Pan, Understanding thermodynamic and kinetic contributions in expanding the stability window of aqueous electrolytes, *Chem*, 2018, **4**, 2872–2882.
38. C. Zhang, J. Holoubek, X. Wu, A. Daniyar, L. Zhu, C. Chen, D. Leonard, I. Rodrı´guez-Pe´rez, J. Jiang, C. Fang and X. Ji, A $ZnCl_2$ water-in-salt electrolyte for a reversible Zn metal anode, *Chem. Commun.*, 2018, **54**, 14097–14099.
39. M. González, O. Borodin, M. Kofu, K. Shibata, T. Yamada, O. Yamamuro, K. Xu, D. Price and M. Saboungi, Nanoscale relaxation in "water-in-salt" and "water-in-bisalt" electrolytes, *J. Phys. Chem. Lett.*, 2020, **11**, 7279–7284.
40. L. Jiang, L. Liu, J. Yue, Q. Zhang, A. Zhou, O. Borodin, L. Suo, H. Li, L. Chen, K. Xu and Y. Hu, High-voltage aqueous Na-ion battery enabled by inert-cation-assisted water-in-salt electrolyte, *Adv. Mater.*, 2020, **32**, e1904427.
41. L. Chen, J. Zhang, Q. Li, J. Vatamanu, X. Ji, T. Pollard, C. Cui, S. Hou, J. Chen, C. Yang, L. Ma, M. Ding, M. Garaga, S. Greenbaum, H. Lee, O. Borodin, K. Xu and C. Wang, A 63 m superconcentrated aqueous electrolyte for high-energy Li-ion batteries, *ACS Energy Lett.*, 2020, **5**, 968–974.
42. M. Harilal, S. Krishnan, B. Vijayan, M. Reddy, S. Adams, A.R.M. Barron and R. Jose, Continuous nanobelts of nickel oxide–cobalt oxide hybrid with improved capacitive charge storage properties, *Mater. Design*, 2017, **122**, 376–384.
43. M. Harilal, S. Krishnan, A. Yar, I. Misnon, M. Reddy, M. Yusoff, J. Dennis and R. Jose, Pseudocapacitive charge storage in single-step-synthesized $CoO–MnO_2–MnCo_2O_4$ hybrid nanowires in aqueous alkaline electrolytes, *J. Phys. Chem. C*, 2017, **121**, 21171–21183.
44. V. Thangadurai, D. Pinzaru, S. Narayanan and A. Baral, Fast solid-state Li ion conducting garnet-type structure metal oxides for energy storage, *J. Phys. Chem. Lett.*, 2015, **6**, 292–299.
45. Y. Cai, B. Zhao, J. Wang and Z. Shao, Non-aqueous hybrid supercapacitors fabricated with mesoporous TiO_2 microspheres and activated carbon electrodes with superior performance, *J. Power Sources*, 2014, **253**, 80–89.
46. M. Cho, M. Kim, H. Kim, K. Kim, J. Yoon and K. Roh, Electrochemical performance of hybrid supercapacitor fabricated using multi-structured activated carbon, *Electrochem. Commun.*, 2014, **47**, 5–8.
47. G.G. Amatucci, F. Badway, A.D. Pasquier and T. Zheng, An asymmetric hybrid non-aqueous energy storage cell, *J. Electrochem. Soc.*, 2001, **148**, A930–A939.
48. S. Pan, M. Yao, J. Zhang, B. Li, C. Xing, X. Song, P. Su, and H. Zhang, Recognition of ionic liquids as high-voltage electrolytes for supercapacitors, *Front. Chem.*, 2020, **8**, 261.
49. H. Peng, Y. Xu, Y. Jiang, X. Wang, R. Zhao, F. Wang, L. Li, and G. Ma, Bread-inspired foaming strategy to fabricate a wine lees-based porous carbon framework for high specific energy supercapacitors, *Sustain. Energy Fuels*, 2021, **5**, 4965–4972.
50. J. Duay, E. Gillette, R. Liu and S. Lee, Highly flexible pseudocapacitor based on freestanding heterogeneous MnO_2/conductive polymer nanowire arrays, *Phys. Chem. Chem. Phys.*, 2012, **14**, 3329–3337.

51. C. Huang, C. Wu, S. Hou, P. Kuo, C. Hsieh and H. Teng, Gel electrolyte derived from poly (ethylene glycol) blending poly (acrylonitrile) applicable to roll-to-roll assembly of electric double layer capacitors, *Adv. Funct. Mater.*, 2012, **22**, 4677–4685.
52. Y. Sudhakar, M. Selvakumar and D. Bhat, $LiClO_4$-doped plasticized chitosan and poly(ethylene glycol) blend as biodegradable polymer electrolyte for supercapacitors, *Ionics*, 2012, **19**, 277–285.
53. C. Ramasamy and J.P.d.v.-M. Anderson, An activated carbon supercapacitor analysis by using a gel electrolyte of sodium salt-polyethylene oxide in an organic mixture solvent, *J. Solid State Electr.*, 2014, **18**, 2217–2223.
54. H. Peng, X. Gao, K. Sun, X. Xie, G. Ma, X. Zhou, and Z. Lei, Physically cross-linked dual-network hydrogel electrolyte with high self-healing behavior and mechanical strength for wide-temperature tolerant flexible, *Chem. Eng. J.*, 2021, **422**, 130353.
55. P. Tamilarasan and S. Ramaprabhu, Stretchable supercapacitors based on highly stretchable ionic liquid incorporated polymer electrolyte, *Mater. Chem. Phys.*, 2014, **148**, 48–56.
56. Y. Lim, J. Yoon, J. Yun, D. Kim, S. Hong, S. Lee, G. Zi and J. Ha, Biaxially stretchable, integrated array of high performance microsupercapacitors, *ACS Nano*, 2014, **8**, 11639–11650.
57. C. Liew, S. Ramesh and A. Arof, Good prospect of ionic liquid based-poly (vinyl alcohol) polymer electrolytes for supercapacitors with excellent electrical, electrochemical and thermal properties, *Int. J. Hydrogen Energ.*, 2014, **39**, 2953–2963.
58. S. Ketabi and K. Lian, Effect of SiO_2 on conductivity and structural properties of PEO-$EMIHSO_4$ polymer electrolyte and enabled solid electrochemical capacitors, *Electrochim. Acta*, 2013, **103**, 174–178.
59. K. Koziol, J. Vilatela, A. Moisala, M. Motta, P. Cunniff, M. Sennett and A. Windle, High-performance carbon nanotube fiber, *Science*, 2007, **318**, 1892–1895.
60. S. Talapatra, S. Kar, S. Pal, R. Vajtai, L. Ci, P. Victor, M. Shaijumon, S. Kaur, O. Nalamasu and P. Ajayan, Direct growth of aligned carbon nanotubes on bulk metals, *Nat. Nanotechnol.*, 2006, **1**, 112–116.
61. D. Futaba, K. Hata, T. Yamada, T. Hiraoka, Y. Hayamizu, Y. Kakudate, O. Tanaike, H. Hatori, M. Yumura and S. Iijima, Shape-engineerable and highly densely packed single-walled carbon nanotubes and their application as super-capacitor electrodes, *Nat. Mater.*, 2006, **5**, 987–994.
62. C. Hu, L. Miao, Q. Yang, X. Yu, L. Song, Y. Zheng, C. Wang, L. Li, L. Zhu, X. Cao and H. Niu, Self-assembly of CNTs on Ni foam for enhanced performance of $NiCoO_2$@ CNT@ NF supercapacitor electrode, *Chem. Eng. J.*, 2021, **410**, 128317.
63. Y. Jiang, J. Ou, Z. Luo, Y. Chen, Z. Wu, H. Wu, X. Fu, S. Luo and Y. Huang, High capacitive antimonene/CNT/PANI free-standing electrodes for flexible supercapacitor engaged with self-healing function, *Small*, 2022, **18**, 2201377.
64. Z. Wu, G. Zhou, L. Yin, W. Ren, F. Li and H. Cheng, Graphene/metal oxide composite electrode materials for energy storage, *Nano Energy*, 2012, **1**, 107–131.
65. H. Zhang, M. Chhowalla and Z. Liu, 2D nanomaterials: Graphene and transition metal dichalcogenides, *Chem. Soc. Rev.*, 2018, **47**, 3015–3017.
66. H. Yang, R. Gao, H. Pu, Y. Yang, Y. Wu, W. Meng and D. Zhao, Hierarchical porous activated graphene nanosheets with an ultra-high potential as electrode material for symmetric supercapacitors, *Micropor. Mesopor. Mat.*, 2020, **306**, 110430.
67. N. Mohamed, M. El-Kady and R. Kaner, Macroporous graphene frameworks for sensing and supercapacitor applications, *Adv. Funct. Mater.*, 2022, **32**, 2203101.
68. G. Ma, H. Peng, J. Mu, H. Huang, X. Zhou and Z. Lei, In situ intercalative polymerization of pyrrole in graphene analogue of MoS_2 as advanced electrode material in supercapacitor, *J. Power Sources*, 2013, **229**, 72–78.
69. Z. Liu, H. Peng, X. Xie, X. Wang, Y. Pu, G. Ma and Z. Lei, One-step self-embedding of CoP nanoparticles in N, P-codoped hard carbon for high-performance lithium ion capacitors, *J. Power Sources*, 2022, **543**, 231831.

70. O. Barbieri, M. Hahn, A. Herzog and R. Kötz, Capacitance limits of high surface area activated carbons for double layer capacitors, *Carbon*, 2005, **43**, 1303–1310.
71. E. Raymundo-Piñero, F. Leroux and F. Béguin, A high-performance carbon for supercapacitors obtained by carbonization of a seaweed biopolymer, *Adv. Mater.*, 2006, **18**, 1877–1882.
72. L. Wang, H. Ji, S. Wang, L. Kong, X. Jiang and G. Yang, Preparation of Fe_3O_4 with high specific surface area and improved capacitance as a supercapacitor, *Nanoscale*, 2013, **5**, 3793–3799.
73. S. Rajasekaran and V. Raghavan, Facile synthesis of activated carbon derived from *Eucalyptus globulus* seed as efficient electrode material for supercapacitors, *Diam. Relat. Mater.*, 2020, **109**, 9.
74. O. Eleri, K. Azuatalam, M. Minde, A. Trindade, N. Muthuswamy, F. Lou and Z. Yu, Towards high-energy-density supercapacitors via less-defects activated carbon from sawdust, *Electrochim. Acta*, 2020, **362**, 137152.
75. Y. Lu, S. Zhang, X. Han, X. Wan, J. Gao, C. Bai, Y. Li, Z. Ge, L. Wei, Y. Chen, Y. Ma and Y. Chen, Controlling and optimizing the morphology and microstructure of 3D interconnected activated carbons for high performance supercapacitors, *Nanotechnology*, 2020, **32**, 085401.
76. Y. Xu, H. Lei, S. Qi, F. Ren, H. Peng, F. Wang, L. Li, W. Zhang and G. Ma, Three-dimensional zanthoxylum leaves-derived nitrogen-doped porous carbon frameworks for aqueous supercapacitor with high specific energy, *J. Energy Storage*, 2020, **32**, 101970.
77. T. Lin, I. Chen, F. Liu, C. Yang, H. Bi, and H. Xu, Nitrogen-doped mesoporous carbon of extraordinary capacitance for electrochemical energy storage, *Science*, 2015, **350**, 7.
78. R. Dash, J. Chmiola, G. Yushin, Y. Gogotsi, G. Laudisio, J. Singer, J. Fischer and S. Kucheyev, Titanium carbide derived nanoporous carbon for energy-related applications, *Carbon*, 2006, **44**, 2489–2497.
79. Y. Chen, C. Jing, X. Fu, M. Shen, K. Li, X. Liu, H. Yao, Y. Zhang and K. Yao, Synthesis of porous NiCoS nanosheets with Al leaching on ordered mesoporous carbon for high-performance supercapacitors, *Chem. Eng. J.*, 2020, **384**, 123367.
80. T. Liu, L. Zhang, W. You and J. Yu, Core–shell nitrogen-doped carbon hollow spheres/Co_3O_4 nanosheets as advanced electrode for high-performance supercapacitor, *Small*, 2018, **14**, 1702407.
81. H. Li, R. Liu, D. Zhao and Y. Xia, Electrochemical properties of an ordered mesoporous carbon prepared by direct tri-constituent co-assembly, *Carbon*, 2007, **45**, 2628–2635.
82. D. Lozano-Castell6, D. Cazorla-Amorós, A. Linares-Solano, S. Shiraishi, H. Kurihara and A. Oya, Influence of pore structure and surface chemistry on electric double layer capacitance in non-aqueous electrolyte, *Carbon*, 2003, **41**, 1765–1775.
83. H.J. Zhou and S. Kitagawa, Metal–organic frameworks (MOFs), *Chem. Soc. Rev.*, 2014, **43**, 5415–5418.
84. J. Liang, H. Peng, Z. Wang, R. Zhao, W. Zhang, G. Ma and Z. Lei, Rare earth metal lanthanum-organic frameworks derived three-dimensional mesoporous interconnected carbon nanosheets for advanced energy storage, *Electrochim. Acta*, 2020, **353**, 136597.
85. P. Sun, Y. Zhang, H. Shi, and F. Shi, Controllable one step electrochemical synthesis of PANI encapsulating 3d-4f bimetal MOFs heterostructures as electrode materials for high-performance supercapacitors, *Chem. Eng. J.*, 2022, **427**, 130836.
86. J. Chmiola, G. Yushin, R. Dash and Y. Gogotsi, Effect of pore size and surface area of carbide derived carbons on specific capacitance, *J. Power Sources*, 2006, **158**, 765–772.
87. M. Rose, Y. Korenblit, E. Kockrick, L. Borchardt, M. Oschatz, S. Kaskel and G. Yushin, Hierarchical micro-and mesoporous carbide-derived carbon as a high-performance electrode material in supercapacitors, *Small*, 2011, **7**, 1108–1117.

88. H.A. Ghaly, A.G. El-Deen, E.R. Souaya and N.K. Allam, Asymmetric supercapacitors based on 3D graphene-wrapped V_2O_5 nanospheres and Fe_3O_4@ 3D graphene electrodes with high power and energy densities, *Electrochim. Acta*, 2019, **310**, 58–69.
89. L. Li, Z. Hu, N. An, Y. Yang, Z. Li and H. Wu, Facile synthesis of MnO2/CNTs composite for supercapacitor electrodes with long cycle stability, *J. Phys. Chem. C*, 2014, **118**, 22865–22872.
90. H. Liu, B. Xu, M. Jia, M. Zhang, B. Cao, X. Zhao and Y Wang, Polyaniline nanofiber/large mesoporous carbon composites as electrode materials for supercapacitors, *Appl. Surf. Sci.*, 2015, **332**, 40–46.
91. J. Zhou, J. Lian, L. Hou, J. Zhang, H. Gou, M. Xia, Y. Zhao, T.A. Strobel, L. Tao and F. Gao, Ultrahigh volumetric capacitance and cyclic stability of fluorine and nitrogen co-doped carbon microspheres, *Nat. Commun.*, 2015, **6**, 8503.
92. H. Peng, S. Qi, Q. Miao, R. Zhao, Y. Xu, G. Ma and Z. Lei, Formation of nitrogen-doped holey carbon nanosheets via self-generated template assisted carbonization of polyimide nanoflowers for supercapacitor, *J. Power Sources*, 2021, **482**, 228993.
93. H. Peng, G. Ma, K. Sun, J. Mu, Z. Zhang and Z. Lei, Facile synthesis of poly(p-phenylenediamine)-derived three-dimensional porous nitrogen-doped carbon networks for high performance supercapacitors, *J. Phys. Chem. C*, 2014, **118**, 29507–29516.
94. P. Simon and Y. Gogotsi, Materials for electrochemical capacitors, *Nat. Mater.*, 2008, **7**, 845–854.
95. C. Largeot, C. Portet, J. Chmiola, P.L. Taberna, Y. Gogotsi and P. Simon, Relation between the ion size and pore size for an electric double-layer capacitor, *J. Am. Chem. Soc.*, 2008, **130**, 2730–2731.
96. H. Yamada, H. Nakamura, F. Nakahara, I. Moriguchi and T. Kudo, Electrochemical study of high electrochemical double layer capacitance of ordered porous carbons with both meso/macropores and micropores, *J. Phys. Chem. C*, 2007, **111**, 227–233.
97. H. Zhou, Y. Peng, H.B. Wu, F. Sun, H. Yu, F. Liu, Q. Xu and Y. Lu, Fluorine-rich nanoporous carbon with enhanced surface affinity in organic electrolyte for high-performance supercapacitors, *Nano Energy*, 2016, **21**, 80–89.
98. J.P. Paraknowitsch and A. Thomas, Doping carbons beyond nitrogen: An overview of advanced heteroatom doped carbons with boron, sulphur and phosphorus for energy applications, *Energy Environ. Sci.*, 2013, **6**, 2839–2855.
99. L. Xie, F. Su, L. Xie, X. Guo, Z. Wang, Q. Kong, G. Sun, A. Ahmad, X. Li, Z. Yi and C. Chen, Effect of pore structure and doping species on charge storage mechanisms in porous carbon-based supercapacitors, *Mater. Chem. Front.*, 2020, **4**, 2610–2634.
100. N. Zhao, L. Deng, D. Luo, S. He and P. Zhang, Oxygen doped hierarchically porous carbon for electrochemical supercapacitor, *Int. J. Electrochem. Sci.*, 2018, **13**, 10626–10634.
101. H. Peng, G. Ma, K. Sun, J. Mu, X. Zhou and Z. Lei, A novel fabrication of nitrogen-containing carbon nanospheres with high rate capability as electrode materials for supercapacitors, *RSC Adv.*, 2015, **5**, 12034–12042.
102. J. Sánchez-González, F. Stoeckli and T.A. Centeno, The role of the electric conductivity of carbons in the electrochemical capacitor performance, *J. Electroanal. Chem.*, 2011, **657**, 176–180.
103. L. Spain, Electronic transport properties of graphite, carbons, and related materials, in *Chemistry & Physics of Carbon*, edited by P. L. Walker, Boca Raton: CRC Press, 2021, pp. 119–304.
104. Z. Li, B. Yang, S. Zhang and C. Zhao, Graphene oxide with improved electrical conductivity for supercapacitor electrodes, *Appl. Surf. Sci.*, 2012, **258**, 3726–3731.

3 Biomass-Derived Materials Toward Low-Carbon Hydrogen Production

Jiale Xie, Pingping Yang and Cheng Huang

1 INTRODUCTION

Hydrogen is a renewable energy carrier in the future zero-carbon society. Nowadays, natural gas is the primary source of hydrogen production, which accounts for 75% of global hydrogen production.[1] Therefore, significant CO_2 emissions were generated during hydrogen production. Another source of hydrogen production is the chlor-alkali industry where hydrogen is a by-product.[2] Hydrogen production via green approaches has attracted increasing interest, such as electrolysis, photocatalysis, and biomass-based technologies. Biomass hydrogen production is believed as a sustainable approach, which is an energy-saving and eco-friendly hydrogen production route.[3] Biomass is renewable organic material that stores chemical energy from the sun through the photosynthesis process, which is one of the largest sources for fuel applications (~100 billion metric tons of carbon per year).[4] As shown in Figure 3.1, biomass can be converted to hydrogen energy through various strategies, including biomass-to-hydrogen direct conversion, and biomass-to-carbon conversion for green hydrogen technologies, such as water electrolysis and solar water splitting.[5–8] Biomass and its derivatives play a critical role in the production of low-carbon hydrogen. Therefore, we should reexamine and evaluate the biomass-derived materials and related hydrogen production techniques. In this

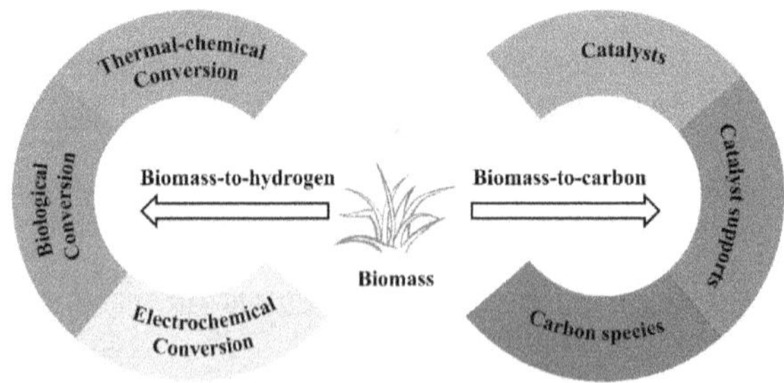

FIGURE 3.1 Strategies of biomass for low-carbon hydrogen production.

DOI: 10.1201/9781003387831-3

chapter, the fundamentals and techniques of low-carbon hydrogen production are comprehensively discussed first. Afterwards, we propose and discuss the above two important strategies for low-carbon hydrogen production by featured examples. At last, we list the remaining challenges and explore the future opportunities in this research hotspot.

2 FUNDAMENTALS AND TECHNIQUES OF LOW-CARBON HYDROGEN PRODUCTION

2.1 BIOMASS GASIFICATION

Gasification converts the biomass into a combustible gas mixture via the partial oxidation at high temperatures (e.g., 800–900°C). The main hydrogen production processes by biomass gasification strategy are shown in Figure 3.2. A sequence of complex thermochemical reactions happened during the gasification processes.[9] There are four zones in the gasifier, including drying, pyrolysis, combustion and reduction. The mixed gaseous product includes CO, H_2, CH_4, and non-combustible gases such as CO_2 and N_2. A high hydrogen and low tar content are expected for hydrogen production. However, the quality of combustible gas was affected by the temperature, pressure, residence time, equivalence ratio, biomass characteristics, and gasifier design.[9] Air gasification of biomass only produces syngas with a low H_2 content and low heating. In steam gasification, the hydrogen yield is far better than pyrolysis and air gasification while the overall conversion efficiency of thermal energy to hydrogen can also reach up to 52%. In addition, fast pyrolysis at the temperature of 800–900°C can convert 60% biomass into gas, such as H_2 and CO, while only leaves 10% biomass as solid char.[10]

2.2 ELECTROLYSIS

The electrolysis process provides a quick and convenient approach to produce pure hydrogen. Compared to water electrolysis, biomass electrolysis generates hydrogen

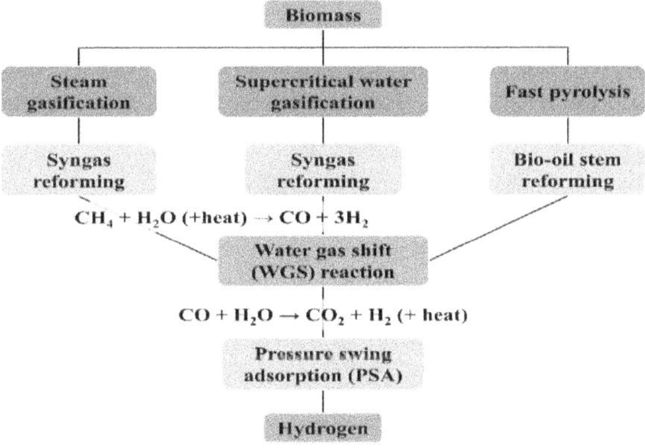

FIGURE 3.2 Main hydrogen production processes by biomass gasification strategy.

by substituting the oxidation of biomass-derived fuels at the anode for the oxygen evolution reaction, which can significantly reduce the electric energy consumption required for hydrogen production.[11] Meanwhile, the value-added chemicals can be coproduced. Electrolysis of renewable biomass for producing hydrogen can be realized by two common technologies: a proton exchange membrane electrolysis cell (PEMEC, Figure 3.3a) and a microbial electrolysis cell (MEC). In a typical PEMEC cell, the proton exchange membrane was sandwiched between a simple graphite-felt anode without coating any catalyst and a carbon cathode coated with a Pt black catalyst for hydrogen evolution. The electrode reactions at anodic and cathodic sides can be expressed respectively as follows (Eqns 3.1–3.2).

$$C_xH_yO_z + H_2O \rightarrow \text{Oxidation products} + H^+ + e^- \tag{3.1}$$

$$H^+ + e^- \rightarrow H_2 \tag{3.2}$$

The hydrogen evolution reaction (HER) has a high Faradaic efficiency of >95%. Due to the multielectron transfer processes, the kinetics of the anodic reaction is sluggish. Therefore, there is a great challenge to develop a highly efficient electrocatalyst for anodic oxidation reaction. Moreover, selectivity is a critical factor during the choosing of electrocatalysts.

2.3 Photocatalysis

Photocatalytic water splitting (without biomass) usually has a very low quantum yield of ~1.8%. However, a quantum yield of >70% can be achieved by phobiorefinery for hydrogen production.[12] In photocatalysis, the energetic band structures of photocatalysts are important for performing the processes of charge transport and transfer at the semiconductor/electrolyte junction (SEJ). When illuminating the aqueous suspension of photocatalysts (Figure 3.3b), electrons on the valence band are excited into the conductive band of photocatalysts. Then, the holes form on the valence band. The photogenerated charges are transported to the SEJ interface with the assistance of the build-in field in photocatalysts. The photogenerated electrons and holes can be used for the reduction and oxidation reactions, respectively. It's noted that the conductive and valence band levels should be lower than the reduction potential, and higher than the oxidation potential of the redox species, respectively. For biomass-based photocatalysis, the biomass oxidation reaction and HER reaction

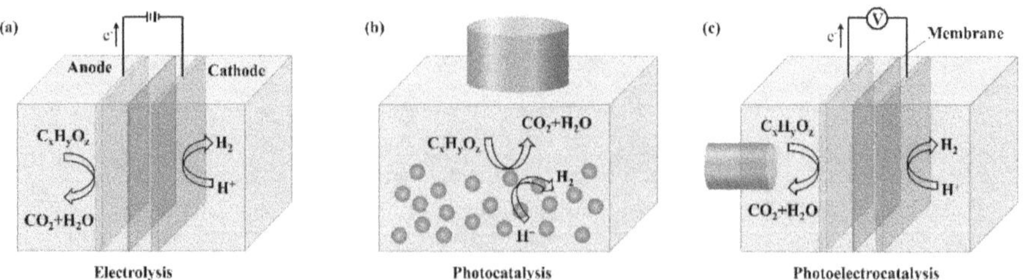

FIGURE 3.3 Schematic illustration of (a) electrolysis, (b) photocatalysis, and (c) photocatalysis for biomass-based hydrogen production.

can simultaneously happen on the surface of photocatalysts. Normally, the cocatalysts are needed to promote the reaction rates.[13]

2.4 Photoelectrocatalysis

As shown in Figure 3.3c, a photoelectrochemical (PEC) cell is made of a *n*-type semiconductive photoanode, a counter electrode (e.g., Pt), an electrolyte, and a membrane. On the photoanodes, the biomass oxidation reactions take place. The requirements for semiconductors are similar to those in photocatalysis to realize the biomass oxidation and HER reactions. The membrane (e.g., Nafion, glass frit, and diaphragm) is used to reduce or/and avoid the effect of the products, which are generated on the photoanode and cathode. Under illumination, the photogenerated electrons transport to the counter electrode through an external circuit, and react with H^+ to generate hydrogen. The photogenerated holes transport to the surface of the photoanode and oxidize the $C_xH_yO_z$ to CO_2 and H_2O. Although the reaction processes are very complex, the chemical reactions can be simply described by the following equations (Eqns 3.3–3.4):

$$C_xH_yO_z + h^+ + OH^- \rightarrow CO_2 + H_2O \tag{3.3}$$

$$H_2O + e^- \rightarrow H_2 + OH^- \tag{3.4}$$

The H-type configuration of a PEC cell is widely used to allow ion motion and gas separation naturally. Additional voltage or bias is usually required to apply to the PEC cell to either reduce the electron-hole recombination rate or help hydrogen evolve if the photoanode is not energetic enough for water splitting.[14]

3 BIOMASS-TO-HYDROGEN

3.1 Thermal-chemical conversion

Thermal-chemical conversion is the main strategy for hydrogen production from biomass. Compared with pyrolysis, gasification is recognized to be more promising in industrial production because of its higher efficiency and H_2 yield (Table 3.1). To decrease the reaction temperature and improve the carbon gasification rate and hydrogen selectivity, the catalysts should be used to promoting tar cracking and reforming of condensable fractions. Typical catalysts are alkaline earth metallic catalysts, metal-based catalysts (e.g., Ni, Ce, and La), and mineral catalysts. Traditional alkali metal catalysts can effectively promote biomass gasification, but some drawbacks exist such as difficult recovery, large amount of loading, easy scaling, and blockage. Ni-based catalysts have been widely used as effective catalysts, and it is better to utilize them in combination with other metals. Metal oxides not only have catalytic activity but also can be used as effective supports for exogenous metal catalysts. For instance, the stability of Ni-based catalysts can be improved by using a metal oxide supporter. Natural mineral catalysts are more suitable for steam gasification of biomass, but their catalytic activity in supercritical water gasification is obviously lower than that of other catalysts. At present, a key step for its development and application is to explore cost-effective catalysts that increase H_2 production, reduce gasification temperature, and promote tar cracking (Table 3.1).

3.2 Biological conversion

There are three kinds of biological conversion, including the biological water gas shift (BWGS) reaction, dark fermentation, and photo-fermentation. Biological conversion happens at low temperatures of 30–60°C and low pressure of 1 atm.

TABLE 3.1
Summary of thermal-chemical conversion for biomass-based hydrogen production

	Steam gasification	Supercritical water gasification	Fast pyrolysis
Main reactions	$C + H_2O \rightarrow H_2 + CO$, $\Delta H_{298K} = 131$ kJ/mol $CO + H_2O \rightarrow H_{2+}CO_2$, $\Delta H_{298K} = \rightarrow 40.9$ kJ/mol $CH_{4+}H_2O \rightarrow 3H_{2+}CO$, $\Delta H_{298K} = 206.3$ kJ/mol $C_aH_{b+}aH_2O \rightarrow$ $aCO + (a + b/2)H_2$	$CH_nO_m + (1 - m)H_2O$ $\rightarrow (n/2 + 1 - \mathrm{m})H_{2+}CO$ $CO + H_2O \rightarrow CO_{2+}H_2$ $CO + 3H_2 \rightarrow$ $CH_4 + H_2O$ $CO_2 + 4H_2 \rightarrow$ $CH_4 + 2H_2O$	$CH_nO_m \rightarrow$ $CH_4 + CO + H_2 +$ Other products $CH_4 + H_2O \rightarrow CO + 3H_2$ $CO + H_2O \rightarrow CO_2 + H_2$
Average H_2 production (wt%, g H_2/100 g biomass)	Without catalyst: 4 g With catalyst: 7 g	Without catalyst: 3 g With catalyst: 5 g	Without catalyst: 2 g With catalyst: 3 g
Reactor	Upper/lower ventilation gasifier; fluidized bed	Batch reactor, continuous reactor	Fluidized bed; fixed bed; rotary furnaces; ablation reactor; drainage bed
Advantage	Suitable for large-scale industrial production with high gasification rate and low-ash production.	High conversion and hydrogen concentration without tar and coke formation and secondary pollution.	Oxygen free and relatively high hydrogen concentration.
Disadvantage	Separation and purification of gas products are difficult.	Difficult to recycle alkaline catalysts and strict operating conditions.	High equipment requirement and energy consumption with low hydrogen yield.
Challenge	Decrease the tar content; optimize the catalyst composition to reduce its deactivation; reduce energetic and material costs	Technological development; reduce energetic costs	Raw bio-oil feeding; scale up of the process; optimize the catalyst composition to reduce its deactivation

Therefore, the energy cost is low. Typically, the enzymes (e.g., nitrogenases and hydrogenases) produced by the microorganisms can be used to synthesize hydrogen. Nitrogenases reduce protons (H^+) from adenosine triphosphate (ATP) and release H_2, while hydrogenases reversibly catalyze the conversion of protons to hydrogen.

In BWGS reaction, photoheterotrophic bacteria can oxidize CO into CO_2, and reduce H_2O into hydrogen. *Rhodospirillum rubrum* and *Carboxydothermus hydrogenoformans* are the most used gram-negative and gram-positive bacteria. The high concentration of CO will limit the activity of bacteria. In dark fermentation, anaerobic microorganisms can generate H_2, CO_2, and small amounts of CH_4, CO, and H_2S. However, the hydrogen yield is lower than that of photolysis. The by-products, pH, hydraulic retention time (HRT), and gas partial pressure can influence the yield of hydrogen. The time-consuming and expensive process impedes its development at the industrial level. Photo-fermentation is catalyzed by nitrogenases in purple non-sulfur bacteria to convert organic acids or biomass into hydrogen from solar energy in a nitrogen-deficient medium. To improve the hydrogen yield, the two-stage fermentation with an initial dark fermentation is used to produce the organic acids first. Then a photo-fermentation process is used to convert the organic acids into hydrogen. However, the operations are difficult, such as the control of different bacteria and parameters.

3.3 Electrochemical conversion

Biomass electrolysis has lower thermodynamic requirements compared to water electrolysis, leading to low ΔE_{eq} and high H_2 production efficiency, however, the kinetics are also sluggish due to the multiple electron transfer process.[11] Moreover, biomass must be fermented and converted into organic acids or alcohols, and some molecules can be converted to hydrogen. Electrocatalyst is a determining factor in reducing the activation energy and enhancing the selectivity toward valuable chemicals. Liu et al. reported an electrolysis approach for directly producing hydrogen from native biomass by using polyoxometalate (POM) solution as both a catalyst and charge carrier (Figure 3.4a).[15] The electric energy consumption could be as low as 0.69 kW h per normal cubic meter of H_2 (Nm^{-3} H_2) at 0.2 A cm^{-2}, which is only 16.7% of the energy consumed for the reported water electrolysis. A novel chemical-electrolysis conversion process was presented. In this process, raw biomass can be effectively decomposed via a chemical conversion by aqueous POM at low temperatures and followed by electrolysis with very low electric energy input to generate hydrogen. Unlike the traditional electrolysis of alcohols, a noble-metal catalyst is not required at the anode.

Zhao et al. reported a membrane-free single-compartment electrolyzer, which can electrooxidize chitin to acetate with >90% yield.[8] The overall energy consumption of electrolysis can be reduced by 15%. In this work, a 3D hierarchical porous nickel (*hp*-Ni, cubic crystal structure) catalyst (Figure 3.4b) was synthesized by template-free cathodic electrodeposition of porous Ni microspheres onto Ni foam (NF). The anode is the activated *hp*-Ni/NF electrode, which was performed at a current density of 50 mA cm^{-2} until a stable OER performance was reached. The *hp*-Ni/NF electrode was transformed to Ni_2P/NF as the cathode. After 4 h of chitin oxidation

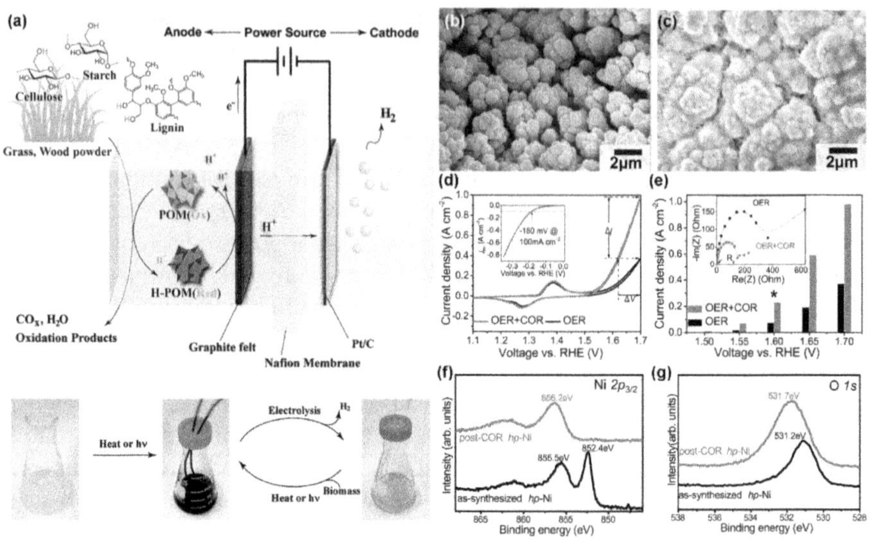

FIGURE 3.4 (a) Schematic illustration of the chemical-electric conversion process.[15] (b) SEM image of an as-synthesized *hp*-Ni sample. (c) SEM image of a post-COR *hp*-Ni sample. (d) Cyclic voltammograms of anodic reactions at a scan rate of 10 mV s^{-1} in 1.0 M KOH before (OER; black curves) and after adding 33.3 mg L^{-1} of chitin (OER + COR; red curves). Δj labels the current density difference at 1.7 V vs. RHE, and ΔV labels the potential difference at 0.4 A cm^{-2}. Inset: Linear sweep voltammograms of HER using Ni_2P/NF catalyst, where the potential for 100 mA cm^{-2} is labeled. (e) Comparison of current density at various potentials in 1.0 M KOH solution with (red bars) and without (black bars) 33.3 mg L^{-1} chitin. Inset: the corresponding electrochemical impedance spectra (EIS) at open-circuit potential. XPS spectra of (f) Ni 2p state, and (g) O 1s state of the samples before (black spectra) and after (red spectra) COR.[8]

reaction (COR), the nanostructure of *hp*-Ni/NF can be retained well (Figure 3.4c). The tiny particles on the surface can be attributed to the undissolved chitin crystals. In the presence of chitin, the current density is $\Delta j \sim 600$ mA cm^{-2} (or 2.5 times) higher at 1.7 V vs. RHE, and the potential is $\Delta V \sim 80$ mV lower at 0.4 A cm^{-2}, than those of the pure OER curves, suggesting COR is kinetically more favorable than OER (Figure 3.4d). The lower charge transfer resistance of the OER + COR curve (Inset of Figure 3.4e) indicates the faster kinetics of COR than that of OER. Figure 3.4e illustrates the COR current is larger than pure OER current in the range of potential used, and selective chitin oxidation (negligible OER) is possible with a potential lower than 1.55 V vs. RHE. The metallic Ni XPS peak (Figure 3.4f) disappearance and the stronger O 1s peak (Figure 3.4g) after COR suggest the enhanced nickel oxidation of the *hp*-Ni/NF electrode, which is beneficial for high COR activity.

3.4 Other Conversion

Thermal-chemical conversion is the most advanced technique for biomass-based hydrogen production. Typically, biomass is converted by thermal gasification

(700–1400°C), pyrolysis (300–1000°C), or supercritical water gasification (above 374°C and 217 atm).[7] The low heat efficiency, high equipment cost, and necessary downstream purification of the produced H_2 significantly limit the applicability of these technologies. Zhang et al. developed a two-step, one-pot process for hydrogen production with a hydrogen yield of 95% from non-food-related biomass and carbohydrate-based waste.[7] Initially, dimethyl sulfoxide (DMSO)-promoted hydrolysis–oxidation of cellulose and hemicellulose in renewable feedstocks to formic acid (FA) takes place (first step). Subsequently, after neutralizing the added catalytic amount of acid in the first step and without purification, H_2 is produced with high selectivity in the presence of a ppm-scale homogeneous iridium catalyst at 90°C (second step). The unwanted side products CO and CH_4 are produced at no more than 22 and 2 ppm, respectively. Because of the low concentration of CO, the H_2 formed is directly applicable in proton exchange membrane (PEM) fuel cells, thereby streamlining the biomass–hydrogen–electricity conversion (Figure 3.5). This procedure has another main advantage. The initially generated FA is a stable and non-toxic H_2 storage material, which conveniently releases H_2 on demand.

4 BIOMASS-TO-CARBON

4.1 Synthetic strategies for biomass-derived carbon

The carbon content of biomass after the oven-drying is around 45–50%.[16] To achieve biomass-derived carbon materials, various carbonization and activation strategies are summarized in Figure 3.6. Pyrolysis and hydrothermal carbonization (HTC) are the two widely used methods to carbonize biomass. Pyrolysis is carried out in an inert or limited oxygen atmosphere at the temperature range of 300–1,200°C, while HTC refers to a thermochemical process used to convert biomass to carbonaceous materials.

The main pyrolysis products obtained from biomass depend on the temperature, temperature ramping rate, particle size, and catalyst used. HTC is performed in a pressurized aqueous environment at a relatively low temperature, typically in the range of 120–250°C, with or without the aid of a catalyst. As a thermochemical conversion technique, the HTC can be influenced by several parameters, such as temperature, residence time, precursor concentration, and catalyst. It uses subcritical

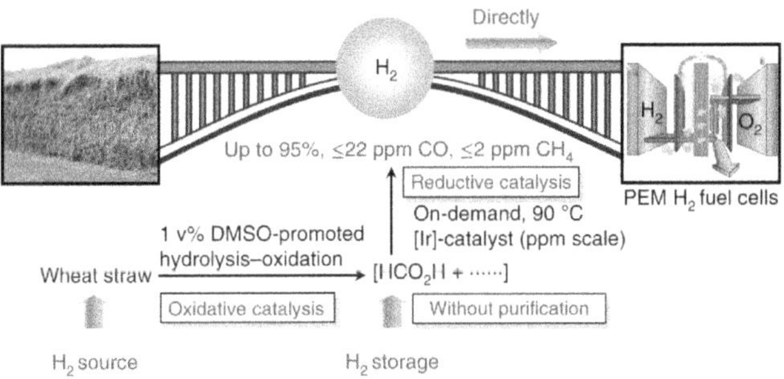

FIGURE 3.5 Streamlining the conversion of non-food-related biomass to electricity via H_2.[7]

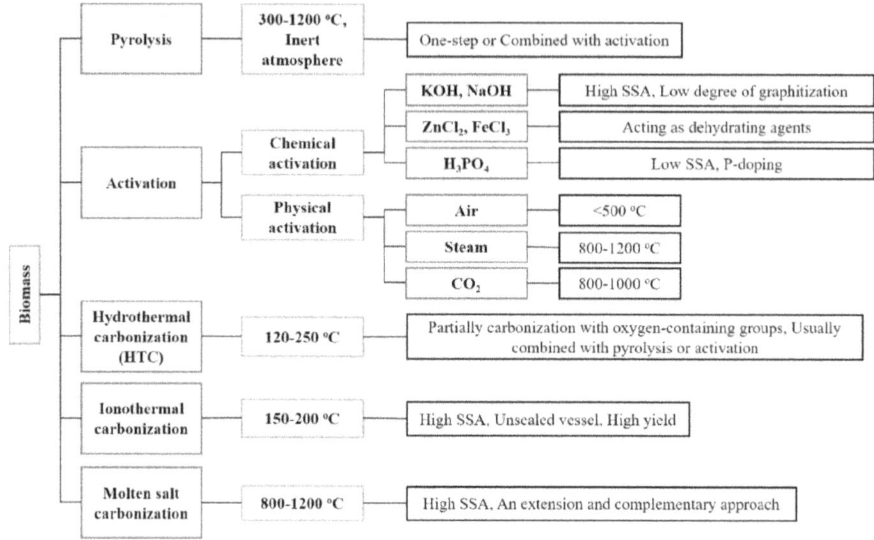

FIGURE 3.6 Strategies for converting biomass-to-carbon materials.

water for the conversion of a biomass to carbonaceous products, resulting in efficient hydrolysis and dehydration of precursors and bestowing hydrochars with a high and tunable content of oxygen-containing functional groups.

Activation is a process of converting carbonaceous materials into active carbon. Both physical and chemical activation means can be used. Physical activation is usually conducted immediately after the pyrolysis step at high temperatures (up to 1,200°C) in an atmosphere of steam or CO_2. Chemical activation is implemented with a chemical agent at a temperature typically ranging from 450 to 900°C. The commonly used chemical activation agents include KOH, NaOH, $ZnCl_2$, $FeCl_3$, H_3PO_4, and K_2CO_3.

4.2 Catalysts for Electrochemical Hydrogen Production

Biomass-derived carbon served as electrocatalysts has been rarely studied for electrochemical hydrogen production. Normally, the noble metal- and transition metal-based materials were applied to catalyze the water splitting. Some typical carbon materials derived from biomass for hydrogen production through water electrolysis are summarized in Table 3.2. Recently, Ding et al. reported a metal-free carbon electrode derived from biomass, which exhibits high electrochemical efficiency.[17] The carbon electrode was made from the annealed hydrothermal carbon on the glassy carbon electrode. As shown in Figure 3.7a, the carbon oxidation reaction has a lower potential of 1.02 V vs. RHE than that of the oxygen evolution reaction of 1.52 V vs. RHE. Therefore, the carbon electrode used is a sacrificial carbon electrode. In a two-electrode cell with carbon pellets, the gas accumulation over a ten-day period is shown in Figure 3.7b. The cathode side produces hydrogen, whereas the anode side produces a marginal amount of oxygen (5%) and the main CO_2 gas

(95%). The total gas production decreased due to the consumption and detachment of small carbon fragments from the pellet. In alkaline, neutral, and acidic electrolytes, the carbon electrodes show significantly different behaviors as shown in Figure 3.7c. The stability of carbon electrodes (Figure 3.7d) is about 90 min until the pellets fall apart at the current density of 120 mA cm^{-2}. Figure 3.7e indicates that carbon electrodes can withstand dynamic fluctuations. After nitrogen doping, the doped carbon electrode delivers a remarkably better performance than the pure carbon electrode (Figure 3.7f). This work provides an industrial clean, scalable, and sustainable idea to obtain "green" hydrogen.

4.3 Catalyst supports for hydrogen production

Biomass carbon materials normally possess a unique structure and heteroatom doping, which facilitate the high mass transfer rate of electrolytes, superior electrical conductivity, high specific surface area, and chemical stability. Qiao et al. chose wild celery as a biomass carbon source and *in situ* grew MoS_2 on/in micrometer scale carbon tube matrix that was carbonized from the prototype keeping (MoS_2/BCTM), which was applied in electrocatalytic hydrogen evolution.[25] Wild celery has a well-defined hexagonal or quasi-hexagonal stem as shown in Figure 3.8a. The diameter of the tube matrix after the carbonization is about 10–40 µm. As shown in Figure 3.8b, the optimal composite of MoS_2/BCTM 1:1 delivers the lowest overpotential of 176 mV

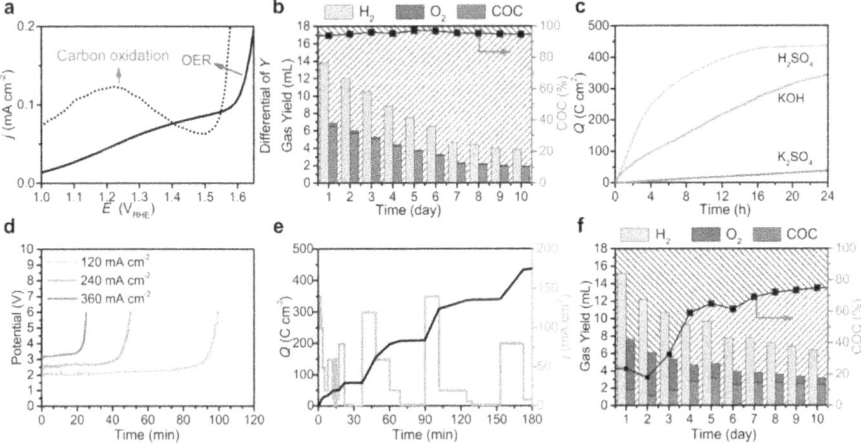

FIGURE 3.7 (a) Current potential curve of a carbon as electrode and the relative differential curve (dashed line). (b) Daily gas collection in a two-electrode cell: hydrogen at the cathode; and oxygen and the carbon oxidation contribution (COC) at the anode. Red-shaded areas demonstrate the hydrogen production from water oxidation (catalytic process), whereas blue-shaded areas are the hydrogen production from carbon oxidation. (c) Charge collection of the carbon pellet in acidic (H_2SO_4, pH 1), alkaline (KOH, pH 13), and neutral (K_2SO_4, pH 7) electrolytes. (d) Stability of the carbon pellets in 1 M KOH at current densities of 120, 240, and 360 mA cm^{-2}. (e) Performance of the carbon pellet under dynamic potential variations to mimic the flexibility of solar-derived electricity. (f) Daily gas collection in a two-electrode cell of the nitrogen-containing carbon pellet (Table 3.2).[17]

TABLE 3.2
Summary of biomass-derived carbon materials for water electrolysis

Biomass	Carbonization method	Activation	Specific surface area (m^2g^{-1}).	Electrolyte	Onset potential (mV)	Overpotential at 10 mA cm^{-2} (mV).	Tafel slope (mV dec^{-1}).	Ref.
Peanut shell	Tow-step thermal treatment	KOH	2,338	0.5 M H_2SO_4	80	400	75.7	18
Spathe-pollen	Two-step thermal treatment	KOH	1,297	0.5 M H_2SO_4	NA	330	63	19
Pine needle	Two-step thermal treatment	KOH	1,931	0.5 M H_2SO_4	4	62	45.9	20
Silk cocoon	One-step thermal treatment	KCl	349.3	0.5 M H_2SO_4	63	137	132	21
Chinese steamed bread flour	Hydrothermal treatment	NA	401.4	0.5 M H_2SO_4	NA	220	77	22
Peanut root nodule	One-step thermal decomposition	$MgCl_2$	513.3	0.5 M H_2SO_4	27	116	67.8	23
Human hair	Two-step pyrolysis	$ZnCl_2$	820	0.5 M H_2SO_4	12	100	57.4	24

at the current density of 10 mA cm^{-2}. In this case, MoS_2 nanosheets are uniform and have orderly dispersion on the surface of BCTM to form nanopore structures, which provide direct channels for mass transport. The Tafel slope for MoS_2/BCTM 1:1 is 51 mV dec^{-1}, which is a little higher than that of Pt/C (30 mV dec^{-1}). Electrochemical impedance study indicates BCTM improves the conductivity of MoS_2 and provides a large number of high-speed channels for charge and mass transport.

Lotus root is another biomass which has a parallel macrochannel structure. Chen et al. prepared one biomass-derived carbon nanosheet coupled with MoO_2/Mo_2C electrocatalyst (MOMC/NC) using lotus root and ammonium molybdate.[26] The synthesis process of MOMC/NC materials is schematically shown in Figure 3.8c. A solid-state reaction was performed at 900°C for 2 h under a nitrogen atmosphere. In 1 M KOH electrolyte, the overpotentials are 138 and 36 mV for MOMC/NC and Pt/C electrodes, when the current density is 10 mA cm^{-2}, respectively. The Tafel slope of MOMC/NC is 56.7 mV dec^{-1}. In 0.5 M H_2SO_4 electrolyte, an overpotential of 169 mV and a Tafel slope of 62.5 mV dec^{-1} can be observed with MOMC/NC-0.20. The kinetic study suggests the performance improvement can be attributed to the small charge transfer resistance and a highly electrochemical surface area of MOMC/NC. Apart from the tube-like biomass, the fiber-like carbon has also been demonstrated as a good catalyst support. As shown in Figure 3.8f, the N-doped biomass carbon fibers (NBCFs) were prepared by the calcination of cattail spikes.[27] Ni nanoparticles were loaded on the surface of NBCFs through simple impregnation and calcination. The in situ

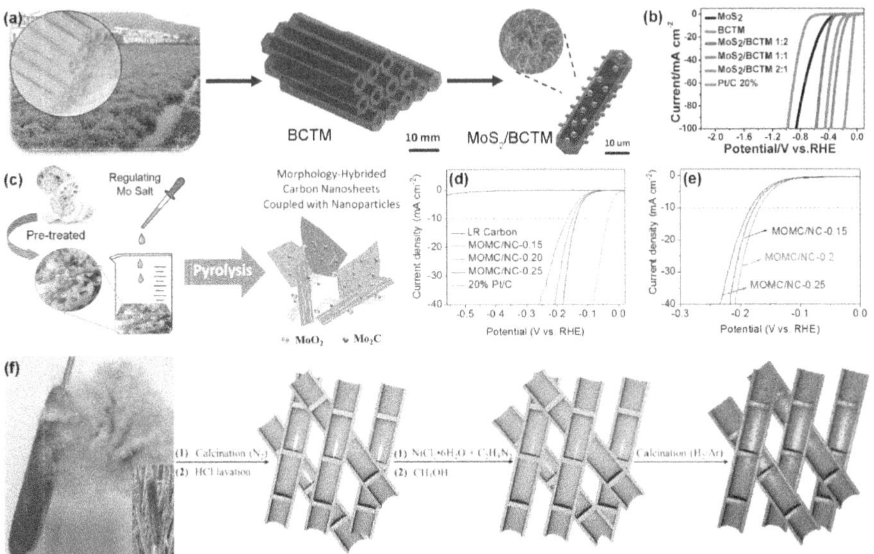

FIGURE 3.8 (a) Schematic of preparation process of MoS_2/BCTM. (b) The polarization curves of MoS_2, BCTM, MoS_2/BCTM 1:2, MoS_2/BCTM 1:1, MoS_2/BCTM 2:1, and Pt/C 20% in 0.5 M H_2SO_4 solution (N_2 saturated). Scan rate: 5 mV s^{-1}.[25] (c) Scheme of the synthesis strategy of MOMC/NC materials. (d) Polarization curves of MOMC/NC materials and 20 wt% Pt/C in 1 M KOH. (e) Polarization curves of MOMC/NC materials in 0.5 M H_2SO_4.[26] (f) Scheme of the preparation of the Ni/NBCF-H_2 samples.[27]

template effect of the Ni-ethylenediamine complex facilitates the formation of Ni-N bonds between the Ni nanoparticles and NBCFs, which not only prevents the aggregation and corrosion of the Ni nanoparticles but also accelerates the electron transfer in the electrochemical reaction.

4.4 Carbon species for solar hydrogen production

Due to the various heteroatoms, biomass materials have been widely used to prepare carbon-based functional materials. Zhang et al. synthesized nitrogen and sulfur-doped carbon nanoparticles (N,S-C) using human hair as carbon sources via a hydrothermal method for photoelectrochemical water splitting.[28] The N,S-doped C and surface plasmon resonance Ag nanoparticles were modified on the surface of TiO_2 nanorods, which were grown on the surface of carbon cloth (CC). The synthesis process of broccoli-like Ag-N,S-C/TiO_2@CC is schematically shown in Figure 3.9. The photocurrent density of Ag-N,S-C/TiO_2@CC is 89.8 μA cm^{-2}, which is higher than Ag/TiO_2@CC, N,SC/TiO_2@CC and TiO_2@CC under AM 1.5G simulated sunlight, at 1.23 V vs. RHE. Under visible light irradiation, the photocurrent density of Ag-N,S-C/TiO_2@CC reaches to 12.6 μA cm^{-2}, which is 21.0 times higher than TiO_2@CC. Under AM 1.5G simulated sunlight, Ag-N,S-C/TiO_2@CC absorb sunlight and are excited to generate electron/hole pairs due to the surface plasmon resonance effect of Ag NPs. At the same time, holes migrate to the surface of TiO_2 nanorods to contact the electrolyte, and water oxidation reaction occurs. Photogenerated electrons migrate along the external circuit to the surface of the counter electrode (Pt), and water reduction reaction occurs. Subsequently, N,S-C NPs further extract holes from the Ag-N,S-C/TiO_2@CC surface through the Ti-N bond, which can reduce the charge depletion layer and promote the separation of photogenerated carriers, extending the lifetime of photogenerated carriers. Therefore, N,S-C NPs and Ag NPs play their respective roles and coordinate to promote charge separation and transfer.

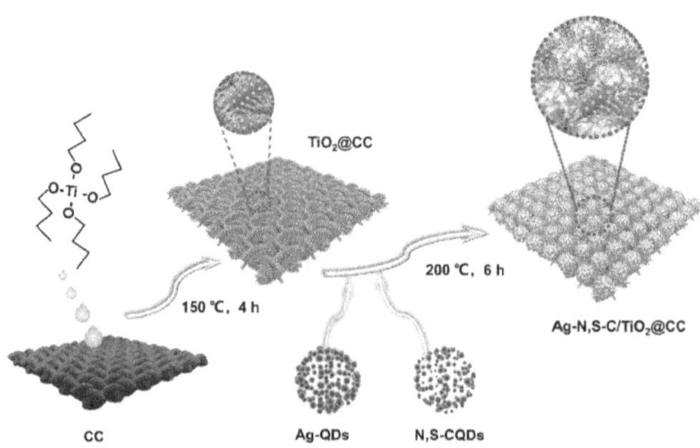

FIGURE 3.9 Synthesis process of broccoli-like Ag-N,S-C/TiO_2@CC using hydrothermal method.[28]

5 PERSPECTIVES AND CONCLUSIONS

Low-carbon hydrogen as a future fuel has been widely accepted. The world's biomass generation is ~100 billion metric tons of carbon per year, which is the largest source of fuel production. Various biohydrogen generation techniques have been studied in depth to improve the yield and volumetric production rates of hydrogen. Thermal-chemical processes are the most common technologies for biomass hydrogen production. Biomass gasification can mainly determine the hydrogen yield through the operation parameters, such as the temperature and steam flow. The stability and activity of the used catalysts can also greatly influence the gasification reaction. Biohydrogen systems have great potential attributed to improvements in bioreactor designs, fast gas separation, and purification, and gene adjustment of enzyme pathways. However, it still requires more advances to render it economically competitive and environmentally benign for large-scale industrial production. For biomass hydrogen production, the urgent issues are to realize the controllability and scalability of the production process, improve the reaction rate and efficiency, save the production costs, and accelerate the industrialization process.

Biomass-to-carbon is another an important strategy to perform low-carbon hydrogen production. This strategy could combine the electrochemical and photocatalytic technologies for hydrogen production. The biomass-derived carbon can be used as catalysts, and catalyst supports during the HERs. The biomass-derived carbon can improve hydrogen production performance due to the self-doping of heteroatoms or/and the catalyst loading of metals and their compounds. In addition, the biomass-derived carbon can protect the metal particles from corrosion and aggregation. However, this strategy is still in the library. The main challenges are the scalable preparation of bio-carbon with low emission of carbon-containing pollutants. Moreover, the method to anchor the catalysts on the surface of biomass-derived carbon should be developed at the industrial level.

REFERENCES

1. C. Bauer, K. Treyer, C. Antonini, J. Bergerson, M. Gazzani, E. Gencer, J. Gibbins, M. Mazzotti, S.T. McCoy, R. McKenna, R. Pietzcker, A.P. Ravikumar, M.C. Romano, F. Ueckerdt, J. Vente and M. van der Spek, On the climate impacts of blue hydrogen production, *Sustain. Energy Fuels*, 2022, **6**, 66–75.
2. D.-Y. Lee, A. Elgowainy and Q. Dai, Life cycle greenhouse gas emissions of hydrogen fuel production from chlor-alkali processes in the United States, *Appl. Energy*, 2018, **217**, 467–479.
3. L. Cao, I.K.M. Yu, X. Xiong, D.C.W. Tsang, S. Zhang, J.H. Clark, C. Hu, Y.H. Ng, J. Shang and Y.S. Ok, Biorenewable hydrogen production through biomass gasification: A review and future prospects, *Environ. Res.*, 2020, **186**, 109547.
4. C.B. Field, M.J. Behrenfeld, J.T. Randerson and P. Falkowski, Primary production of the biosphere: Integrating terrestrial and oceanic components, *Science*, 1998, **281**, 237–240.
5. R.D. Cortright, R.R. Davda and J.A. Dumesic, Hydrogen from catalytic reforming of biomass-derived hydrocarbons in liquid water, *Nature*, 2002, **418**, 964–967.
6. H.G. Cha and K.-S. Choi, Combined biomass valorization and hydrogen production in a photoelectrochemical cell, *Nat. Chem.*, 2015, **7**, 328–333.

7. P. Zhang, Y.-J. Guo, J. Chen, Y.-R. Zhao, J. Chang, H. Junge, M. Beller and Y. Li, Streamlined hydrogen production from biomass, *Nat. Catal.*, 2018, **1**, 332–338.
8. H. Zhao, D. Lu, J. Wang, W. Tu, D. Wu, S.W. Koh, P. Gao, Z.J. Xu, S. Deng, Y. Zhou, B. You and H. Li, Raw biomass electroreforming coupled to green hydrogen generation, *Nat. Commun.*, 2021, **12**, 2008.
9. B. Dou, H. Zhang, Y. Song, L. Zhao, B. Jiang, M. He, C. Ruan, H. Chen and Y. Xu, Hydrogen production from the thermochemical conversion of biomass: Issues and challenges, *Sustain. Energy Fuels*, 2019, **3**, 314–342.
10. Y. Kalinci, A. Hepbasli and I. Dincer, Biomass-based hydrogen production: A review and analysis, *Int. J. Hydrogen Energy*, 2009, **34**, 8799–8817.
11. H. Luo, J. Barrio, N. Sunny, A. Li, L. Steier, N. Shah, I.E.L. Stephens and M.-M. Titirici, Progress and perspectives in photo- and electrochemical-oxidation of biomass for sustainable chemicals and hydrogen production, *Adv. Energy Mater.*, 2021, **11**, 2101180.
12. K.A. Davis, S. Yoo, E.W. Shuler, B.D. Sherman, S. Lee and G. Leem, Photocatalytic hydrogen evolution from biomass conversion, *Nano Convergence*, 2021, **8**, 6.
13. X. Wang, X. Zheng, H. Han, Y. Fan, S. Zhang, S. Meng and S. Chen, Photocatalytic hydrogen evolution from biomass (glucose solution) on Au/CdS nanorods with Au^{3+} self-reduction, *J. Solid State Chem.*, 2020, **289**, 121495.
14. X. Lu, S. Xie, H. Yang, Y. Tong and H. Ji, Photoelectrochemical hydrogen production from biomass derivatives and water, *Chem. Soc. Rev.*, 2014, **43**, 7581–7593.
15. W. Liu, Y. Cui, X. Du, Z. Zhang, Z. Chao and Y. Deng, High efficiency hydrogen evolution from native biomass electrolysis, *Energy Environ. Sci.*, 2016, **9**, 467–472.
16. J. Deng, M. Li and Y. Wang, Biomass-derived carbon: Synthesis and applications in energy storage and conversion, *Green Chem.*, 2016, **18**, 4824–4854.
17. Y. Ding, M. Greiner, R. Schlögl and S. Heumann, A metal-free electrode: From biomass-derived carbon to hydrogen, *ChemSusChem*, 2020, **13**, 4064–4068.
18. K.R.A. Saravanan, N. Prabu, M. Sasidharan and G. Maduraiveeran, Nitrogen-self doped activated carbon nanosheets derived from peanut shells for enhanced hydrogen evolution reaction, *Appl. Surf. Sci.*, 2019, **489**, 725–733.
19. N. Prabu, R.S.A. Saravanan, T. Kesavan, G. Maduraiveeran and M. Sasidharan, An efficient palm waste derived hierarchical porous carbon for electrocatalytic hydrogen evolution reaction, *Carbon*, 2019, **152**, 188–197.
20. G. Zhu, L. Ma, H. Lv, Y. Hu, T. Chen, R. Chen, J. Liang, X. Wang, Y. Wang, C. Yan, Z. Tie, Z. Jin and J. Liu, Pine needle-derived microporous nitrogen-doped carbon frameworks exhibit high performances in electrocatalytic hydrogen evolution reaction and supercapacitors, *Nanoscale*, 2017, **9**, 1237–1243.
21. X. Liu, M. Zhang, D. Yu, T. Li, M. Wan, H. Zhu, M. Du and J. Yao, Functional materials from nature: Honeycomb-like carbon nanosheets derived from silk cocoon as excellent electrocatalysts for hydrogen evolution reaction, *Electrochim. Acta*, 2016, **215**, 223–230.
22. A. Nsabimana, F. Wu, J. Lai, Z. Liu, R. Luque and G. Xu, Simple synthesis of nitrogen-doped porous carbon from Chinese steamed bread flour and its catalytic application for hydrogen evolution reaction, *Electrochim. Acta*, 2018, **290**, 30–37.
23. Y. Zhou, Y. Leng, W. Zhou, J. Huang, M. Zhao, J. Zhan, C. Feng, Z. Tang, S. Chen and H. Liu, Sulfur and nitrogen self-doped carbon nanosheets derived from peanut root nodules as high-efficiency non-metal electrocatalyst for hydrogen evolution reaction, *Nano Energy*, 2015, **16**, 357–366.
24. X. Liu, W. Zhou, L. Yang, L. Li, Z. Zhang, Y. Ke and S. Chen, Nitrogen and sulfur co-doped porous carbon derived from human hair as highly efficient metal-free electrocatalysts for hydrogen evolution reactions, *J. Mater. Chem. A*, 2015, **3**, 8840–8846.

25. S. Qiao, J. Zhao, B. Zhang, C. Liu, Z. Li, S. Hu and Q. Li, Micrometer–scale biomass carbon tube matrix auxiliary MoS_2 heterojunction for electrocatalytic hydrogen evolution, *Int. J. Hydrogen Energy*, 2019, **44**, 32019–32029.
26. X. Chen, J. Sun, T. Guo, R. Zhao, L. Liu, B. Liu, Y. Wang, J. Li and J. Du, Biomass-derived carbon nanosheets coupled with MoO_2/Mo_2C electrocatalyst for hydrogen evolution reaction, *Int. J. Hydrogen Energy*, 2022, **47**, 30959–30969.
27. S. Hong, N. Song, E. Jiang, J. Sun, G. Chen, C. Li, Y. Liu and H. Dong, Nickel supported on nitrogen-doped biomass carbon fiber fabricated via in-situ template technology for pH-universal electrocatalytic hydrogen evolution, *J. Colloid Interface Sci.*, 2022, **608**, 1441–1448.
28. Y. Zhang, Y. Lei, T. Zhu, Z. Li, S. Xu, J. Huang, X. Li, W. Cai, Y. Lai and X. Bao, Surface plasmon resonance metal-coupled biomass carbon modified TiO_2 nanorods for photoelectrochemical water splitting, *Chin. J. Chem. Eng.*, 2022, **41**, 403–411.

4 Biomass-Derived Carbon Materials for Microbial Fuel Cell Anodes

Long Zou

1 INTRODUCTION

Energy shortage, climate warming, and environmental pollution have brought a series of serious problems to human beings, while the rapidly growing worldwide population, urbanization, and industrialization are exacerbating the situation. Currently, fossil fuels, including coal, crude oil, and natural gas, account for about 85% of the world's total energy consumption,[1] but their limited reserves will soon be unable to meet the growing demand of mankind. In addition, various undesirable environmental issues associated with the use of fossil fuels, such as greenhouse gas emissions and acid rain,[2] also demonstrate that they are no longer the ideal energy source for human sustainable development. Because of this, the exploration of sustainable and clean energy sources including solar, wind, hydro, and biomass energy, as well as efforts to develop technologies for energy storage and conversion, have aroused great interest in academia and industry over the past few decades.[3, 4]

Microbial electrochemical systems, as a new energy conversion route, which use electroactive microorganisms to catalyze various electrochemical reactions, are making surprising strides in capturing energy from wastes and producing high-value chemicals economically.[5–7] Among them, microbial fuel cell (MFC) is the earliest and most common prototype, accounting for over 80% of microbial electrochemical systems reported in the literature.[8, 9] Meanwhile, a plethora of derivatives, such as microbial electrolysis cell,[10] microbial desalination cell[11] and microbial electrosynthesis,[12] have also sprung up, showing a remarkable scalability and adaptability of microbial electrochemical systems. However, despite significant improvements in recent years, MFCs still suffer from some significant challenges, including low energy output density, high capital costs associated with manufacturing and operational processes, and difficulties in scaling up, which have hindered their commercialization.[13] For example, the power densities of liter-scale MFCs are usually in the order of several W m^{-3}, almost three to four orders of magnitude lower than those of commercial chemical fuel cells, meaning that the current MFCs are not competitive as an energy production unit.[14] Considering the fundamentals of microbial electrochemistry, the sluggish interfacial electron transfer between the abiotic electrode and the microbial biocatalysts with poor conductivity is the most critical limiting factor.[15] Thus, beyond doubt, improving the interfacial electron transfer rate is an important

DOI: 10.1201/9781003387831-4

issue to be solved at present, and it is the key route to solve the above bottlenecks. As a key component of the MFC device, the anode not only interacts with microbial cells as an electron acceptor but also provides physical support for microbial biofilm growth. The development of electrode materials with low cost, high activity, and excellent biocompatibility will have far-reaching significance in promoting MFCs from proof-of-concept to real-world application.

Carbon materials are the most widely studied and mature MFC anodes with abundant reserves, high specific surface area, good mechanical and chemical stability, corrosion resistance, and no biological toxicity.[16, 17] However, the commercial monolithic carbon materials such as carbon cloth (CC), carbon paper (CP), carbon felt (CF), graphite rod (GR), graphite felt (GF) and graphite plate (GP) still suffer from relatively low performance when used in MFCs.[16] The imperfections in available pore structure and surface electrochemical activity, leading to poor biofilm formation and depressed interfacial electron transfer, are counted as the main limiting factors. In addition, nanoscale carbon materials, including graphene, carbon nanotubes, carbon quantum dots, etc., have also been developed for direct use as MFC anodes, or as surface modifiers, often achieving much higher performance than those monolithic materials.[18] Although currently commercially available, the relatively high manufacturing cost and potential biotoxicity to microorganisms limit their large-scale applications.[19] Moreover, these nanostructured carbons are products of fossil fuels, often requiring very harsh synthesis conditions and facing the challenges of environmental degradation and unsustainability.[20] Therefore, the search for alternative carbon electrode materials that are competitive in terms of performance, economy, and environment is a high-priority item. With respect to this, the synthesis and utilization of carbon materials from the richest natural biomass, including plants, animals, and microorganisms,[21] have attracted considerable interest from academia and industry because of the abundant resource, easy processability, and inherent biocompatibility. Biomass-derived carbon (BDC) materials can inherit the natural porous or layered structures of precursors, with high specific surface area and rich heteroatomic doping, which can promote microbial colonization, electrolyte diffusion, and interfacial electrochemical activity.[15, 22] Moreover, it is estimated that the anode material cost could account for 20%–50% of the overall MFC operating cost,[23] the fabrication of cost-effective, environment-friendly anode materials from biomass waste seems to be a promising approach in the framework of a circular economy.

This chapter briefly describes the basic principle of MFCs and summarizes the recent progress in the application of different types of BDC materials in MFC anodes. The significant factors affecting the performance of BDC anodes are discussed from aspects of elemental composition, structural characteristics, and surface chemistry, which will guide the further rational design of BDC materials for high-performance MFC anodes. Finally, the current challenges and outlooks in the future development of BDC anodes are also highlighted.

2 FOUNDATION OF MFCS

The first reported prototype of MFCs dates back to 1911 when the English botanist M. C. Potter succeeded in using microorganisms to produce electricity.[24] After a long period of stagnation, interest in MFCs has grown exponentially since the beginning

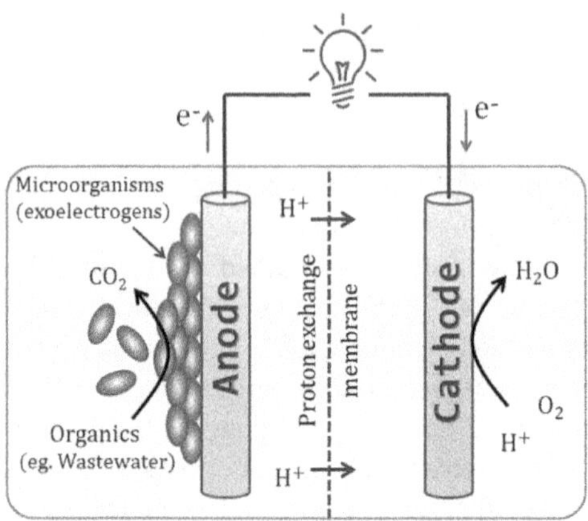

FIGURE 4.1 Schematic diagram of a typical H-type dual-chamber MFC.

of the 21st century, as their promising potential for complementing the existing energy-intensive wastewater treatment systems with a technology that can be energy self-sustainable or even have a net positive energy output while waste organic matters are cleaned.[25] The most remarkable feature of this technology is the direct conversion of chemical energy stored in organic substrates, from simple molecules (e.g., glucose, acetate, lactate, and carbohydrates) to complex compounds (e.g., cellulose, molasses, industrial, and agricultural wastes), into directly usable electricity through electrode-assisted microbial oxidation. Up to now, a variety of different configurations of MFCs, including H-type, cube-type, flat-plate type, horizontal and vertical tubular types, have been developed.[26] As illustrated in Figure 4.1, a typical H-type dual-chamber MFC is usually separated by a proton exchange membrane into an anodic chamber and a cathodic chamber. In the anodic chamber, microorganisms that are electrochemically active degrade organic substrates through intracellular oxidative metabolism, then export the produced electrons outside the cells for delivery to a solid electrode (anode) through a specific pathway called extracellular electron transfer (EET), accompanied by the release of protons into the electrolyte. The collected electrons by the anode are then transported via an external wire to another solid electrode (cathode) placed in the cathodic chamber, where they are used as an energy source for the reduction of terminal electron acceptors (such as oxygen that react with the protons diffused from the anode chamber to form water), usually with the involvement of suitable catalysts such as carbon supported platinum.

Compared with chemical fuel cells, the use of live microbial catalysts in the anodic chamber to generate electrons and transfer them to the anode is the distinguishing feature. Accordingly, the electrical interaction between microbial cells and the anode electrode, especially the interfacial electron transport, is one of the key factors determining the performance of MFCs. A series of electroactive microorganisms are known to be capable of interacting with electrodes, the most representative of which are *Geobacter* and *Shewanella* genera, with others belonging to the genera *Pseudomonas, Citrobacter, Klebsiella, Arcobacter, Comamonas, Desulfuromonas,* and *Enterobacter*.[27] Given such a broad range of electroactive microorganisms, it is not surprising that the types of pathways they use for EET tend to be diverse.

In general, EET pathways are mainly divided into three types (Figure 4.2): direct EET through direct physical contact between the out-membrane cytochromes and the electrode surface, long-distance EET by means of electrically conductive nanowires (pili and membrane extensions), and mediated EET through soluble electron shuttles. Both *G. sulfurreducens* and *S. oneidensis* have been documented to have a substantial diversity of cytochromes, whose integral role in the EET of these two typical electroactive microorganism strains is well established. For example, the deletion of out-membrane cytochrome OmcZ seriously impaired the EET ability of *G. sulfurreducens*, resulting in a great loss of current output. For *S. oneidensis*, a well-known Mtr route encoded by *mtrDEF-omcA-mtrCAB* gene cluster is responsible for the direct EET. Its mutants lacking the cytochrome-encoding genes generated 20% less current, while a 35% increase in current generation was observed when the cytochrome MtrC was overexpressed in the wild-type strain.[28] *G. sulfurreducens* also allows for longer electron transfer via electroconductive pili,[29] which facilitates the transfer of electrons to the electrode from those cells on the outer side of a thick biofilm although they cannot make direct contact with the electrode surface. *S. oneidensis* has a similar long-distance EET mechanism but utilizes cytochrome-rich extracellular extensions instead of pili.[30] In addition, the production and use of soluble electron shuttles by electroactive microorganisms appears to be a more common strategy for achieving long-distance electron transfer, which enables planktonic cells to transfer electrons to electrodes they do not touch. Such as, *Shewanella* genera performs indirect EET by producing flavins, which are reduced by cytochromes on the cell surface or in periplasmic space and diffused to the electrode to reduce the latter. Flavin-mediated electron transfer was shown to account for about 75% of the total EET capacity of *Shewanella* genera.[31] Another representative case of electroactive microorganisms making use of an indirect EET mechanism is *Pseudomonas* genera, which are capable of producing phenazine derivatives such as pyocyanin as electron shuttles.[32] Regardless of the EET mode, increasing the cell load of electroactive microorganisms on the anode electrodes, thus improving the probability of direct EET or reducing the electron shuttle diffusion distance, is obviously an effective means to enhance the electricity generation of MFCs.

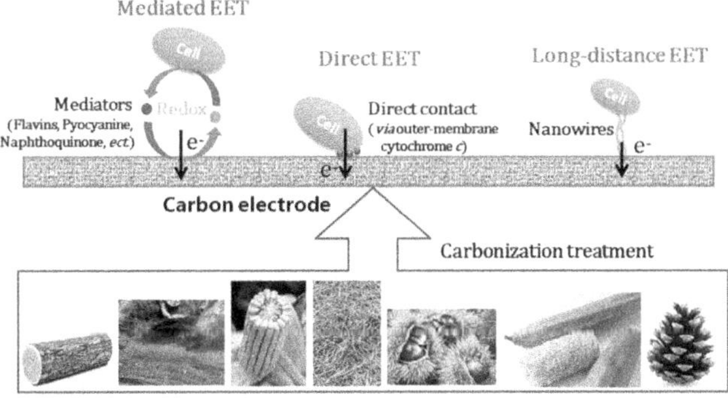

FIGURE 4.2 Schematic diagram of EET pathways from electrogenic bacteria to carbon electrode derived from natural biomass.

3 APPLICATION OF BIOMASS-DERIVED CARBON (BDC) MATERIALS IN MFC ANODES

The anode is the key component of an MFC, which largely determines the overall performance and operating cost. With the rapid development of environmentally friendly manufacturing technologies including pyrolysis and hydrothermal carbonization,[33] BDC materials have been widely used as candidates for MFC anode materials in recent years (Table 4.1). This section will focus on summarizing their application in different forms and the improved MFC performance.

3.1 BDCs for Free-Standing Anodes

Monolithic BDCs prepared from bulk biomass precursors, usually natural plants, by high-temperature pyrolysis process can be directly used as free-standing MFC anodes, which is convenient and economical. For example, tubular bamboo charcoal was prepared through the direct carbonization of a bamboo tube as a binder-free MFC anode, which achieved a 50% improvement in the maximum power density over a conventional graphite tube.[35, 71] The power density delivered by this tubular bamboo charcoal anode was related to its inner diameter and could be increased by simple electrode stacking. Various types of natural wood are also commonly used precursors for producing monolithic BDCs. A high-performance three-dimensional (3D) carbon anode was prepared from a piece of basswood through hydrothermal treatment for surface activation and followed by pyrolysis, which delivered 8.3 times higher power output than a conventional CC anode because its hierarchically porous structure guaranteed high surface area and multi-transportation-pathways.[36] Cedar wood pieces were also used as precursors to prepare binder-free MFC anodes through pyrolysis at different temperatures (400–900°C), and their performance varied due to the temperature-dependent conductivity and pore structure.[38] Corncob is a representative crop residue for preparing porous biochar due to its high volume-to-mass ratio. Recently, a 3D N-doped microporous carbon foam anode originated from natural corncob was reported to achieve a then-record power density of 4.99 W m^{-2} and current density of 12.30 A m^{-2}.[41] The plant tissues with inherent unique structures, such as chestnut shell,[39, 40] pinecone,[44] and loofah sponge,[43] were favored for the preparation of binder-free MFC anode materials because these specific structures (e.g., well-ordered macropores and high porosity) are expected to effectively promote microbial colonization and electrolyte diffusion. For similar reasons, some everyday table foods including bread[37] and cake[47] were also being tried to produce MFC anodes with attractive ability in electricity generation. In addition to these plant-originated biomass, animal-originated, and microbial ones, such as silk cocoon[46] and algae,[34] have also been developed for the preparation of free-standing anodes for MFCs. Furthermore, from the point of view of resource recovery and economy, it is worth advocating making high-performance anode materials from waste biomass. This strategy has been successfully practiced in the recycling of waste paper[49] and sewage sludge,[45] and the resulting BDCs are significantly superior in performance and cost to commercial carbon materials.

TABLE 4.1
Summary of biomass-derived carbons used for MFC anodes and their performance

Biomass precursor	Feature	MFC type	Microorganism	MFC performance	Ref
Biomass-derived carbons for free-standing anodes					
Algae	3D non-woven interconnected macroporous networks	Dual-chamber	*S. oneidensis* MR-1	2,385 mW m^{-2} (vs. 110 m W m^{-2} for carbon felt)	34
Bamboo tube	Tubular structure	Dual-chamber	Effluent from a matured MFC	1,652 ± 18 mW m^{-2} (vs. 1,102 ± 8 mW m^{-2} for graphite tube)	35
Basswood chip	3D hierarchically porous structure	Dual-chamber	*S. oneidensis* MR-1	483 mW m^{-2} (vs. 52 mW m^{-2} for carbon cloth)	36
Bread	3D macroporous carbon foam	Dual-chamber	Pre-acclimated bacteria from activated anaerobic sludge	3,134 mW m^{-2} (vs. 1218 mW m^{-2} for carbon brush)	37
Cedar wood piece	Ordered honeycomb structure	Single-chamber	Primary wastewater effluent and dehydrated sludge	9.9 ± 0.6 mW m^{-2} (vs. 1.9 ± 0.2 mW m^{-2} for carbon felt)	38
Chestnut shell	Hierarchically structured urchin shape	Single-chamber	Effluent of matured MFC anolyte	759 ± 38 mW m^{-2} (vs. 830 ± 42 mW m^{-2} for carbon brush)	39
Chestnut shell	Macroporous structure with rich mesopores and micropores	Single-chamber	Municipal wastewater	23.58 W m^{-3} (vs. 10.4 W m^{-3} for carbon cloth)	40
Corncob	3D N-doped macroporous foam	Dual-chamber	Anaerobic sludge	4,990 ± 20 mW m^{-2} (vs. 1,160 ± 30 mW m^{-2} for carbon felt)	41
Kapok fiber	Hollow structure	Single-chamber	Anaerobic sludge	104.1 mW g^{-1} (vs. 5.5 mW g^{-1} for solid fiber)	42

(Continued)

TABLE 4.1
(Continued)

Biomass precursor	Feature	MFC type	Microorganism	MFC performance	Ref
Loofah sponge	3D structure decorated with carbon-coated TiO_2 core-shell nanoparticles	Single-chamber	Pre-acclimated bacteria from a well-running MFC	2,590 ± 120 mW m^{-2} (vs. 860 ± 30 mW m^{-2} for graphite)	43
Pinecone	3D macroporous structure	Dual-chamber	Effluent from well-started MFC and anaerobic sludge	10.88 W m^{-3} (vs. 4.94 W m^{-3} for carbon felt)	44
Sewage sludge	Carbon micro-wires on surface	Dual-chamber	Effluent from a matured MFC	2,228 mW m^{-2} (vs. 462 mW m^{-2} for graphite plate)	45
Silk cocoon	Enriched nitrogen contents and hierarchical pores	Single-chamber	Pre-acclimated bacteria from activated anaerobic sludge	5 mW g^{-1} (vs. 2 mW g^{-1} for carbon cloth)	46
Steamed cake	3D foam with surface anchored carbon nanoparticles	Single-chamber	Effluent from a matured MFC	1,307 mW m^{-2} (vs. 707 mW m^{-2} for carbon felt)	47
Sugarcane	3D structure embedded with N-doped Fe nanoparticles	Dual-chamber	Activated anaerobic sludge	3,012.7 mW m^{-2} (vs. 1,518.3 mW m^{-2} for carbon cloth)	48
Waste paper	3D carbon aerogel integrated with CeO_2/N-doped graphene	Dual-chamber	Pre-acclimated bacteria from a matured MFC	1,468 mW m^{-2} (vs. 152 mW m^{-2} for bare carbon aerogel)	49
Biomass-derived carbons for anode modifiers					
Almond shell	Powders on carbon cloth	Dual-chamber	A mixture of aerobic and anaerobic sludge	4,346 ± 12 mW m^{-2} (vs. 252 ± 31 mW m^{-2} for carbon cloth)	50
Bacterial cell	Combined with FeS nanoparticles on carbon cloth	Dual-chamber	Mixed microbes	793 mW m^{-2} (vs. 5,542 mW m^{-2} for carbon cloth)	51

Biomass precursor	Feature	MFC type	Microorganism	MFC performance	Ref
Cedar Wood	Powders on carbon felt	Single-chamber	Primary wastewater effluent and dehydrated sludge	9.0 ± 0.6 mW m^{-2} (vs. 1.9 ± 0.2 mW m^{-2} for carbon felt)	38
Cellulose	Powders on carbon cloth	Dual-chamber	*S. putrefaciens* CN32	446 mW m^{-2}	52
Cellulose	Derived graphene oxide (GO) combined with polyaniline on graphite rod	Dual-chamber	Synthetic wastewater	15.13 mA m^{-2}	53
Coffee waste	Powders on carbon cloth	Single-chamber	*E. coli* DH5α	3,927 mW m^{-2} (vs. 975 mW m^{-2} for activated carbon)	54
Goose feather	N-doped activated carbon on carbon paper	Single-chamber	*S. oneidensis* MR-1	960 ± 30 mA m^{-2} (vs. 480 ± 50 mW m^{-2} for carbon paper)	55
Lignin	Derived GO combined with metal oxides (ZnO and TiO_2) on graphite rod	Dual-chamber	Synthetic wastewater	75.43 mA m^{-2} for GO-ZnO and 67.54 mA m^{-2} for GO-TiO_2	56
Mango wood	Spraying powders on carbon felt	Single-chamber	Bacteria from a matured MFC	589.8 mW m^{-2} (vs. 272.2 mW m^{-2} for carbon cloth)	57
Municipal sludge	Rich in N- and O-containing functional groups, synergistic with polypyrrole (PPy)	Dual-chamber	*S. oneidensis* MR-1	568.5 mW m^{-2} (1.9 times larger than that of PPy/activated carbon-modified graphite felt)	58
Plant peel	Powders on graphite sheet	Single-chamber	*P. aeruginosa*	584 mA m^{-2}	59
Plant polyphenols (tannic acid)	Nanoparticles on carbon cloth	Single-chamber	Anaerobic granular sludge and bacteria from a matured MFC	1.72 W m^{-2} (1.82 times higher than the undecorated carbon cloth)	60
Sargassum	As dopant for electrochemical polymerization of pyrrole on stainless-steel sponge (SS)	Dual-chamber	*G. sulfurreducens*	45.2 W m^{-3} (vs. 15.4 W m^{-3} for PPy/SS)	61

(Continued)

TABLE 4.1 (Continued)

Biomass precursor	Feature	MFC type	Microorganism	MFC performance	Ref
Silver grass	Powders on carbon cloth	Dual-chamber	*E. coli*	963 mW cm^{-2} (vs. 258 mW cm^{-2} for Vulcan XC-72)	62
Sludge	Porous structure	Single-chamber	Pre-acclimated anaerobic granular sludge	2,165 ± 21 mW m^{-2} (vs. ~717 mW m^{-2} for carbon cloth)	63
Biomass-derived carbons for anode amendments					
Coconut shell	Graphite plate as current collector	Sediment-MFC	Sediment from the Pearl River	Improved power generation by 2–10 folds and enhanced TOC removal rate 1.7–4 folds	64
Coconut shell	Mixed with metal compounds (Si, Zn and Cu), graphite as current collector	Dual-chamber	Soak effluent of a tannery industry	38.72 ± 0.5 mW m^{-2} (vs. 30.42 ± 0.5 mW m^{-2} for graphite particles)	65
Corncob	Carbon rod as current collector	Single-chamber	Microbes from cow dung	49.92 ± 0.5 W m^{-2} (vs. 2.05 ± 0.06 W m^{-2} for no addition)	66
Corncob	Biochar as wetland matrix	Sediment-MFC	Soil microbes	102.08 mW m^{-2} (vs. 78.24 mW m^{-2} for no addition)	67
Maize straw	Carbon cloth as current collector	Single-chamber	Domestic wastewater	2,400 ± 280 mW m^{-3} (vs. 1,310 ± 240 mW m^{-3} for no addition)	68
Pomelo peel	Graphite felt as current collector	Dual-chamber	Sludge acclimated in synthetic swine wastewater	Enhance efficiency in removing antibiotics	69
Sugarcane cube	Graphite rod as current collector	Dual-chamber	Effluent from a matured MFC	59.94 ± 2.81 W m^{-3} (vs. 16.54 ± 1.44 W m^{-3} for granular activated carbon)	70

3.2 BCDs for Anode Modifiers

Powdered BDCs, especially nanostructured ones, are often prepared to modify conventional monolithic carbon electrodes with the assistance of polymer-based adhesive, rather than being used directly as MFC anodes. In general, such BDCs can significantly improve the surface roughness, biocompatibility, and availability of pore structures of conventional carbon electrodes, thereby enhancing the microbial colonization capacity and electrocatalytic reaction kinetics for efficient MFC anodes. The biochar derived from cedar wood[38] and mango wood[57] was adopted to modify conventional CF electrodes, and obtained about five and two times higher power density when used as the MFC anode, respectively. Herbs are also common natural biomass for biochar production, such as silver grass-derived activated carbon with a superhigh specific surface area of 3,027 m^2g^{-1} achieved a high power output density of 963 mW cm^{-2} when used as the surface modifier of a CF anode.[62] Interestingly, it was found that direct high-temperature pyrolysis of lignocellulosic biomass including cellulose[53] and lignin[56] could produce graphene-like BDCs, which were well-behaved candidates for decorating MFC anodes. In addition, the gelatinous cellulose membrane produced by bacteria was proved to be an excellent precursor for preparing carbon nanofiber aerogel with a 3D open hole structure that significantly enhanced the performance of MFC anode.[52, 72] Activated sludge, a by-product of wastewater treatment, is abundant and expensive to dispose of, so the conversion of it into high-value-added products such as biochar has attracted increasing attention. Recently, it was reported that the areal power density of CC anode was about three fold higher after modification with sludge-derived carbon.[63] This improvement was attributed to the enhanced microbial activity, optimized microbial community structure, and promoted cytochrome expression. Ma et al. developed an interesting approach for loading the sludge-derived carbon powders into the GF substrate electrode through pyrrole electro-polymerization rather than using polymeric binders. When used as the MFC anode, the PPy/SC-modified electrode with a higher surface N/C ratio and higher relative contents of O=C-NH_2 and O=C-O functional groups delivered a power density increased by 1.7 times as compared to the PPy/activated carbon-modified GF anode, which substantially demonstrated the superiority of sludge-derived carbon materials. In addition, domestic waste, such as almond shell,[50] coffee residue[54] and poultry feather,[55] have also been used as the precursors for the production of BDCs for modification of MFC anodes.

3.3 BCDs for Anode Amendments

Powdered or granular BDCs are common amendments for MFCs, especially sediment MFCs integrated with constructed wetland systems for environmental pollutant removal.[73] Macroporous BDCs were prepared from sugarcane and packed into the anode chamber of a dual-chamber MFC, which achieved a maximum power density of 59.94 W m^{-3}, about 2.62 times higher than that filled with the granular active carbon, due to an improved amount of biomass growth and the abundance of electroactive microorganisms.[70] The BDC derived from coconut shells was reported to improve the electricity generation by 2–10 folds and enhance the total organic carbon

(TOC) removal rate by 1.7–4 folds when used as the amendment in a sediment MFC.[64] Similarly, the sediment MFC amended by corncob biochar obviously exceeded that without biochar in terms of chemical oxygen demand (COD) removal efficiency for textile wastewater treatment (83% vs. 66%), decolorization efficiency (90% vs. 65%), total dissolved solid (TDS) reduction (84% vs. 67%), and power density (102.08 mW m^{-2} vs. 78.24 mW m^{-2}).[67] In addition, the sulfuric acid-modified corncob biochar was assessed for the effect on the performance of MFC for dye wastewater treatment, which achieved an improved power density of 49.92 W m^{-2}, current density of 0.28 A m^{-2}, decolorization efficiency of 81.6%, and TDS reduction of 84.4% when a dose of 0.5 g was used. It has been proved that the biochar amendments derived from other biomass precursors, such as maize straw[68] and pomelo peel,[69] could significantly improve the performance of MFCs, regardless of their application.

4 FACTORS AFFECTING THE PERFORMANCE OF BDC ANODES

Given the necessity of electrical interaction, the BDC materials used as MFC anodes must have several specific characteristics for promoting interactions between the electroactive microorganisms and the electrode surface. The surface morphology and chemistry, conductivity, and elemental composition of the anode materials greatly influence the aforementioned interactions, which will be discussed in detail in this section.

4.1 Pore structure and surface area

There is no doubt that the pore structure and surface area of an anode are the critical factors affecting the MFC performance.[15] On the one hand, open and orderly macropores larger than several microns are necessary for microbial implantation and growth, which will greatly promote the electroactive biofilm formation for high electricity production. On the other hand, nanoscale pores are beneficial for significantly increasing surface area to provide more electrochemical reaction sites, as well as facilitating electrolyte diffusion to rapidly feed substrates. As shown in Table 4.1, various 3D macroporous BDCs derived from different biomass precursors have been documented with different degrees of improvement in MFC anodes. In general, these macroporous structures directly originated from the intrinsic structure of the used biomass precursors through the facile pyrolysis approach. The pore size is an important parameter to determine the power generation capacity and stability of the MFC anode. It was found that the bamboo charcoal tube anodes with different inner pore diameters delivered various voltage outputs, and large-sized anodes could run stably for more than 30 days, while the small-sized ones could not.[71] Notably, the stability of macroporous structure depends on the pyrolysis temperature, which of course is usually different for different biomass. Bataillou et al.[38] recently investigated the relationship between pyrolysis temperature and physicochemical features of BDC derived from cedar woody biomass and found that when the pyrolysis temperature increased from 400 to 700°C, the produced BDC showed honeycomb-like macroporous structure with increased specific surface area, but this structure collapsed at 900°C.

In addition, the pyrolysis temperature also has a great influence on the nanoscale pore structure of BDCs, given the loss of volatile components in the biomass precursors. In our previous work,[72] we found that the nanoscale pore of BDC derived from bacterial cellulous pellicle could be tuned by controlling the pyrolysis temperature, in which the micropores were enlarged into mesopores as the increase in temperature from 600 to 1,000°C. As consequence, the BDC with abundant mesopores showed greatly improved microbial electrocatalysis and much higher power output density.

In the preparation of BDCs, the chemical activation approach is often used to create a more abundant pore structure (usually micro- and meso-pores) and a larger specific surface area. To achieve this, the biomass precursor materials are mixed with activation agents (e.g., KOH, $ZnCl_2$, NaOH, K_2CO_3, H_3PO_4, and $FeCl_3$) and then treated at a high temperature under an inert gas atmosphere.[74] In this case, the activation agents can react with the carbon skeleton and etch it, resulting in a porous microstructure. For example, the dried goose feather was mixed with KOH at a mass ratio of 1:3, followed by thermal treatment at 650°C for 3 h in a nitrogen atmosphere, producing activated BDC powder with plentiful microporous structure.[55] Besides, the post-activation of produced BDC powders is also common. The specific surface area of chestnut shell-derived BDC powder increased from 342 to 881 m^2g^{-1} after post-activation using KOH as an activation agent assisted by high-temperature treatment, which was due to the formation of an abundant microporous structure.[40] The combination of the inherent macropores and nanoscale pores enabled this BDC-based anode to 2.3 times higher power density than the conventional CC anode. Likewise, the silver grass-derived BDC post-activated by the KOH agent possessed an unprecedented specific surface area of over 3,000 m^2g^{-1} with the coexistence of macro-, meso-, and micropores.[62] Due to their functions on hosting bacterial cells to form biofilm and facilitating internal mass transfer, favoring fast electron transfer, and promoting nutrient transport to the biofilm, respectively, it achieved a high power output of 963 mW cm^{-2} when used as an *E. coli*-inoculated MFC anode. Apparently, the regulation of pore structure of BDCs at different scales during the preparation process is of great importance to their application in MFC anode.

4.2 Electrical Conductivity

The electrical conductivity of electrode materials is a crucial parameter affecting any electrochemical energy system. For a BDC, its electrical conductivity depends mainly on the degree of graphitization, which is mainly related to the pyrolysis temperature and time.[74] Generally, the higher the pyrolysis temperature and the longer the pyrolysis time, the better the electrical conductivity of the products. Woody-derived biochar pyrolyzed at 400°C was not conducting, while its electrical resistivity decreased by several orders of magnitude as the pyrolysis temperature increased from 500 to 900°C.[38] It was also found that the conductivity and carbon content of sewage sludge-derived carbon prepared through methane chemical vapor method had positive correlation with the treat temperature, the MFC using the biocarbon anode processed at 1,200°C obtained five times higher electricity production than that equipped with a graphite anode.[45] Therefore, enhancing the conductivity of

BDCs without damaging the pore structure by increasing the pyrolysis temperature is generally beneficial to the performance of MFC.

4.3 Surface accessibility

The surface of the MFC anode is where electroactive microorganisms interact directly. It is conceivable that electrode surface chemistry including wettability, charge, roughness, and redox activity will affect microbial metabolism and growth and reaction kinetics.[15, 75] The wettability of a BDC electrode, which usually deteriorates with increasing pyrolysis temperature, directly influences the attachment of microbial cells and other biological components such as proteins and electron shuttles. It has been proved that the hydrophilic anode surface favored MFC performance in terms of shorter start-up time and higher power generation than the hydrophobic one because of better biofilm growth.[76] While the effect of surface wettability of BDCs on the MFC anode performance has not been fully elucidated. Positive-charged anode surface is apparently favorable to microbial adhesion and biofilm growth, given the fact of a negative-charged cell envelope. However, the surface of BDCs is usually negatively charged because of the ubiquity of hydroxyl and carboxyl functional groups.[77] Surface ammonification with positively charged active molecules is an effective method to overcome this bottleneck. Many studies have shown that positive-charged polyaniline (PANI) has significant advantages in this respect because of its excellent conductivity and biocompatibility. Our previous study demonstrated that the surface functionalization of mesoporous BDC derived from green tree leaves extractive with chemically polymerized PANI not only significantly improved the microbial biofilm growth but also enhanced the interaction between cells and electrode surface, thus resulting in a power output density ten times that of the control anode (CC electrode).[78] Recently, Yaqoob et al. also proved the evident superiority of PANI-modified graphene-derived cellulose in terms of power generation and the remediation efficiency of heavy metals (Cd^{2+} and Pb^{2+}) compared to the unmodified graphene anode.[53] Evidently, surface modification and functionalization of BDCs are effective means to improve their biological accessibility for high-performance MFCs, which is worthy of further study.

4.4 Heteroatom self-doping

Ever-growing attention has been paid to heteroatom-doped carbon materials as the fascinating advantages in regulating electronic structure, optimizing intrinsic activity, and improving surface wettability for heterogeneous electrocatalysis. Since biomass precursors are rich in various biogenic elements (e.g., N, P, S) because of the presence of proteins and nucleic acids, heteroatomic doping of BDCs is common,[79] which has been found to be beneficial to improve the performance of MFC anodes. Particularly, the N doping has been widely reported to promote the contact between electrogenic and anodes, and enhance the electrochemical reaction kinetics at the anode interface.[41, 44, 80–83] Recently, an N-doped macroporous carbon foam derived from the fermented wheat flour that naturally contains N element, was proved to be high-performance MFC anode after the modification with PANI.[83] The macroporous carbon foams with N doping prepared from naturally abundant nitrogenous corncobs were also documented to achieve high power density and current density when used as MFC anodes.[41] Moreover, the N-doped bioanodes with good biocompatibility resulted in not only the enrichment of electroactive *Geobacter* cells but also the improvement in the EET of the *Geobacter* because of the presence of pyrrolic N atoms. The N, P-codoped porous BDC prepared by one-step carbonization of pinecone was found to facilitate the electroactive bacterial adhesion and accelerate the MFC start-up.[44] Recently, N, P, S, Co-codoped carbon nanoparticles derived from plant polyphenols (tannic acid) were presented as a highly efficient BDC modifier of traditional CC electrode, and the decorated CC electrode when used as MFC anode exhibited a power density increased by 1.82 times and a current density increased by 1.44 times compared with the undecorated one.[60]

5 SUMMARY AND PERSPECTIVES

This chapter focuses on the latest progress of BDCs from various natural biomasses as anode materials for MFCs and highlights the effects of pore structure, electrical conductivity, surface chemistry, and heteroatom self-doping on their bioelectrochemical performance. Evidently, the use of sustainable and cheap biomass as raw precursors to prepare valuable carbon electrodes can reduce the dependence on non-renewable resources, such as expensive carbon materials from fossil sources, and bring great benefits to the commercialization and applications of MFCs to a certain extent. Due to the need to undergo a high-temperature carbonization process, BDC materials usually have rich porous structure and high specific surface area, which is very conducive to biocatalyst loading and electrolyte diffusion, while providing more electrochemical reaction sites. Moreover, the inherent heteroatom self-doping of BDC materials gives them surface chemical features that are easier for bacteria and biomolecules to get close to, as well as easily regulated electronic structure and catalytic activities, leading to the enhanced bioelectrocatalytic kinetics of MFC anodes. Thus, the development of BDC materials for MFC anodes is greatly promising and is in full swing.

However, there are still some challenges along this avenue. For example, the structure and properties of BDCs depend to a large extent on the raw biomass precursors, which often vary significantly with the location, growing environment, and harvest time. Therefore, the reproducibility of the prepared BCD anodes is a tricky challenge, as it is critical for the stable operation of the MFC systems. Moreover, due to the inherent structure of the raw precursors, the pore size and structure of BCDs are generally widely distributed and disordered, so many of them are not feasible for the bacterial electrocatalytic process in MFCs. In addition, the degradation of BCD anodes during MFC operation due to biotic and abiotic corrosion is also a challenge that cannot be ignored.

To address these challenges, some perspectives are highlighted to develop high-performance BCD materials for MFC anodes. First, considering the economic and environmental benefits, the use of various waste biomass rather than economically valuable ones (such as starch and its derivatives, cocoon silk) is preferred for the preparation of BDC materials. Second, it is a significant direction to modulate the pore structure of BDCs reasonably according to the demand, of course, this needs a clear understanding of the effect of the pore structure on the anodic bioelectrocatalysis as the premise. It has been recognized that the selection of appropriate biomass precursors to directly obtain a wealth of macro pores larger than a few microns in size is satisfactory, considering that these macro pores significantly promote the colonization of electrogenic bacteria. The cutting of nano-sized pores is also necessary, perhaps by introducing soft/hard templates or controlling carbonization parameters. Third, in view of the significant contribution of heteroatom doping, the effective controlling of concentration of different doping species and their ratio the preparation process of BDCs is worth a lot of effort. Finally, it is urgent to elucidate the key factors leading to the degradation of BDC anodes in the process of MFC operation, which will provide a basis for ensuring the stability of MFCs and promoting their applications.

REFERENCES

1. J. Barber, Photosynthetic energy conversion: Natural and artificial, *Chem. Soc. Rev.*, 2009, **38**, 185–196.
2. N.S. Weliwatte and S.D. Minteer, Photo-bioelectrocatalytic CO_2 reduction for a circular energy landscape, *Joule*, 2021, **5**, 2564–2592.
3. Q. Zhang, L. Suresh, Q. Liang, Y. Zhang, L. Yang, N. Paul and S.C. Tan, Emerging technologies for green energy conversion and storage, *Adv. Sustain. Syst.*, 2021, **5**, 2000152.
4. C. Senthil and C.W. Lee, Biomass-derived biochar materials as sustainable energy sources for electrochemical energy storage devices, *Renew. Sustain. Energ. Rev.*, 2021, **137**, 110464.
5. H. Wang and Z.J. Ren, A comprehensive review of microbial electrochemical systems as a platform technology, *Biotechnol. Adv.*, 2013, **31**, 1796–1807.
6. B.E. Logan and K. Rabaey, Conversion of wastes into bioelectricity and chemicals by using microbial electrochemical technologies, *Science*, 2012, **337**, 686–690.
7. H. Chen, F.Y. Dong and S.D. Minteer, The progress and outlook of bioelectrocatalysis for the production of chemicals, fuels and materials, *Nat. Catal.*, 2020, **3**, 225–244.
8. T. Cai, L. Meng, G. Chen, Y. Xi, N. Jiang, J. Song, S. Zheng, Y. Liu, G. Zhen and M. Huang, Application of advanced anodes in microbial fuel cells for power generation: A review, *Chemosphere*, 2020, **248**, 125985.
9. U. Schroder, F. Harnisch and L.T. Angenent, Microbial electrochemistry and technology: Terminology and classification, *Energ. Environ. Sci.*, 2015, **8**, 513–519.
10. E. Yang, H.O. Mohamed, S.-G. Park, M. Obaid, S.Y. Al-Qaradawi, P. Castan, K. Chon and K.-J. Chae, A review on self-sustainable microbial electrolysis cells for electro-biohydrogen production via coupling with carbon-neutral renewable energy technologies, *Bioresour. Technol.*, 2021, **320**, 124363.
11. A. Al-Mamun, W. Ahmad, M.S. Baawain, M. Khadem and B.R. Dhar, A review of microbial desalination cell technology: Configurations, optimization and applications, *J. Clean. Prod.*, 2018, **183**, 458–480.
12. K. Rabaey and R.A. Rozendal, Microbial electrosynthesis – Revisiting the electrical route for microbial production, *Nat. Rev. Microbiol.*, 2010, **8**, 706–716.
13. J.V. Boas, V.B. Oliveira, M. Simões and A.M.F.R. Pinto, Review on microbial fuel cells applications, developments and costs, *J. Environ. Manage.*, 2022, **307**, 114525.

14. J. Hu, Q. Zhang, D.-J. Lee and H.H. Ngo, Feasible use of microbial fuel cells for pollution treatment, *Renew. Energ.*, 2018, **129**, 824–829.
15. L. Zou, Y. Qiao and C.M. Li, Boosting microbial electrocatalytic kinetics for high power density: Insights into synthetic biology and advanced nanoscience, *Electrochem. Energy Rev.*, 2018, **1**, 567–598.
16. X. Fan, Y. Zhou, X. Jin, R.-B. Song, Z. Li and Q. Zhang, Carbon material-based anodes in the microbial fuel cells, *Carbon Energy*, 2021, **3**, 449–472.
17. S. Li, C. Cheng and A. Thomas, Carbon-based microbial-fuel-cell electrodes: From conductive supports to active catalysts, *Adv. Mater.*, 2017, **29**, 1602547.
18. A.G. Olabi, T. Wilberforce, E.T. Sayed, K. Elsaid, H. Rezk and M.A. Abdelkareem, Recent progress of graphene based nanomaterials in bioelectrochemical systems, *Sci. Total Environ.*, 2020, **749**, 141225.
19. M. Maas, Carbon nanomaterials as antibacterial colloids, *Materials*, 2016, **9**, 617.
20. J. Deng, M. Li and Y. Wang, Biomass-derived carbon: Synthesis and applications in energy storage and conversion, *Green Chem.*, 2016, **18**, 4824–4854.
21. W. Long, B. Fang, A. Ignaszak, Z. Wu, Y.-J. Wang and D. Wilkinson, Biomass-derived nanostructured carbons and their composites as anode materials for lithium ion batteries, *Chem. Soc. Rev.*, 2017, **46**, 7176–7190.
22. X. Tang, D. Liu, Y.-J. Wang, L. Cui, A. Ignaszak, Y. Yu and J. Zhang, Research advances in biomass-derived nanostructured carbons and their composite materials for electrochemical energy technologies, *Prog. Mater Sci.*, 2021, **118**, 100770.
23. A.A. Yaqoob, M.N.M. Ibrahim and S. Rodríguez-Couto, Development and modification of materials to build cost-effective anodes for microbial fuel cells (MFCs): An overview, *Biochem. Eng. J.*, 2020, **164**, 107779.
24. M.C. Potter, Electrical effects accompanying the decomposition of organic compounds, *Proc. R. Soc. B-Biol. Sci.*, 1911, **84**, 260–276.
25. C. Santoro, C. Arbizzani, B. Erable and I. Ieropoulos, Microbial fuel cells: From fundamentals to applications. A review, *J. Power Sources*, 2017, **356**, 225–244.
26. H. Gul, W. Raza, J. Lee, M. Azam, M. Ashraf and K.-H. Kim, Progress in microbial fuel cell technology for wastewater treatment and energy harvesting, *Chemosphere*, 2021, **281**, 130828.
27. B.E. Logan, R. Rossi, A.A. Ragab and P.E. Saikaly, Electroactive microorganisms in bioelectrochemical systems, *Nat. Rev. Microbiol.*, 2019, **17**, 307–319.
28. O. Bretschger, A. Obraztsova, C.A. Sturm, I.S. Chang, Y.A. Gorby, S.B. Reed, D.E. Culley, C.L. Reardon, S. Barua, M.F. Romine, J. Zhou, A.S. Beliaev, R. Bouhenni, D. Saffarini, F. Mansfeld, B.-H. Kim, J.K. Fredrickson and K.H. Nealson, Current production and metal oxide reduction by Shewanella oneidensis MR-1 wild type and mutants, *Appl. Environ. Microbiol.*, 2007, **73**, 7003–7012.
29. R.J. Steidl, S. Lampa-Pastirk and G. Reguera, Mechanistic stratification in electroactive biofilms of Geobacter sulfurreducens mediated by pilus nanowires, *Nat. Commun.*, 2016, **7**, 12217.
30. S. Pirbadian, S.E. Barchinger, K.M. Leung, H.S. Byun, Y. Jangir, R.A. Bouhenni, S.B. Reed, M.F. Romine, D.A. Saffarini, L. Shi, Y.A. Gorby, J.H. Golbeck and M.Y. El-Naggar, Shewanella oneidensis MR-1 nanowires are outer membrane and periplasmic extensions of the extracellular electron transport components, *Proc. Natl Acad. Sci. U. S. A.*, 2014, **111**, 12883–12888.
31. N.J. Kotloski and J.A. Gralnick, Flavin electron shuttles dominate extracellular electron transfer by shewanella oneidensis, *mBio*, 2013, **4**, e00553–00512.
32. Y. Qiao, Y.-J. Qiao, L. Zou, C.-X. Ma and J.-H. Liu, Real-time monitoring of phenazines excretion in Pseudomonas aeruginosa microbial fuel cell anode using cavity microelectrodes, *Bioresour. Technol.*, 2015, **198**, 1–6.

33. B. Zhang, Y. Jiang and R. Balasubramanian, Synthesis, formation mechanisms and applications of biomass-derived carbonaceous materials: A critical review, *J. Mater. Chem. A*, 2021, **9**, 24759–24802.
34. Y.-T. Shi, Y.-Y. Yu, Z.-A. Xu, J. Lian and Y.-C. Yong, Superior carbon belts from Spirogyra for efficient extracellular electron transfer and sustainable microbial energy harvesting, *J. Mater. Chem. A*, 2019, **7**, 6930–6938.
35. J. Zhang, J. Li, D. Ye, X. Zhu, Q. Liao and B. Zhang, Tubular bamboo charcoal for anode in microbial fuel cells, *J. Power Sources*, 2014, **272**, 277–282.
36. Z.-A. Xu, X.-M. Ma, Y.-T. Shi, J.M. Moradian, Y.-Z. Wang, G.-F. Sun, J. Du, X.-M. Ye and Y.-C. Yong, Surface activated natural wood biomass electrode for efficient microbial electrocatalysis: Performance and mechanism, *Int. J. Energ. Res.*, 2022, **46**, 8480–8490.
37. L. Zhang, W. He, J. Yang, J. Sun, H. Li, B. Han, S. Zhao, Y. Shi, Y. Feng, Z. Tang and S. Liu, Bread-derived 3D macroporous carbon foams as high performance free-standing anode in microbial fuel cells, *Biosens. Bioelectron.*, 2018, **122**, 217–223.
38. G. Bataillou, C. Lee, V. Monnier, T. Gerges, A. Sabac, C. Vollaire and N. Haddour, Cedar wood-based biochar: Properties, characterization, and applications as anodes in microbial fuel cell, *Appl. Biochem. Biotechnol.*, 2022, **194**, 4169–4186.
39. S. Chen, J. Tang, X. Jing, Y. Liu, Y. Yuan and S. Zhou, A hierarchically structured urchin-like anode derived from chestnut shells for microbial energy harvesting, *Electrochim. Acta*, 2016, **212**, 883–889.
40. Q. Chen, W. Pu, H. Hou, J. Hu, B. Liu, J. Li, K. Cheng, L. Huang, X. Yuan, C. Yang and J. Yang, Activated microporous-mesoporous carbon derived from chestnut shell as a sustainable anode material for high performance microbial fuel cells, *Bioresour. Technol.*, 2018, **249**, 567–573.
41. Y. Wang, C. He, W. Li, W. Zong, Z. Li, L. Yuan, G. Wang and Y. Mu, High power generation in mixed-culture microbial fuel cells with corncob-derived three-dimensional N-doped bioanodes and the impact of N dopant states, *Chem. Eng. J.*, 2020, **399**, 125848.
42. H. Zhu, H. Wang, Y. Li, W. Bao, Z. Fang, C. Preston, O. Vaaland, Z. Ren and L. Hu, Lightweight, conductive hollow fibers from nature as sustainable electrode materials for microbial energy harvesting, *Nano Energy*, 2014, **10**, 268–276.
43. J. Tang, Y. Yuan, T. Liu and S. Zhou, High-capacity carbon-coated titanium dioxide core-shell nanoparticles modified three dimensional anodes for improved energy output in microbial fuel cells, *J. Power Sources*, 2015, **274**, 170–176.
44. R. Wang, D. Liu, M. Yan, L. Zhang, W. Chang, Z. Sun, S. Liu and C. Guo, Three-dimensional high performance free-standing anode by one-step carbonization of pinecone in microbial fuel cells, *Bioresour. Technol.*, 2019, **292**, 121956.
45. H. Feng, Y. Jia, D. Shen, Y. Zhou, T. Chen, W. Chen, Z. Ge, S. Zheng and M. Wang, The effect of chemical vapor deposition temperature on the performance of binder-free sewage sludge-derived anodes in microbial fuel cells, *Sci. Total Environ.*, 2018, **635**, 45–52.
46. M. Lu, Y. Qian, C. Yang, X. Huang, H. Li, X. Xie, L. Huang and W. Huang, Nitrogen-enriched pseudographitic anode derived from silk cocoon with tunable flexibility for microbial fuel cells, *Nano Energy*, 2017, **32**, 382–388.
47. H. Yuan, G. Dong, D. Li, L. Deng, P. Cheng and Y. Chen, Steamed cake-derived 3D carbon foam with surface anchored carbon nanoparticles as freestanding anodes for high-performance microbial fuel cells, *Sci. Total Environ.*, 2018, **636**, 1081–1088.
48. B. Song, J. Li, Z. Wang, J. Ali, L. Wang, Z. Zhang, F. Liu, E.M. Glebov, J. Zhang and X. Zhuang, N-doped Fe nanoparticles anchored on 3D carbonized sugarcane anode for high power density and efficient chromium(VI) removal, *J. Environ. Chem. Eng.*, 2022, **10**, 108751.

49. N. Senthilkumar, M.A. Aziz, M. Pannipara, A.T. Alphonsa, A.G. Al-Sehemi, A. Balasubramani and G. Gnana Kumar, Waste paper derived three-dimensional carbon aerogel integrated with ceria/nitrogen-doped reduced graphene oxide as freestanding anode for high performance and durable microbial fuel cells, *Bioproc. Biosyst. Eng.*, 2020, **43**, 97–109.
50. M. Li, S. Ci, Y. Ding and Z. Wen, Almond shell derived porous carbon for a high-performance anode of microbial fuel cells, *Sustain. Energ. Fuels*, 2019, **3**, 3415–3421.
51. Z. Li, P. Zhang, Y. Qiu, Z. Zhang, X. Wang, Y. Yu and Y. Feng, Biosynthetic FeS/BC hybrid particles enhanced the electroactive bacteria enrichment in microbial electrochemical systems, *Sci. Total Environ.*, 2021, **762**, 143142.
52. D. Wang, Y. Wang, J. Yang, X. He, R.-J. Wang, Z.-S. Lu and Y. Qiao, Cellulose aerogel derived hierarchical porous carbon for enhancing flavin-based interfacial electron transfer in microbial fuel cells, *Polymers*, 2020, **12**, 664.
53. A.A. Yaqoob, M.N. Mohamad Ibrahim, K. Umar, S.A. Bhawani, A. Khan, A. M. Asiri, M.R. Khan, M. Azam and A.M. AlAmmari, Cellulose derived graphene/polyaniline nanocomposite anode for energy generation and bioremediation of toxic metals via benthic microbial fuel cells, *Polymers*, 2021, **13**, 135.
54. Y.-H. Hung, T.-Y. Liu and H.-Y. Chen, Renewable coffee waste-derived porous carbons as anode materials for high-performance sustainable microbial fuel cells, *ACS Sustain. Chem. Eng.*, 2019, **7**, 16991–16999.
55. Y.-X. Wang, W.-Q. Li, C.-S. He, G.-N. Zhou, H.-Y. Yang, J.-C. Han, S.-Q. Huang and Y. Mu, Efficient bioanode from poultry feather wastes-derived N-doped activated carbon: Performance and mechanisms, *J. Clean. Prod.*, 2020, **271**, 122012.
56. A.A. Yaqoob, M.N.M. Ibrahim, A.S. Yaakop and M. Rafatullah, Utilization of biomass-derived electrodes: A journey toward the high performance of microbial fuel cells, *Appl. Water Sci.*, 2022, **12**, 99.
57. M. Li, Y.-W. Li, Q.-Y. Cai, S.-Q. Zhou and C.-H. Mo, Spraying carbon powder derived from mango wood biomass as high-performance anode in bio-electrochemical system, *Bioresour. Technol.*, 2020, **300**, 122623.
58. X. Ma, C. Feng, W. Zhou and H. Yu, Municipal sludge-derived carbon anode with nitrogen- and oxygen-containing functional groups for high-performance microbial fuel cells, *J. Power Sources*, 2016, **307**, 105–111.
59. A. Vempaty, A. Kumar, S. Pandit, M. Gupta, A.S. Mathuriya, D. Lahiri, M. Nag, Y. Kumar, S. Joshi and N. Kumar, Evaluation of the Datura peels derived biochar-based anode for enhancing power output in microbial fuel cell application, *Biocatal. Agric. Biotechnol.*, 2023, **47**, 102560.
60. K. Zhu, S. Wang, H. Liu, S. Liu, J. Zhang, J. Yuan, W. Fu, W. Dang, Y. Xu, X. Yang and Z. Wang, Heteroatom-doped porous carbon nanoparticle-decorated carbon cloth (HPCN/CC) as efficient anode electrode for microbial fuel cells (MFCs), *J. Clean. Prod.*, 2022, **336**, 130374.
61. G. Wu, H. Bao, Z. Xia, B. Yang, L. Lei, Z. Li and C. Liu, Polypyrrole/sargassum activated carbon modified stainless-steel sponge as high-performance and low-cost bioanode for microbial fuel cells, *J. Power Sources*, 2018, **384**, 86–92.
62. M. Rethinasabapathy, J.H. Lee, K.C. Roh, S.-M. Kang, S.Y. Oh, B. Park, G.-W. Lee, Y.L. Cha and Y.S. Huh, Silver grass-derived activated carbon with coexisting micro-, meso- and macropores as excellent bioanodes for microbial colonization and power generation in sustainable microbial fuel cells, *Bioresour. Technol.*, 2020, **300**, 122646.
63. K. Zhu, Y. Xu, X. Yang, W. Fu, W. Dang, J. Yuan and Z. Wang, Sludge derived carbon modified anode in microbial fuel cell for performance improvement and microbial community dynamics, *Membranes*, 2022, **12**, 120.

64. S.S. Chen, J.H. Tang, L. Fu, Y. Yuan and S.G. Zhou, Biochar improves sediment microbial fuel cell performance in low conductivity freshwater sediment, *J. Soils Sed.*, 2016, **16**, 2326–2334.
65. M. Naveenkumar and K. Senthilkumar, Microbial fuel cell for harvesting bio-energy from tannery effluent using metal mixed biochar electrodes, *Biomass Bioenerg.*, 2021, **149**, 106082.
66. K. Sonu, M. Sogani, Z. Syed, A. Dongre and G. Sharma, Effect of corncob derived biochar on microbial electroremediation of dye wastewater and bioenergy generation, *ChemistrySelect*, 2020, **5**, 9793–9798.
67. K. Sonu, M. Sogani and Z. Syed, Integrated constructed wetland-nicrobial fuel cell using biochar as wetland matrix: Influence on power generation and textile wastewater treatment, *ChemistrySelect*, 2021, **6**, 8323–8328.
68. J. Dong, Y. Wu, C. Wang, H. Lu and Y. Li, Three-dimensional electrodes enhance electricity generation and nitrogen removal of microbial fuel cells, *Bioproc. Biosyst. Eng.*, 2020, **43**, 2165–2174.
69. D. Cheng, N. Huu Hao, W. Guo, S.W. Chang, N. Dinh Duc, J. Li, L. Quang Viet, N. Thi An Hang and T. Van Son, Applying a new pomelo peel derived biochar in microbial fell cell for enhancing sulfonamide antibiotics removal in swine wastewater, *Bioresour. Technol.*, 2020, **318**, 123886.
70. Y. Zhou, G. Zhou, L. Yin, J. Guo, X. Wan and H. Shi, High-performance carbon anode derived from sugarcane for packed microbial fuel cells, *ChemElectroChem*, 2017, **4**, 168–174.
71. J. Li, J. Zhang, D. Ye, X. Zhu, Q. Liao and J. Zheng, Optimization of inner diameter of tubular bamboo charcoal anode for a microbial fuel cell, *Int. J. Hydrogen Energy*, 2014, **39**, 19242–19248.
72. L. Zou, Y. Qiao, Z.-Y. Wu, X.-S. Wu, J.-L. Xie, S.-H. Yu, J. Guo and C.M. Li, Tailoring unique mesopores of hierarchically porous structures for fast direct electrochemistry in microbial fuel cells, *Adv. Energ. Mater.*, 2016, **6**, 1501535.
73. T. Saeed, M. J. Miah and A.K. Yadav, Development of electrodes integrated hybrid constructed wetlands using organic, construction, and rejected materials as filter media: Landfill leachate treatment, *Chemosphere*, 2022, **303**, 135273.
74. H. Zhang, Y. Zhang, L. Bai, Y. Zhang and L. Sun, Effect of physiochemical properties in biomass-derived materials caused by different synthesis methods and their electrochemical properties in supercapacitors, *J. Mater. Chem. A*, 2021, **9**, 12521–12552.
75. C. Li and S. Cheng, Functional group surface modifications for enhancing the formation and performance of exoelectrogenic biofilms on the anode of a bioelectrochemical system, *Crit Rev. Biotechnol.*, 2019, **39**, 1015–1030.
76. C. Santoro, M. Guilizzoni, J P. C. Baena, U. Pasaogullari, A. Casalegno, B. Li, S. Babanova, K. Artyushkova and P. Atanassov, The effects of carbon electrode surface properties on bacteria attachment and start up time of microbial fuel cells, *Carbon*, 2014, **67**, 128–139.
77. Z. Tan, S. Yuan, M. Hong, L. Zhang and Q. Huang, Mechanism of negative surface charge formation on biochar and its effect on the fixation of soil Cd, *J. Hazard, Mater.*, 2020, **384**, 121370.
78. L. Zou, Y. Qiao, C. Zhong and C.M. Li, Enabling fast electron transfer through both bacterial outer-membrane redox centers and endogenous electron mediators by polyaniline hybridized large-mesoporous carbon anode for high-performance microbial fuel cells, *Electrochim. Acta*, 2017, **229**, 31–38.
79. A. Gopalakrishnan and S. Badhulika, Effect of self-doped heteroatoms on the performance of biomass-derived carbon for supercapacitor applications, *J. Power Sources*, 2020, **480**, 228830.

80. X. Wu, Y. Qiao, C. Guo, Z. Shi and C.M. Li, Nitrogen doping to atomically match reaction sites in microbial fuel cells, *Commun. Chem.*, 2020, **3**, 68.
81. G. Massaglia, V. Margaria, M.R. Fiorentin, K. Pasha, A. Sacco, M. Castellino, A. Chiodoni, S. Bianco, F.C. Pirri and M. Quaglio, Nonwoven mats of N-doped carbon nanofibers as high-performing anodes in microbial fuel cells, *Mater. Today Energy*, 2020, **16**, 100385.
82. H.-R. Yuan, L.-F. Deng, X. Qian, L.-F. Wang, D.-N. Li, Y. Chen and Y. Yuan, Significant enhancement of electron transfer from Shewanella oneidensis using a porous N-doped carbon cloth in a bioelectrochemical system, *Sci. Total Environ.*, 2019, **665**, 882–889.
83. D. Jiang, H. Xie, H. Chen, K. Cheng, L. Li, K. Xie and Y. Wang, Polyaniline@N-doped macroporous carbon foam as self-supporting anodes for microbial fuel cells, *Int. J. Hydrogen Energy*, 2022, **47**, 35458–35467.

5 Biomass-Derived Graphene-Like Carbon Materials for Supercapacitor Applications

Gaojie Li, Biao Gao, Paul K. Chu and Kaifu Huo

1 INTRODUCTION

Fossil fuels are causing environmental problems and also getting depleted. Therefore, researchers have proposed renewable energy sources such as solar, wind, tide, and waste heat, which can provide energy without CO_2 emissions and pollutants. However, since renewable energy sources are often difficult to use on a large scale due to their geographical, intermittent, and unstable characteristics, the development of efficient energy conversion and storage technologies is highly required. The common electrochemical energy storage devices are rechargeable batteries and supercapacitors (SCs). SCs have attracted a lot of attention due to desirable characteristics such as the higher power density (1–103 kW kg^{-1}), faster charging/discharging capability, and longer cycling life (>106 cycles) compared to conventional lead-acid batteries and lithium-ion batteries.[1] In addition, SCs can be used in a wide temperature range (−70 to 85°C), whereas the use of conventional batteries is limited to a narrow temperature range. Owing to these excellent properties, SCs have great potential and value in future energy systems. Figure 5.1 depicts the Ragone plots of various energy storage systems.[2] Batteries can provide high energy densities of 150–500 Wh kg^{-1}, but their power density is limited due to the sluggish movement of electrons and ions. In order to maintain a high energy output, they are usually discharged for more than 10 min or even longer. In contrast, SCs with higher power densities can fully discharge in 10 s with an output power between 10 and 20 kW kg^{-1}. However, it is difficult for SCs to achieve energy densities of 30 Wh kg^{-1} or higher. As the key component in SCs, electrode materials have a decisive influence on the energy/power densities of SCs and therefore, the development of high-performance electrode materials is the hotspot for the SCs.

Carbonaceous materials are very important to electrochemical energy storage on account of their outstanding chemical stability, tunable porosity, large specific surface area (SSA), and abundant electroactive surface sites. Several carbon isomers

DOI: 10.1201/9781003387831-5

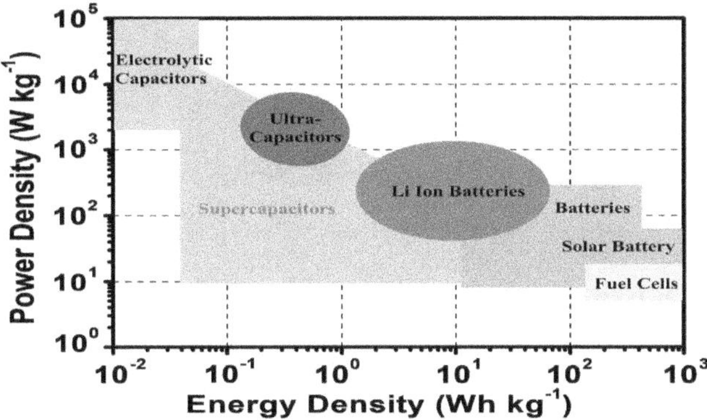

FIGURE 5.1 Ragone plots of various energy storage systems[2] (reproduced with permission: Copyright 2022, Royal Society of Chemistry).

have been utilized in electrodes and in particular, graphene has recently attracted great interest due to its distinct properties in terms of the electron conductivity (200 S m^{-1}), large SSA (2,630 $m^2 g^{-1}$), and high intrinsic carrier mobility (10,000 $cm^2 V^{-1} s^{-1}$).[3] With sp^2-hybridized carbon atoms arranged in a honeycomb-like network in a single plane, graphene is an ultrathin substance with high conductivity.[4] In addition, graphene with the long-range π–π conjugation possesses many attractive properties including chemical stability, thermal conductivity, and mechanical strength. However, there are some limitations for SCs application. First, large-scale synthesis is difficult and costly, and second, graphene layers undergo irreversible sheet stacking during electrode preparation, consequently reducing the active surface area and capacitance. Biomass-derived graphene-like carbon (BDGC) materials are excellent for SCs due to many virtues such as sustainability, reproducibility, abundant natural sources, and low cost.

In this chapter, recent progress in the preparation and capacitive properties of BDGC from biomass precursors are summarized and the common synthetic methods are described. The impact of the morphology, SSA, porous structure, and degree of graphitization on the electrochemical properties are discussed and future research directions are proposed.

1.1 Fundamentals of Supercapacitors

SCs also known as electrochemical capacitors or ultracapacitors have attracted considerable attention for electrochemical energy storage by bridging the gap between conventional dielectric capacitors and lithium-ion batteries. According to the charge storage mechanism, SCs are divided into three types: (1) Electric double-layer capacitors (EDLCs), which electrostatically store charges by adsorbing ions on the electrode surface; (2) Pseudocapacitors which electrochemically store energy *via* the rapid surface-controlled redox reactions with the Faraday types; (3) Hybrid capacitors which combine the electrodes with capacitive properties and battery-like performances to produce higher energy and power densities.

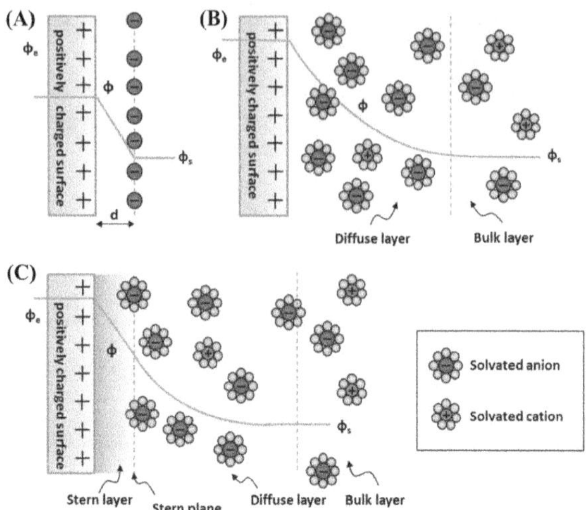

FIGURE 5.2 EDLC models: (A) Helmholtz, (B) Gouy–Chapman, and (C) Stern (reproduced with permission: Copyright 2014, Wiley-Blackwell).[5]

1.2 Charge Storage Mechanism

EDLCs store charge at the interface between the electrode and electrolyte via electrostatic ion adsorption/desorption. By applying a voltage between the electrodes, charges accumulate on the surface of the electrode, which is balanced by the adsorption of oppositely charged ions from the electrolyte. To avert short circuits, the two electrodes of EDLCs are separated by a separator. At the interface between the electrode and electrolyte, two layers of reverse charges are established during charging. EDLCs are comparable to parallel plate capacitors when an electrode of surface area S (m^2) is polarized, as shown in Figure 5.2.[5] Originally, the storage mechanism of EDLCs follows the Helmholtz model which describes the charge distribution at the interface (Figure 5.2A). The double layers (DL) of opposite charges are formed at the interface with an effective thickness (d) equal to the atomic distance. The potential in the vicinity of the electrode is gradually lowered and the simplified Helmholtz double layer can be envisioned as an electrical capacitor with a capacitance of C_H defined by Eqn 5.1:

$$C_H = \frac{\varepsilon_r \varepsilon_0 S}{d} \tag{5.1}$$

where ε_r is the dielectric constant of the electrolyte, ε_0 is the relative dielectric constant of vacuum ($\varepsilon_0 = 8.854 \times 10^{-12}$ F m^{-1}), and S is the surface area of the active materials.

Gouy and Chapman have considered the reality of a diffused layer in the electrolyte resulting from the accumulation of ions on the surface of the electrode, as shown in Figure 5.2B. The potential decreases exponentially from the electrode surface to the electrolyte. Stern has integrated the Helmholtz and Gouy–Chapman models into one model because the Helmholtz and Gouy–Chapman models are not sufficient to explain the actual phenomenon. The diffusion of ions in the electrolyte as well as the connection between the dipole moment of the electrolyte and active substance cannot be explained by the Helmholtz model.[6] In the Stern model, two ion distribution regions are involved: the diffusion layer (outside region) and Helmholtz layer,[6] as shown in Figure 5.2C. These two layers correspond to two series capacitors with capacitances of C_H (Helmholtz layer) and C_D (diffusion layer) contributing to the total capacitance of the electrode (C_{DL}), which can be evaluated by Eqn 5.2[5]:

$$\frac{1}{C_{DL}} = \frac{1}{C_H} + \frac{1}{C_D} \tag{5.2}$$

In order to provide the basis for the comparison between different electrode materials, it is a common practice to carry out galvanostatic charging-discharging (GCD) test. In a symmetrical system, the total capacitance (C, F g^{-1}) of the devices (two electrodes equal in mass, thickness, size, and materials) is calculation by Eqn 5.3:

$$C = 4\frac{It}{\Delta Vm} \tag{5.3}$$

where I (A) is the current, t (s) is the discharging time, ΔV (V) is the voltage window, and m (g) is the total mass of active materials. The stored specific energy E (Wh kg^{-1}) in a supercapacitor, also known as the energy density, is given by Eqn 5.4:

$$E = \frac{1}{2}CV^2 \tag{5.4}$$

where m (kg) is the mass of the supercapacitor and V is the maximum voltage window. Hence, a larger E can be achieved by increasing C or V. The maximum specific power (W kg^{-1}) also depends on the maximum voltage and is given by Eqn 5.5:

$$P = \frac{U_{max}^2}{4 \cdot ESR \cdot m} \tag{5.5}$$

where ESR (Ω) is the equivalent series resistance which is the sum of the ion resistance of the electrolyte impregnated in the separator, resistance of the electrode, and interface resistance between the electrode and current collector.

The pseudocapacitor is based on a Faradaic charging process in which the redox electron transfer process occurs on the electrode surface, as in the case of a battery. Unlike batteries, the capacitance stems from the formation of a specific relationship between the degree of charge acceptance and voltage change. The chemical reaction in pseudocapacitors is a fast and reversible process involving electron transfer, and the electrode materials do not undergo any phase change and not involve the creation or destruction of chemical bonds. Pseudocapacitors can provide a higher specific capacitance than EDLCs, thereby making them attractive to applications involving large energy densities. The pseudocapacitance is given by the derivative of the charge stored (Δq) to the changing potential (ΔV) (Eqn 5.6):

$$C = \frac{d(\Delta q)}{d(\Delta V)} \tag{5.6}$$

Hybrid capacitors can achieve both high energy and power densities. The charge storage process depends on both the capacitive and Faraday characteristics such as ion adsorption and desorption at the electrode-electrolyte interface, reversible surface redox reactions on the electrode, and reversible reactions of the entire electrode.[7] The integration opens new opportunities for further improvement of energy storage and bridging the gap between batteries and EDLCs.

1.3 Supercapacitors

Supercapacitor cells have two different assembled structures, namely two-electrode cells and three-electrode cells. The two-electrode cells are used for research and commercial applications, while the three-electrode cells are mainly used for research because of their higher accuracy. The two-electrode cells consist of two electrodes, a metal current collector, and a separator. As for the three-electrode configuration, a reference electrode is added to make the measurement more accurate. Supercapacitor cells with the two and three-electrode configurations are schematically depicted in Figure 5.3A,B.[8]

1.3.1 EDLCs

EDLCs consist of two porous carbon-based materials as the electrodes, electrolyte, and separator. The charges stored are an electric double-layer electrostatic form and transferred between the electrode and electrolyte. Since the difference in potentials causes opposite charge attraction, there is no buildup of charges on the surface of the electrode when a voltage is applied. Inversely, electrolyte ions traverse the separator and diffuse to the oppositely charged electrodes, as shown in Figure 5.4A (EDLCs) which is illustrated with porous carbon.[9] The porous carbon with large SSA shortens the electrolyte ion transport channels and furthermore, the mechanism of EDLCs enables rapid energy adsorption and desorption, efficient energy transfer, and better power delivery.

1.3.2 Pseudocapacitors

The energy storage of pseudocapacitors involves three different mechanisms: adsorption-desorption (underpotential deposition), reduction oxidation (redox pseudocapacitor), and intercalation-deintercalation (intercalation pseudocapacitor).[11] The schematic diagram of a redox supercapacitor with metal oxide as the electrode and KOH as the electrolyte is shown in Figure 5.4B. When ions are inserted into the electrode, ion insertion occurs not only on the surface but also in the pores and layers of the redox active or conducting polymers. Furthermore, pseudocapacitive electrodes exhibit the capacitor-like behavior because their cyclic voltammograms resemble bilayer rectangles and linear hydrostatic discharge. Nonetheless, pseudocapacitors have a high capacity but

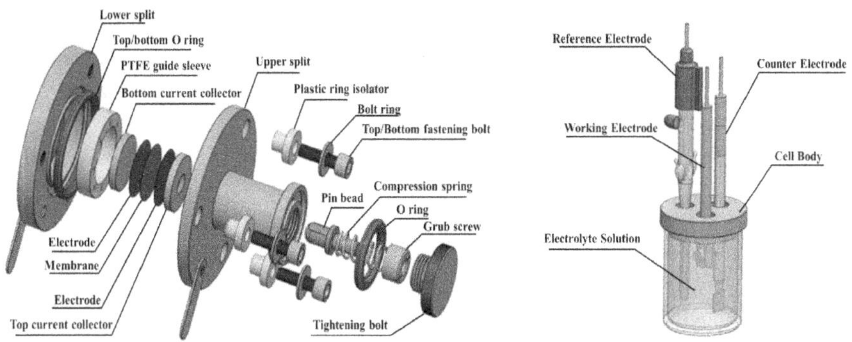

FIGURE 5.3 Schematic diagrams of two and three-electrode cells: (A) two-electrode configuration and (B) three-electrode configuration (reproduced with permission: Copyright 2022, Elsevier).[8]

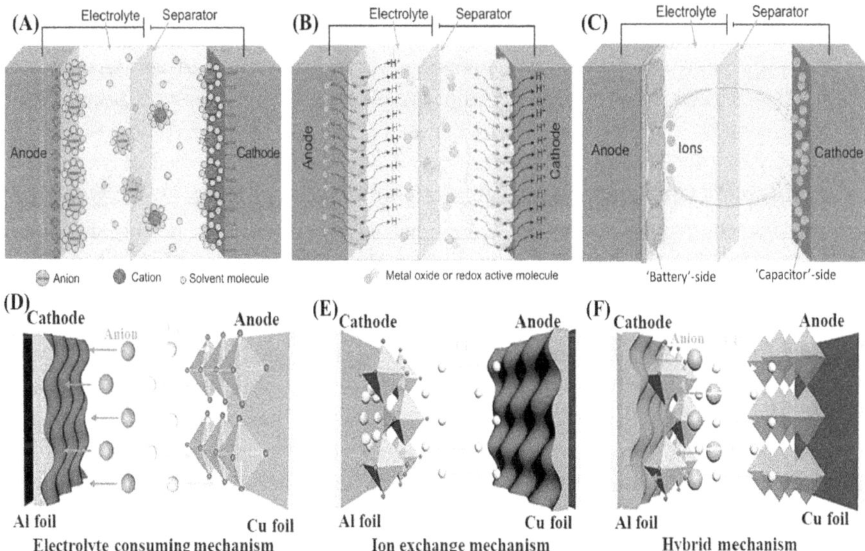

FIGURE 5.4 Schematic of the charge storage mechanisms: (A) EDLCs, (B) pseudocapacitors, and (C) battery-type electrodes (reproduced with permission: Copyright 2021, Elsevier)[9]; (D), (E), and (F) schematic of three energy storage mechanisms for MICs (reproduced with permission: Copyright 2021, Elsevier).[10]

poor stability and short cycling life due to the easy expansion and contraction during charging-discharging cycles. The active materials form defects due to the repeated insertion/extraction of redox-reactive ions on the surface of the active materials, leading to severe expansion and collapse of the electrode structure. Therefore, the cyclic stability of pseudocapacitors is slightly worse than that of EDLCs.

1.3.3 Metal-ion capacitors

Metal-ion capacitors (MICs) have been used in advanced hybrid electrochemical energy storage systems to solve the problems of low energy of capacitors and low power of LIBs. MICs consist of a battery-type anode (electrochemical insertion or conversion) and a capacitor-type cathode (physisorption) in an electrolyte containing metal ions, which have many characteristics of LIBs and SCs (Figure 5.4C). During charging and discharging, the cathode and anode undergo physical (adsorption/desorption) and chemical (intercalation/delamination or conversion) processes to store and release the energy, respectively. Due to the integration of two energy storage mechanisms in one device, MICs have several advantages including: (1) higher energy density than SCs; (2) higher power density than LIBs; (3) wider working temperature range from −50 to 85°C; and (4) better self-discharging performance than SCs.

Depending on whether the electrolyte is consumed during the electrochemical process, the energy storage mechanisms of MICs can be classified into three types: electrolyte consumption mechanisms, ion exchange mechanisms, and hybrid energy storage mechanisms[10]:

1. Electrolyte consumption mechanisms. In this system, the battery materials use alkali metal ion de/intercalation compounds as the anode (positive) and capacitive materials such as activated carbon, graphene, and carbon nanotubes extracted from biomass as the cathode (negative). During charging, the cations and anions move toward the anode and cathode electrodes, respectively, consistent with the mechanism of SCs. However, instead of physical adsorption, alkali metal ions can insert into the compounds containing alkali metal ions or undergo reduction reactions at the anode. During discharging, the alkali metal ions desorb from the anode back to the electrolyte, and the anions desorb from the cathode to reach charge equilibrium (Figure 5.4D).[10]
2. Ion exchange mechanisms. In this system, the cathode is the battery material that provides the alkali metal ions, while the anode is the capacitive material. During charging and discharging, the electrolyte concentration is kept constant and it only serves to transfer alkali metal ions, similar to a rocking chair-type alkali metal ion battery. Unlike the battery, the alkali metal ions are de-embedded from the cathode and adsorb on the anode surface when MICs are charged and vice versa (Figure 5.4E).
3. Hybrid energy storage mechanisms. The distinctive feature of these MICs is that one or both electrodes contain both the battery and capacitor materials. During charging, the all alkali metal ions are removed from the cathode into the electrolyte, and the alkali metal ions supplied by the electrolyte are inserted into the anode and vice versa (Figure 5.4F).

2 BIOMASS-DERIVED GRAPHENE-LIKE CARBON MATERIALS FOR SUPERCAPACITORS

Graphene-based materials are promising electrode materials for SCs on account of the large theoretical surface area, high electronic conductivity, and electrochemical stability. Nevertheless, their large-scale applications are often restricted by irreversible aggregation and stacking caused by the strong van der Waals forces between the substrate surfaces of graphene nanosheets. This significantly restricts the diffusion pathways of ions within some narrow channels, thus preventing the electrolyte ions from entering the surface of the graphene nanosheets, leading to a significant loss of the ion-accessible surfaces and making the specific capacitance much smaller than the theoretical value (ca. 550 F g^{-1}).[12] BDGC materials are an effective solution because they have a large surface area, porous structure, high graphitization degree, and chemically stable surface by controlling the preparation method. What's more, BDGC materials have interconnected pore networks, good electrical conductivity, and electrochemically stable surface boding well for SCs.

2.1 Chemical composition of biomass

The chemical components in biomass are covalently bonded carbohydrate monomers which consist of starch, lignocellulose, triglycerides, phenols, terpenes, fatty acids, and trace amounts of iron, calcium, potassium, silicon, and other elements.[11] For example, lignocellulosic biomass consists of mainly cellulose, hemicellulose, and lignin (Figure 5.5).[11, 13] There is also biomass composed of carbon, hydrogen, and oxygen together with trace amounts of nitrogen, sulfur, and phosphorus. The components in plant-based biomass can be converted into carbon by heat treatment or carbonization. The components in lignocellulose interact strongly with covalent or noncovalent forces. Cellulose is a linear polysaccharide consisting of a large molecular weight D-glucose linked uniformly by a β-1-4 glucosidic bond and accounts for 40%–50% of lignocellulosic biomass.[13] Hemicellulose with branching and amorphous characteristics as well as a small degree of polymerization makes up 25%–35% of the total biomass. Thus, hemicellulose tends to decompose thermally. Lignin is a component in lignocellulosic biomass and amorphous polymers accounting for 15%–20% of the biomass. In addition, lignin has a number of aromatic functional groups consisting of p-coumarol, pinacol, and sinapyl alcohol.

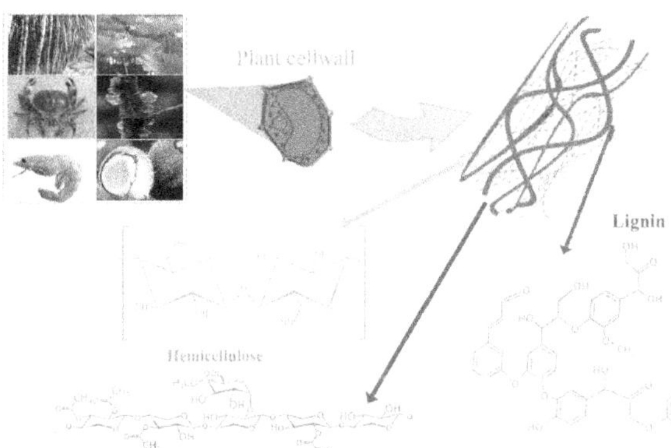

FIGURE 5.5 Structure of lignocellulosic biomass (reproduced with permission: Copyright 2023, Elsevier).[11]

2.2 Graphene-like carbon materials derived from biomass

The biomass mainly composed of cellulose and hemicellulose is broadly found on earth and can be used to synthesize porous carbon due to the unique natural structure as well as high carbon and low ash contents. Graphene-like carbon (GLC) produced from biomass generally has a porous structure that provides connectivity for different pore sizes, shortens the ion pathways, and lowers the resistance in ion transport. For example, coconut shells can easily be converted into porous GLCs by a chemical treatment with $ZnCl_2$ and $FeCl_3$ activation.[14] The resulting GLCs have a large SSA (1,874 $m^2 g^{-1}$) and high graphitization degree ($I_G/I_D \approx 1.81$). The carbon atoms with sp^2 covalent bonds provide high electrical conductivity to reduce the resistance of the system. Similarly, three-dimensional GLCs (3D GLCs) with graphene-like sheets, large SSA (1,503 $m^2 g^{-1}$), and high electrical conductivity (32.14 S cm^{-1}) formed from coconut shells can be obtained by K_2CO_3 catalytic graphitization.[15] SEM shows that the honeycomb-like structure is composed of ultrathin graphene sheets (Figure 5.6A) and well-developed interconnected pores and pore sizes in the range of micrometers (Figure 5.6B). The graphene nanosheets are curved and wrinkled and composed of about three to four layers, as revealed by TEM. The Raman spectrum of 3D GLCs shows low-intensity D-band, sharp G–band, and 2D-band, with an $I_{D/G}$ of only 0.088, while the ratio of $I_{2D/G}$ is as high as 0.855 (Figure 5.6C) confirming high-quality graphene sheets. Hence, 3D GLCs have outstanding electrochemical properties. Without conductive additives, the capacitance of 3D GLCs is 91.15 F g^{-1} at 0.2 A g^{-1} and retention is 85.1% after 5,000 cycles at 0.1 A g^{-1} in an organic electrolyte.

The nitrogen-doped (N-doped) GLC nanosheets from pine nut shells can be obtained by combination with melamine and KOH activation. The GLCs are composed of ultrathin nanosheets and irregular nanosheets and the thickness of the nanosheets is about 5–8 nm (Figure 5.6D–E).[16] The GLC nanosheets have outstanding electrochemical properties due to the large SSA (2,090 $m^2 g^{-1}$) and the inherent hierarchical porous sheets structure. Hence, the GLCs exhibit a superior specific capacitance of 324 F g^{-1} at 0.05 A g^{-1}, high rate capacitance of 258 F g^{-1} at 20 A g^{-1}, and excellent cyclic stability of 94.6% after 10,000 cycles at 2 A g^{-1} in 6 M KOH (Figure 5.6F).

2.2.1 GLCs from forest plants and residues

The preparation of BDGCs from forest plants and residues by different activations provides a green and economical possibility for developing 2D carbon or 3D skeleton carbon consisting of 2D carbon nanosheets. For instance, porous GLC sheets with an ultrathin thickness (3.8 nm) and a hierarchically porous structure with a large meso-pore ratio from pine bark can be obtained by potassium acetate activation (Figure 5.7A).[17] The GLC sheets have a high capacitance (128.1 F g^{-1} at 1 A g^{-1}), good cycling stability, and satisfactory energy density (32.4 Wh kg^{-1}) and power density (30.375 kW kg^{-1}) in the $TEABF_4$/AN electrolyte.[17] Similarly, GLCs with large amounts of interconnected nanosheets with a thickness of 10 nm and large SSA (979.89 $m^2 g^{-1}$) are prepared from Enteromorpha by high-temperature melting of K_2CO_3 (Figure 5.7B).[18] The GLCs have wide pore distributions and a suitable

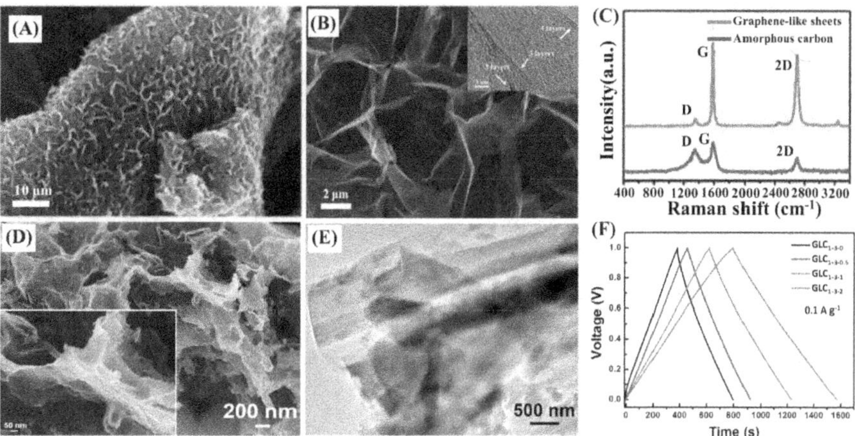

FIGURE 5.6 (A) SEM image and (B) enlarged SEM image; (C) Raman scattering spectra of 3D GLCs (reproduced with permission: Copyright 2018, Royal Society of Chemistry)[15]; (D) SEM image, (E) TEM image, and (F) GCD curves of GLC nanosheets from pine nut shells (reproduced with permission: Copyright 2019, American Chemical Society).[16]

amount of oxygen, thus enabling the electrolyte ions to enter the surface and interior of the electrode (Figure 5.7C). In addition, the GLC nanosheets with abundant porosity, well-developed porous structure, large SSA, and high conductivity are prepared from fresh lotus receptacles by "drilling" holes with H_2O_2 and exfoliating into graphene-like nanosheets with HAc, followed by carbonization (Figure 5.7D).[19] The GLC nanosheets have very high specific capacitance (340 F g^{-1} at 0.5 A g^{-1}), large capacitance retention of 98% after 10,000 cycles, and outstanding energy density (23.33 Wh kg^{-1}).

3D GLCs with the 3D interconnected graphene nanosheet structure can be prepared from spruce bark hydrothermally and KOH activation (Figure 5.7E).[12] The 3D GLCs are composed of intertwined vertically aligned graphene nanosheets, which link to each other and form a 3D honeycomb-like structure to offer porous space between the contiguous layers (Figure 5.7F). The Raman spectrum of 3D GLCs shows low-intensity D-band, sharp G-band, and 2D-band at around 1,350 cm^{-1}, 1,584 cm^{-1} and 2,700 cm^{-1} (Figure 5.7G). The 2D band is a feature of the two-phonon resonance second order, and its width and position are sensitive to the number of layers in the graphene sheet. For instance, single-layered graphene displays a sharp 2D peak, while two-layered graphene shows a relatively wide 2D band. The D peak at 1,350 cm^{-1} is attributed to carbon atom breathing vibration which is related to the six-membered sp^2 carbon rings. The D peak indicates the existence of defects in the materials, which are absent from defect-free graphene. The intensity ratio of the G- and D-band (I_G/I_D) is used to measure the defects in materials. The 3D GLCs possess a 3D interconnected structure, large SSA (2,385 $m^2 g^{-1}$), many hierarchical pores (1.68 $cm^3 g^{-1}$), and open surfaces with graphene nanosheets, thus showing an outstanding capacitance of 239 F g^{-1} at 1 A g^{-1} and high energy density of 74.4 Wh kg^{-1} in the $TEABF_4$/AN electrolyte.

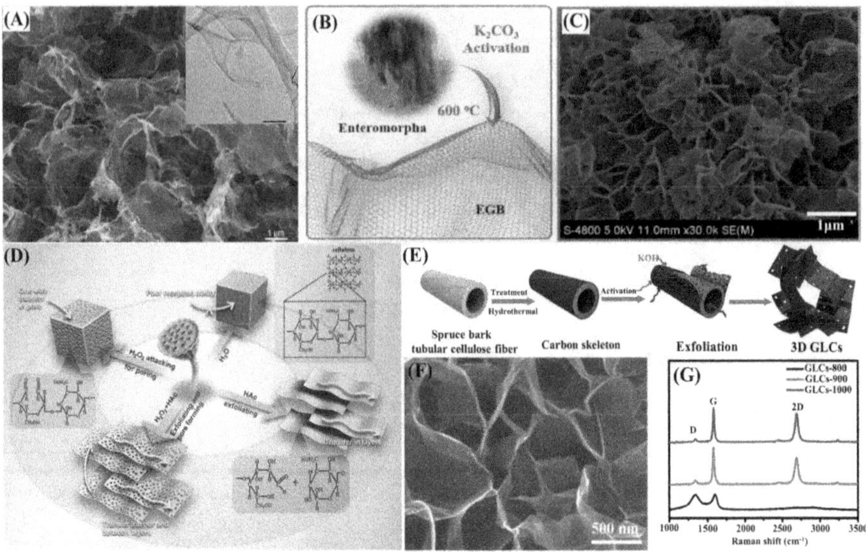

FIGURE 5.7 (A) SEM and TEM (inset) images of GLCs (reproduced with permission: Copyright 2019, American Chemical Society)[17]; (B) preparation of GLCs from Enteromorpha, (C) SEM image of GLCs (reproduced with permission: Copyright 2020, Elsevier)[18]; (D) preparation process of GLC nanosheets from the lotus receptacle (reproduced with permission: Copyright 2017, Wiley-VCH Verlag)[19]; (E) preparation of 3D GLCs from spruce bark, (F) SEM image and (G) Raman scattering spectra of 3D GLCs (reproduced with permission: Copyright 2017, Elsevier).[12]

2.2.2 GLCs from agricultural products

The preparation of high-quality GLC materials from agricultural products for SCs has advantages such as the low cost, abundant raw materials, and sustainable nature. For example, GLC nanosheets can be prepared from hemp bast fibers by hydrothermal carbonization and KOH activation (Figure 5.8A).[20] The GLC nanosheets are made of highly interconnected carbon and ultrathin carbon nanosheets (10–30 nm in thickness) and have high meso-porosity (58%). Hence, the GLC nanosheets with a large SSA (2,287 m^2g^{-1}) and electrical conductivity (226 S m^{-1}) show a high specific capacitance of 142 F g^{-1} with excellent capacitance retention as well as a high energy density of 19 Wh kg^{-1} in the ionic liquid electrolyte. Similarly, the porous GLCs are prepared from fungus by hydrothermal treatment in KOH and carbonization (Figure 5.8B).[21] The porous GLCs with a hierarchically interconnected porous framework and wide pore distributions have a large SSA (1,103 m^2g^{-1}) and bulk density (0.96 g cm^{-3}) thus providing more storage sites and shorter transport paths for the electrolyte ions. As a result, the porous GLCs exhibit a high volumetric capacitance of 360 F cm^{-3} and retention of 99% after 10,000 cycles. The N-doped porous GLC sheets formed from layered peanut seed coats with triethanolamine by thermal exfoliation and pyrolysis have graphene-like sheets with a thickness of ~4 nm and large SSA (887 m^2g^{-1}), which improve the electrical conductivity (8.1 S cm^{-1}) and wettability (Figure 5.8C).[22] Consequently, the GLC sheets show ultra-high rate capabilities

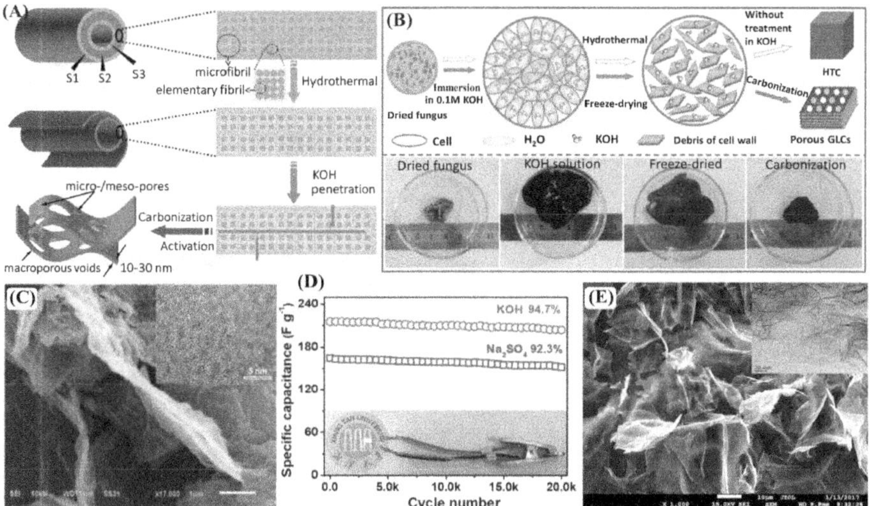

FIGURE 5.8 (A) Preparation of GLCs from hemp (reproduced with permission: Copyright 2013 American Chemical Society)[20]; (B) preparation of porous GLCs from fungus (reproduced with permission: Copyright 2015, Elsevier)[21]; (C) SEM and HR-TEM (inset) images of the porous GLC sheets, (D) cycling stability of the porous GLC sheets (reproduced with permission: Copyright 2020, Elsevier)[22]; (E) SEM image of the porous GLCs formed from sugarcane bagasse pith (reproduced with permission: Copyright 2017, Elsevier).[23]

(72.5% capacity at 200 A g^{-1}) and cycling stability (92.3% of capacity retention after 20,000 cycles at 10 A g^{-1}) in 6.0 M KOH and 1 M Na_2SO_4 electrolytes, respectively (Figure 5.8D). Similarly, the N-doped porous GLC sheets synthesized by KOH activation (Figure 5.8E)[23] have a large-size multiscale wrinkled nanosheet structure, large SSA (1,786.1 m^2g^{-1}), and appropriate amount of N. The N-doped GLC sheets show a large specific capacitance (339 F g^{-1} at 0.25 A g^{-1}), long-term cycling stability (97.9% capacitance retention after 10,000 cycles), and high energy density (11.77 Wh kg^{-1}) in 6 M KOH.[24]

2.2.3 GLCs from marine sources

Marine products are abundant and produced from the ocean that can be eaten or used such as fish, shellfish, and raw plants (seaweed, nori, kelp, etc.). Shellfish mainly includes crab, shrimp, lobster shells, and oysters. Shellfish harbors useful proteins, calcium carbonate, and chitin. Chitin is a linear polymer (poly b-(1, 4)-*N*-acetyl-D-glucosamine) containing nitrogen, which is the second most abundant natural biopolymer on earth (after cellulose).[25] Shellfish is a renewable porous carbon precursor for the direct preparation of graphene carbon nanosheets with N dopant and tunable porosity.

The N-doped GLC sheets are prepared from naturally layered shrimp shells by simultaneous carbonization and auto-activation followed through ultrasonic liquid exfoliation (Figure 5.9A).[26] The GLC sheets possess the graphene sheet structure (5 nm thickness), large SSA (1,946 m^2g^{-1}), and big nitrogen content (8.75 wt%) for

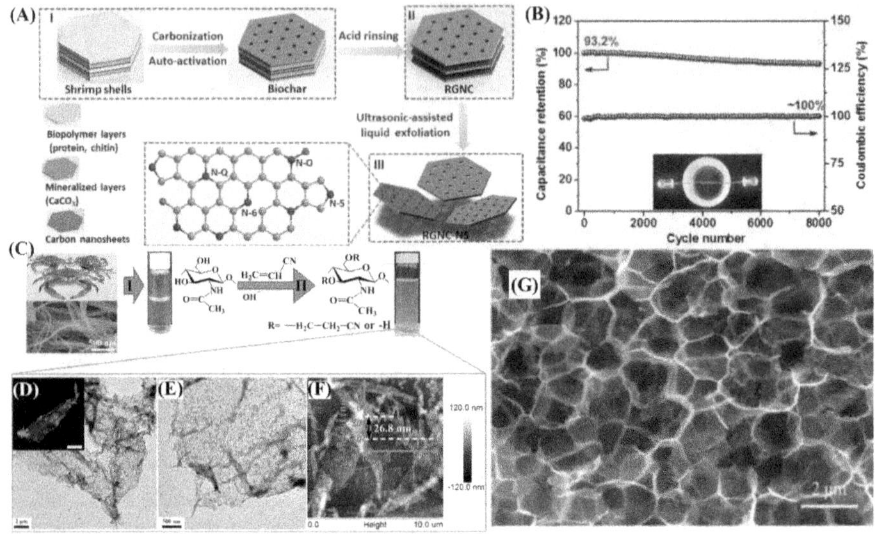

FIGURE 5.9 (A) Synthetic pathway of N-doped GLC sheets; (B) cyclic stability (two blue LEDs in inset) of N-doped GLC sheets (reproduced with permission: 2016, Royal Society of Chemistry)[26]; (C) synthetic pathway and microscopic of GLC sheets, (D and E) TEM and (F) AFM images of GLC sheets (reproduced with permission: 2017 WILEY-VCH Verlag GmbH & Co. KGaA, Weinheim)[27]; (G) SEM images of N-doped GLCs (reproduced with permission: 2016, Royal Society of Chemistry).[28]

improved conductivity (7.8 S cm^{-1}) and wettability with the electrolyte. The GLC sheets have an excellent rate capability with a high capacitance of 322 F g^{-1} at 0.5 A g^{-1}, long cyclic stability (93.2% capacitance retention after 20,000 cycles), and high energy-power density (30 Wh kg^{-1} and 64,000 W kg^{-1}) in ionic liquid electrolytes (Figure 5.9B). Similarly, the N-doped porous GLC sheets formed from chitosan show a large SSA of 1,321.3 m^2g^{-1}, large pore volume of 4.21 cm^3g^{-1}, ultrathin nanosheets structure, and appropriate porosity.[29] The GLC sheets with a thickness of 27 nm are fabricated from crab chitin *via* hydrophobization-induced interfacial assembly followed by the aqueous NaOH/urea treatment (Figure 5.9C).[27] The layers possess structural flexibility and robustness, thin chitin nanosheets (26.8 nm), and a large SSA of 724.2 m^2g^{-1} (Figure 5.9D–F). When the hybrid films of chitin and GLC sheets are used in the flexible device, the ultrathin soft film has a good specific capacitance of 162.4 F g^{-1} at 0.5 A g^{-1} and high cycling stability (capacity retention up to 95% after 10,000 cycles at 5 A g^{-1}) in 6 M KOH. In addition, the 3D honeycomb-like N-doped GLC films formed by mixing chitin and graphene oxide are synthesized by dissolution and coagulation of chitin and graphene oxide in the NaOH/urea aqueous solution using a repeated freezing-thawing process followed by carbonization under Ar.[28] The honeycomb-like N-doped GLC films have a homogeneous interconnected open-cell framework with the average pore size of about 2 μm (Figure 5.9G). The porous N-doped GLC films inherit the high conductivity of graphene and have abundant porosity and rich active sites resulting in excellent electrochemical properties.

2.3 BIOMASS-DERIVED CARBON-BASED COMPOSITES

Compared to pure carbon materials, pseudocapacitor materials are based on transition metal oxides or hydroxides, such as MnO_2, Fe_2O_3, Co_2O_3, and $Ni(OH)_2$, which have high capacity and good redox activity. The redox activity of the materials is mainly derived from the multivalent nature of the transition metals. However, the poor inherent conductivity and small SSA lead to a slow charge transfer and suboptimal cycling stability at high current densities.

BDGCs can provide abundant space for the deposition of transition metal oxides or hydroxides and reduced agglomeration. The superior electrical conductivity of BDGCs can expedite charge transfer to improve the electrochemical performance. In addition, the good structural stability and proper porous structure of BDGCs can buffer the volume expansion caused by transition metal oxides or hydroxides. Therefore, the combination of transition metal oxides or hydroxides with BDGCs improves the electrochemical properties. For example, the multifunctional heteroatom (Fe, N, S) Co-doped GLC sheets (Fe-N-S/GLC) with special 2D layer structures are formed from silkworm chrysalides shells by liquid-phase exfoliation to delaminate the multilayered biochars (Figure 5.10A).[30] Owing to the unique integration of the graphene-like structures with a thickness of 4.3 nm, large SSA (2,491 m^2g^{-1}), hierarchical pores, homogenous co-doping, and high electronic conductivity (7.6 S cm^{-1}), the Fe-N-S/GLC sheets have excellent super-capacitive properties such as a large capacitance of 173 F g^{-1} at 1 A g^{-1} and energy density of 29.1 Wh kg^{-1} in an ionic liquid electrolyte.

$Ni(OH)_2$ is an attractive material for SCs due to the high theoretical specific capacitance (2,358 F g^{-1}), specific redox behavior, high redox activity, and environmental friendliness. However, the inherently low electrical conductivity (10^{-17} S cm^{-1}) of $Ni(OH)_2$ leads to redox reactions occurring only on the surface and low utilization. To overcome these problems, $Ni(OH)_2$ is combined with GLCs to enhance the electrochemical performance. In this respect, GLCs with the nanosheet-like structure formed from peach gum with $Ni(OH)_2$ to $Ni(OH)_2$/GLCs have excellent electrochemical properties (Figure 5.10B).[31] In the process, GLCs are first obtained by hydrothermal carbonization and magnesium acetate activation to form $Ni(OH)_2$. The synergistic effects of $Ni(OH)_2$ and GLCs produce $Ni(OH)_2$/GLCs with a large energy density of 36.9 Wh kg^{-1} and power density of 400 W kg^{-1}.

Other pseudocapacitive materials, such as MnO_2, Co_3O_4, Fe_2O_3, NiO, and $NiMn_2O_4$ can also be combined with GLCs to promote capacitance. The N-doped 3D porous GLCs (3DPGLCs) are formed from luffa complexes with $NiMn_2O_4$ nanocrystals ($NiMn_2O_4$/3DPGLCs) by carbonization and a hydrothermal method to form an efficient binder-free electrode (Figure 5.10C).[32] The 3DPGLCs with a porous structure and defective surface can adjust *in situ* growth of $NiMn_2O_4$ to form the $NiMn_2O_4$/3DPGLCs composite. Because the $NiMn_2O_4$/3DPGLCs composite has a high conductive network and abundant pores, it not only is conducive to electrolyte and $NiMn_2O_4$ infiltration but also transfers electrons quickly for the electrochemical reaction. The $NiMn_2O_4$/3DPGLCs electrode has a high specific capacitance of 1,308.2 F g^{-1} at 1 A g^{-1} and rate performance of 77.9% at 15 A g^{-1}. The GLCs formed from cleaned agaric are combined with MnO_2 nanosheets (MnO_2/GLCs) by a

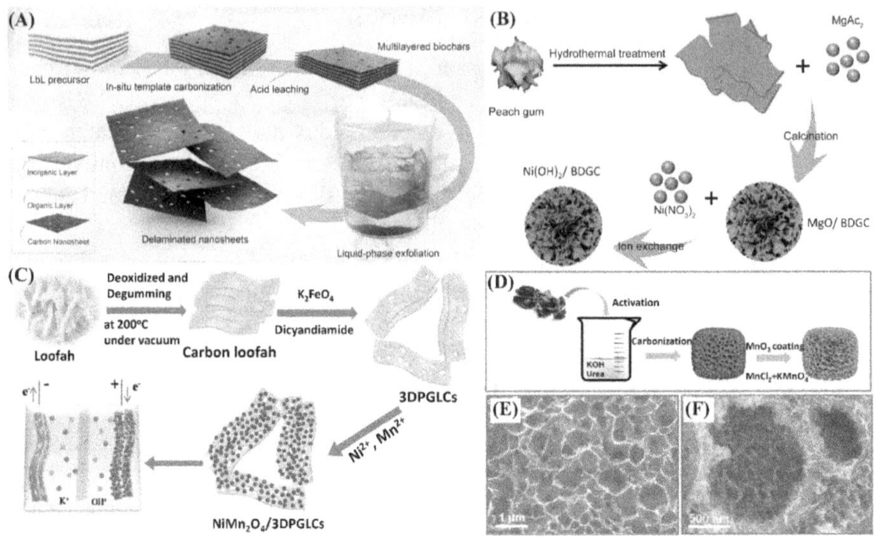

FIGURE 5.10 (A) Preparation of the Fe-N-S/GLC sheets (reproduced with permission: Copyright 2021, Elsevier)[30]; (B) preparation of $Ni(OH)_2$/GLCs composites (reproduced with permission: Copyright 2019, Elsevier)[31]; (C) preparation of $NiMn_2O_4$/3DPGLCs (reproduced with permission: Copyright 2020, Elsevier)[32]; (D) schematic diagram of the MnO_2/N-GLCs composite. (E) SEM image of GLCs; (F) SEM image of the MnO_2/GLCs composite (reproduced with permission: Copyright 2020, Elsevier).[33]

two-step process (Figure 5.10D).[33] The first step consists of a direct heat treatment of the cleaned agaric and KOH activation is performed to obtain the GLCs. The second step includes *in situ* chemical deposition and dense MnO_2 nanosheets are grown on porous carbon (Figure 5.10E,F). The conductive frame of the N-doped porous carbon and synergetic effects between GLCs and MnO_2 yields MnO_2/GLCs with a large SSA (2,250 $m^2 g^{-1}$) produce excellent electrochemical performance such as a specific capacitance of 330 F g^{-1} at 1 A g^{-1}. Similarly, the GLC sheets are formed from salvia splendens petals coupled with MnO_2 (MnO_2/GLCs) by carbonization and hydrothermal processes.[34] The MnO_2/GLCs composite possesses a large SSA (483 $m^2 g^{-1}$) and suitable pore size distribution (2–5 nm). Because of the synergistic effects of MnO_2 and GLC, the MnO_2/GLCs composite electrode shows a high specific capacitance of 438 F g^{-1} at 0.5 A g^{-1} and large rate capability (67.8% capacity at 50 A g^{-1}) in the Na_2SO_4 electrolyte. The composites containing different GLCs and transition metal oxides or hydroxides are prepared from kapok fiber/NiO, fruit/Co_3O_4, and wheat straw/Fe_2O_3 and have been used as electrode materials in SCs.

3 PREPARATION METHODS OF GLCS FROM BIOMASS

In this section, the preparation of graphene-like materials from biomass is classified into physical activation, chemical activation, microwave-assisted activation, template method, and other methods.

3.1 Physical Activation

Physical activation mainly utilizes water steam and CO_2 as the activators to prepare BDGCs. The common preparation process involves carbonization at 400–800°C followed by activation at 700–1,500°C, using the chemical reactions between carbon and H_2O (Eqn 5.7) or CO_2 (Eqn 5.8)[35]:

$$C+H_2O \rightarrow H_2+C \quad H=+132\ \text{kJ mol}^{-1} \tag{5.7}$$

$$C+CO_2 \rightarrow 2CO \quad H=+173\ \text{kJ mol}^{-1} \tag{5.8}$$

The activation process involves the controlled combustion of activated carbon atoms and the elimination of volatile species. The quality of GLCs depends on the conditions, such as temperature, time, gas flow rate, and activator. In addition, the different biomass materials require the appropriate activators to obtain GLCs with large SSA and suitable pores. For instance, GLCs with SSA of 1,700 $m^2 g^{-1}$ and pore volume of 1.135 $cm^3 g^{-1}$ are prepared from coconut shell char by CO_2 activation.[36] In addition to CO_2 activation, water steam is widely used as an effective activator due to its low cost, cleanliness, and safety. For example, GLCs prepared from shrimp shells by water steam activation at 900°C show an SSA of 560.6 $m^2 g^{-1}$ and a high degree of graphitization (I_D/I_G=0.7686).[37]

3.2 Chemical Activation

Chemical activation is a process in which the precursor is first mixed with certain chemicals and then activated to obtain BDGCs at a temperature of 500–1,000°C. Compared to physical activation, chemical activation has several advantages[38]: (i) The pores are well developed; (ii) The activation time is shorter; (iii) GLCs have large SSA; (iv) There is a high degree of graphitization. Various activating regents have been used, for instance, KOH, NaOH, H_3PO_4, $ZnCl_2$, $FeCl_3$, and K_2CO_3. KOH is a popular alkali activator that produces the best effects. The specific reactions are as follows (Eqns 5.9–5.13)[39]:

$$6KOH+2C \leftrightarrow 2K+2K_2CO_3+3H_2 \tag{5.9}$$

$$K_2CO_3 \leftrightarrow K_2O+CO_2 \tag{5.10}$$

$$K_2CO_3+2C \leftrightarrow 2K+3CO \tag{5.11}$$

$$K_2O+C \leftrightarrow 2K+CO \tag{5.12}$$

$$CO_2+C \rightarrow 2CO \tag{5.13}$$

In the chemical activation stage, the pores are developed with the consumption or redevelopment of carbon. KOH reacts with carbon to form metallic K, K_2CO_3, and

H_2 gas at a temperature lower than 570°C (Eqn 5.9). K_2CO_3 is decomposed to form K_2O and CO_2 pyrolysis occurs at around 600°C (Eqn 5.10). KOH is completely consumed at 700°C. K_2CO_3 and K_2O start to decompose and are absent from the system at above 800°C (Eqns 5.11 and 5.12).[38] The activated products of CO_2 may also take part in the pore-forming process due to the physical activation effects. The produced CO_2 reacts with carbon to produce CO at a high temperature (Eqn 5.13). At a higher activation temperature (>750°C, Eqns 5.9–5.11), almost all the biochar is activated by the reaction between KOH and carbon. The K^+ diffuses into the interior of the carbon skeleton to expand or create new pores during the activation process.

Different types of biomass have different SSA, pore volume, and degree of graphitization after chemical activation. For example, the GLC sheets formed from gelatin by KOH activation have a large SSA of 3,106 $m^2 g^{-1}$, pore volume of 1.70 $cm^3 g^{-1}$, and thin sheets of graphene.[40] The N-doped hierarchical porous GLCs aerogels formed from chitosan by carbonization at 800°C and KOH activation have large SSA of 2,435.2 $m^2 g^{-1}$ and pore volume of 1.65 $cm^3 g^{-1}$.[41] Other GLCs such as petroleum pitch (2,132 $m^2 g^{-1}$) and sorghum (4,092 $m^2 g^{-1}$) also have enhanced SSA after KOH activation.

Compared to KOH, NaOH activation has the preponderant in terms of merits such as less corrosion, lower weight dosage, and low cost. NaOH is also extensively employed to fabricate porous GLCs. NaOH reacts with carbon to form pores and the enlargement increases graphitization in the product. There are some possible reactions between the reactive intermediates and carbon surface producing H_2, CO_2, and CO, which give rise to the development of pores according to the following equations (Eqns 5.14–5.18)[42]:

$$6NaOH+2C \leftrightarrow 2Na+2Na_2CO_3+3H_2 \tag{5.14}$$

$$Na_2CO_3 \leftrightarrow Na_2O + CO_2 \tag{5.15}$$

$$Na_2CO_3+2C \leftrightarrow 2Na+3CO \tag{5.16}$$

$$Na_2O+C \leftrightarrow 2Na+CO \tag{5.17}$$

$$CO_2+C \rightarrow 2CO \tag{5.18}$$

For example, GLCs obtained from chitin by NaOH activation at 800°C show a large SSA of 294 $m^2 g^{-1}$ and a pore size of 25 nm.[43] Similarly, GLCs obtained from guava seeds have a large SSA of 2,573.6 $m^2 g^{-1}$ and a total pore volume of 1.260 $cm^3 g^{-1}$ after NaOH activation.[44]

3.3 Microwave-Assisted Activation

Microwave-assisted activation is a new technology for the construction of BDGCs. In the microwave system, heat is generated within the substance. The microwave transmitted to biological tissues causes molecules and polar side chains of proteins to oscillate at high frequencies due to electromagnetic oscillations of molecules and high-speed molecular motion, leading to the generation of heat.[45] The heat on the molecular level increases the temperature rapidly and uniformly, which is not possible by conventional heating. Therefore, microwave-induced activation is carried out in a closed space, thus solving most of the problems associated with conventional pyrolytic activation, consequently giving rise to higher efficiency, resource saving, shorter cycles, and low costs. Li et al. have combined microwave heating and steam activation to extract high-quality porous GLC nanosheets from black sesame.[46] The GLC nanosheets have high and uniform porosity, large SSA (2,414.5 $m^2 g^{-1}$), and a high degree of graphitization. In short, owing to the many advantages offered by microwave-induced physical activation, it has been used to prepare GLCs with some good results.

3.4 Template Methods

Template synthesis is one of the effective methods to fabricate porous BDGCs with controlled structures and higher efficiency. In this technique, the template simply acts as a bracket around which the other types of materials are synthesized. The template reacts with the carbon precursor to increase the degree of graphitization of the carbon structure during carbonization and then removes the template with chemicals or heating to guide pore formation. Currently, the two types of templates are hard templates and soft templates.

Hard templates, such as MgO, $CaCO_3$, ZnO, Fe_2O_3, and SiO_2 are commonly used to fabricate porous structures of GLCs. The structural features of porous GLCs prepared from hard templates depend on the physical/chemical properties of the template. The preparation of porous GLCs by the hard template method involves four steps: (i) synthesis of the ideal hard template; (ii) effective mixing of carbon sources and template; (iii) high-temperature pyrolysis under certain atmosphere, and (iv) hard template removal by acid or alkali rinsing.[47] For instance, corrugated GLC nanosheets prepared from tar pitch with the plate-like MgO template followed by KOH activation have a large SSA of 2,132 $m^2 g^{-1}$ and pore volume of 1.23 $cm^3 g^{-1}$. Corrugated GLC nanosheets have micropores, mesopores, and macropores with a controllable pore size distribution.[48]

ZnO templates have received much attention due to their unique physicochemical properties. The carbon precursors tend to react with ZnO and produce gases during high-temperature processing, which plays a role in physical activation. ZnO has the bifunctional hard templating and physical activation and the GLCs with well-developed porous structures and wide pore size distributions can be obtained. For example, 3D flower-like and hierarchical of GLCs are obtained from pitch with a ZnO template and chemical activation. The GLCs have not only 3D interconnected porous graphene sheets and wide pore size distribution but also a large SSA of 761.5 $m^2 g^{-1}$ and total pore volume of 0.49 $cm^3 g^{-1}$.[49] Compared to MgO and ZnO, $CaCO_3$ is a reliable template and this kind of dual template is used to prepare GLCs. $CaCO_3$ preferentially decomposes to form CaO particles and CO_2 gas at 500°C. The generated CO_2 gas is a reactant with carbon to produce different porous structures. When CaO is removed, hierarchical porous GLCs are formed. For instance, hierarchical porous GLCs with an interconnected pore structure are obtained from cornstalk without pith using the $CaCO_3$ template and $K_2C_2O_4$ activation. The hierarchical porous GLCs have a large SSA (1,910 $m^2 g^{-1}$) and well-developed hierarchical porosity.[50]

Soft templates are also effective in constructing graphene. They are usually organic molecules or super-molecules with functional groups. In certain solvents, the functional groups of the soft templates can provide hydrogen bonding, hydrophobic and hydrophilic interactions, and electrostatic interactions. When a suitable solvent is added, the soft template becomes micelles that interact with the molecules in the carbon precursors. During carbonization, the micelles break down to form the carbon source and biomass charcoal to produce GLCs with a unique porous structure. The 3D interconnected porous GLCs are synthesized using cheap coal tar pitch as the carbon source, $BMIMBF_4$ ionic liquid as the soft template, and KOH activation.[51] The 3D GLCs composed of thin carbon nanosheets with well-developed pores possess a large SSA of 1,593 $m^2 g^{-1}$ and a total pore volume of 0.85 $cm^3 g^{-1}$.

3.5 Other Methods

Other activation methods like hydrothermal carbonization, salt-based method, and physicochemical activation have also been applied to the preparation of BDGCs. The BDGCs with well-developed and interconnected porous structures are formed from natural reed membranes by hydrothermal and carbonization treatments.[52] The BDGCs with a hierarchical porosity and graphene structure formed from kitchen waste hydrolyzed residues with NaCl–KCl molten salts at 800°C have a large SSA (818.65 $m^2 g^{-1}$).[53] Self-activation allows the conversion of biomass to BDGCs without additional activators, since biomass has inherent inorganic salts or other metal ions (K, Na, Fe, and Ca) that can etch the carbonization precursors or react with gases emitted during carbonization (CO_2, H_2O, and H_2). This self-activating pathway has been applied to the preparation of BDGCs from coconut shells, almond kernels, and wood chips.

4 CONCLUSIONS AND OUTLOOK

BDGC is a promising electrode material in energy storage applications due to its abundant natural resources, renewability, low cost, adjustable porous structure, and good physicochemical stability. The rational design and preparation of BDGCs with adjustable SSA, pores, surface chemistry, and degree of graphitization are important for the application in SCs. The quality and performance of the BDGCs depend on the precursors and fabrication techniques. Although various biomass feedstock precursors used for physical activation, chemical activation, microwave-assisted activation, and template methods for the synthesis of BDGC have great potential for SCs, there are still some challenges for practical applications.

First, the carbon source plays an important role in the quality of the BDGC materials. Therefore, it is imperative to establish a unified technical standard or grading system for the carbon source. Second, comprehensive and standardized characterization methods should be established to evaluate the microscopic morphology of BDGC. Specifically, each 3D graphene structure has a unique structure and sheet quality. Although the 3D structures are typically characterized by SEM, TEM, and AFM, some features pertaining to the microstructure of BDGC materials are still not well understood. Third, a standardized protocol is needed to determine the conductivity. The total conductivity should be calculated based on the actual cross-sectional area for a more reliable comparison. Alternatively, at least detailed information about the mass density and the actual contact area should be provided. Fourth, the relationship between the 3D microstructure of graphene and its main properties is not well understood. In practical applications, a database containing information about the structure and properties of various 3D graphene structures should be established. Such a database will allow not only systematic analysis and comparison of different 3D graphene structures but also easy selection of the proper 3D graphene structures for special applications.

In addition to the aforementioned challenges, the application of BDGC should be expanded. The diversity of biomass in terms of resource, type, structure, and composition bodes well for a wide range of applications including rechargeable batteries, desalination, catalysts, and oil/water separation. For example, BDGC materials with a large SSA, suitable pore size, and pore size distribution are desirable for seawater desalination, water treatment, and oil-water separation. Furthermore, these materials can be used as catalysts or substrates in metal-based catalytic applications.

REFERENCES

1. W. Zuo, R. Li, C. Zhou, Y. Li, J. Xia and J. Liu, Battery-supercapacitor hybrid devices: Recent progress and future prospects, *Adv. Sci.*, 2017, **4** (7), 1600539.
2. L. Sun, Y. Gong, D. Li and C. Pan, Biomass-derived porous carbon materials: Synthesis, designing, and applications for supercapacitors, *Green Chem.*, 2022, **24** (10), 3864–3894.
3. M. Athanasiou, S.N. Yannopoulos and T. Ioannides, Biomass-derived graphene-like materials as active electrodes for supercapacitor applications: A critical review, *Chem. Eng. J.*, 2022, **446**, 137191.
4. Z. Sun, S. Fang and Y. Hu, 3D graphene materials: From understanding to design and synthesis control, *Chem. Rev.*, 2020, **120** (18), 10336–10453.
5. R. Beguin, A. Balducci and E. Frackowiak, Carbons and electrolytes for advanced supercapacitors, *Adv. Mater.*, 2014, **26** (14), 2219–2251.
6. K.O. Oyedotun, J. Conradie and K.A. Adegoke, Advances in supercapacitor development: Materials, processes, and applications, *J. Electron. Mater.*, 2022, **52** (1), 96–129.
7. J. Zhao and A.F. Burke, Electrochemical capacitors: Materials, technologies and performance, *Energy Stor. Mater.*, 2021, **36**, 31–55.
8. A.G. Olabi, Q. Abbas, A. Al Makky and M.A. Abdelkareem, Supercapacitors as next generation energy storage devices: Properties and applications, *Energy*, 2022, **248**, 123617.
9. S. Saini, P. Chand and A. Joshi, Biomass derived carbon for supercapacitor applications: Review, *J. Energy Storage*, 2021, **39**, 102646.
10. D. Zhang, L. Li and Y. Zhang, Metal chalcogenides-based materials for high-performance metal ion capacitors, *J. Alloys Comd*, 2021, **869**, 159352.
11. D.S. Priya, L.J. Kennedy and G.T. Anand, Emerging trends in biomass-derived porous carbon materials for energy storage application: A critical review, *Mater. Today Sustain.*, 2023, **21**, 100320.

12. Z. Sun, M. Zheng, H. Hu, H. Dong, B. Lei and Y. Liu, From biomass wastes to vertically aligned graphene nanosheet arrays: A catalyst-free synthetic strategy towards high-quality graphene for electrochemical energy storage, *Chem. Eng. J.*, 2018, **336**, 550–561.
13. X. Luo, S. Chen, T. Hu, Y. Chen and F. Li, Renewable biomass-derived carbons for electrochemical capacitor applications, *SusMat*, 2021, **1** (2), 211–240.
14. L. Sun, C. Tian, M. Li, R. Wang, J. Yin and H. Fu, From coconut shell to porous graphene-like nanosheets for high-power supercapacitors, *J. Mater. Chem. A*, 2013, **1** (21), 6462.
15. J. Xia, N. Zhang, S. Chong, Y. Chen and C. Sun, Three-dimensional porous graphene-like sheets synthesized from biocarbon via low-temperature graphitization for a supercapacitor, *Green Chem.*, 2018, **20** (3), 694-700.
16. L. Guan, L. Pan, Z. Yang, H. Hu and M. Wu, Synthesis of biomass-derived nitrogen-doped porous carbon nanosheests for high-performance supercapacitors, *ACS Sustain. Chem. Eng.*, 2019, **7** (9), 8405–8412.
17. D. Wang, J. Nai, L. Xu and T. Sun, A potassium formate activation strategy for the synthesis of ultrathin graphene-like porous carbon nanosheets for advanced supercapacitor applications, *ACS Sustain. Chem. Eng.*, 2019, **7** (23), 18901–18911.
18. Y. Qi, B. Ge, Y. Zhang, M. Akram and X. Xu, Three-dimensional porous graphene-like biochar derived from Enteromorpha as a persulfate activator for sulfamethoxazole degradation: Role of graphitic N and radicals transformation, *J. Hazard. Mater.*, 2020, **399**, 123039.
19. S. Lu, M. Jin, Y. Zhang, Y. Niu, J. Gao and C. Li, Chemically exfoliating biomass into a graphene-like porous active carbon with rational pore structure, good conductivity, and large surface area for high-performance supercapacitors, *Adv. Energy Mater.*, 2018, **8** (11), 1702545.
20. H. Wang, Z. Xu, J.K. Tak, D. Harfield, A.O. Anyia and D. Mitlin, Interconnected carbon nanosheets derived from hemp for ultrafast supercapacitors with high energy, *ACS Nano*, 2013, **7** (6), 5131–5141.
21. C. Long, X. Chen, L. Jiang, L. Zhi and Z. Fan, Porous layer-stacking carbon derived from in-built template in biomass for high volumetric performance supercapacitors, *Nano Energy*, 2015, **12**, 141–151.
22. B. Liu, M. Yang, D. Yang, H. Li, Graphene-like porous carbon nanosheets for ultra-high rate performance supercapacitors and efficient oxygen reduction electrocatalysts, *J. Power Sources*, 2020, **456**, 227999.
23. Q. Niu, K. Gao, Q. Tang, Y. Zhang, S. Wang and L. Wang, Large-size graphene-like porous carbon nanosheets with controllable N-doped surface derived from sugarcane bagasse pith/chitosan for high performance supercapacitors, *Carbon*, 2017, **123**, 290–298.
24. A. Gopalakrishnan and S. Badhulika, Ultrathin graphene-like 2D porous carbon nanosheets and its excellent capacitance retention for supercapacitor, *J. Ind. Eng. Chem.*, 2018, **68**, 257–266.
25. B. Duan, F. Liu, M. He and L. Zhang, Ag-Fe_3O_4 nanocomposites@chitin microspheres constructed by in situ one-pot synthesis for rapid hydrogenation catalysis, *Green Chem.*, 2014, **16** (5), 2835–2845.
26. W. Tian, Q. Gao, L. Zhang, C. Yang, Z. Li, Y. Tan, W. Qian and H. Zhang, Renewable graphene-like nitrogen-doped carbon nanosheets as supercapacitor electrodes with integrated high energy-power properties, *J. Mater. Chem. A*, 2016, **4** (22), 8690–8699.
27. J. You, M. Li, B. Ding, X. Wu and C. Li, Crab chitin-based 2D soft nanomaterials for fully biobased electric devices, *Adv. Mater.*, 2017, **29** (19), 1606895.
28. B. Wang, S. Li, X. Wu, J. Liu and J. Chen, Biomass chitin-derived honeycomb-like nitrogen-doped carbon/graphene nanosheet networks for applications in efficient oxygen reduction and robust lithium storage, *J. Mater. Chem. A*, 2016, **4** (30), 11789–11799.

29. L. Kong, L. Su, W. Yang, G. Shao and X. Qin, Graphene-like nitrogen-doped porous carbon nanosheets as both cathode and anode for high energy density lithium-ion capacitor, *Electrochim. Acta*, 2020, **349**, 136303.
30. W. Tian, Q. Gao, A. VahidMohammadi, J. Dang, Z. Li, X. Liang, M.M. Hamedi and L. Zhang, Liquid-phase exfoliation of layered biochars into multifunctional heteroatom (Fe, N, S) co-doped graphene-like carbon nanosheets, *Chem. Eng. J.*, 2021, **420**, 127601.
31. D. Yu, X. Zheng, M. Chen and X. Dong, Large-scale synthesis of $Ni(OH)_2$/peach gum derived carbon nanosheet composites with high energy and power density for battery-type supercapacitor, *J. Colloid Interf. Sci.*, 2019, **557**, 608–616.
32. M. Zhang, Z. Song, H. Liu and T. Ma, Biomass-derived highly porous nitrogen-doped graphene orderly supported $NiMn_2O_4$ nanocrystals as efficient electrode materials for asymmetric supercapacitors, *Appl. Surf. Sci.*, 2020, **507**, 145065.
33. D. Li, J. Lin, Y. Lu, Y. Huang, X. He, C. Yu, J. Zhang and C. Tang, MnO_2 nanosheets grown on N-doped agaric-derived three-dimensional porous carbon for asymmetric supercapacitors, *J. Alloys Comd*, 2020, **815**, 152344.
34. B. Liu, Y. Liu, M. Yang and H. Li, MnO_2 nanostructures deposited on graphene-like porous carbon nanosheets for high-rate performance and high-energy density asymmetric supercapacitors, *ACS Sustain. Chem. Eng.*, 2019, **7** (3), 3101–3110.
35. J. Yin, W. Zhang, N.A. Alhebshi, N. Salah and H.N. Alshareef, Synthesis strategies of porous carbon for supercapacitor applications, *Small Methods*, 2020, **4** (3), 1900853.
36. S. Guo, J. Peng, W. Li, K. Yang, L. Zhang, S. Zhang and H. Xia, Effects of CO_2 activation on porous structures of coconut shell-based activated carbons, *Appl. Surf. Sci.*, 2009, **255** (20), 8443–8449.
37. X. Liu, C. He, X. Yu, Y. Bai, L. Ye, B. Wang and L. Zhang, Net-like porous activated carbon materials from shrimp shell by solution-processed carbonization and H_3PO_4 activation for methylene blue adsorption, *Powder Technol.*, 2018, **326**, 181–189.
38. Z. Zhu and Z. Xu, The rational design of biomass-derived carbon materials towards next-generation energy storage: A review, *Renew. Sust. Energy Rev.*, 2020, **134**, 110308.
39. M. Usha Rani, T.N. Rao and A.S. Deshpande, Corn husk derived activated carbon with enhanced electrochemical performance for high-voltage supercapacitors, *J. Power Sources*, 2020, **471**, 228387.
40. X. Fan, C. Yu, J. Yang, M. Zhang and J. Qiu, A layered-nanospace-confinement strategy for the synthesis of two-dimensional porous carbon nanosheets for high-rate performance supercapacitors, *Adv. Energy Mater.*, 2015, **5** (7), 1401761.
41. P. Hao, Z. Zhao, Y. Leng, H. Liu and B. Yang, Graphene-based nitrogen self-doped hierarchical porous carbon aerogels derived from chitosan for high performance supercapacitors, *Nano Energy*, 2015, **15**, 9–23.
42. V. Yang, R.A. Senthil, J. Pan, A. Khan, S. Osman, L. Wang, W. Jiang and Y. Sun, Highly ordered hierarchical porous carbon derived from biomass waste mangosteen peel as superior cathode material for high performance supercapacitor, *J. Electroanal. Chem.*, 2019, **855**, 113616.
43. B. Duan, X. Zheng, Z. Xia, X. Fan, L. Guo, J. Liu, Y. Wang, Q. Ye and L. Zhang, Highly biocompatible nanofibrous microspheres self-assembled from chitin in NaOH/urea aqueous solution as cell carriers, *Angew. Chem. Int. Ed.*, 2015, **54** (17), 5152–5156.
44. O. Pezoti, A.L. Cazetta, O.O. Santos Júnior, J.V. Visentainer and V.C. Almeida, NaOH-activated carbon of high surface area produced from guava seeds as a high-efficiency adsorbent for amoxicillin removal: Kinetic, isotherm and thermodynamic studies, *Chem. Eng. J.*, 2016, **288**, 778–788.
45. Z. Li, D. Guo, Y. Liu, H. Wang and L. Wang, Recent advances and challenges in biomass-derived porous carbon nanomaterials for supercapacitors, *Chem. Eng. J.*, 2020, **397**, 125418.

46. T. Li, R. Ma, X. Xu, S. Sun and J. Lin, Microwave-induced preparation of porous graphene nanosheets derived from biomass for supercapac itors, *Micropor. Mesopor. Mat.*, 2021, **324**, 111277.
47. W. Zhang, R. Cheng, L. Ma and X. He, A review of porous carbons produced by template methods for supercapacitor applications, *New Carbon Mater.*, 2021, **36** (1), 69–81.
48. X. He, N. Zhang, X. Shao, M. Wu, M. Yu and J. Qiu, A layered-template-nanospace-confinement strategy for production of corrugated graphene nanosheets from petroleum pitch for supercapacitors, *Chem. Eng. J.*, 2016, **297**, 121–127.
49. Q. Wang, J. Yan, T. Wei, M. Zhang, X. Jing and Z. Fan, Three-dimensional flower-like and hierarchical porous carbon materials as high-rate performance electrodes for supercapacitors, *Carbon*, 2014, **67**, 119–127.
50. J. Li, Q. Jiang, L. Wei, L. Zhong and X. Wang, Simple and scalable synthesis of hierarchical porous carbon derived from cornstalk without pith for high capacitance and energy density, *J. Mater. Chem. A*, 2020, **8** (3), 1469–1479.
51. X. Xie, X. He, F. Wei, N. Xiao and J. Qiu, Interconnected sheet-like porous carbons from coal tar by a confined soft-template strategy for supercapacitors, *Chem. Eng. J.*, 2018, **350**, 49–56.
52. C. Ban, Z. Xu, D. Wang, Z. Liu and H. Zhang, Porous layered carbon with interconnected pore structure derived from reed membranes for supercapacitors, *ACS Sustain. Chem. Eng.*, 2019, **7** (12), 10742–10750.
53. P. Li, H. Xie, W. Hu, T. Xie, Y. Wang and Y. Zhang, Molten salt and air induced nitrogen-containing graphitic hierarchical porous biocarbon nanosheets derived from kitchen waste hydrolysis residue for energy storage, *J. Power Sources*, 2019, **439**, 227096.

6 Biomass-Derived Porous Carbon Materials for Supercapacitors

Bei Liu, Yong Ye and Ting Li

1 INTRODUCTION

Biomass is currently attracting significant attention as an alternative precursor to fossil-based feedstock in the production of carbon materials for supercapacitors (SCs), due to the advantages of naturally renewable, abundant, low cost, easily accessible, and environmentally friendly .[1, 2] In recent years, various forms of biomass such as plants, animals, and microorganisms have been widely utilized to prepare carbon materials.[3, 4]

The biomass-derived carbon materials (BCMs) offer numerous benefits: (1) cost-effective and abundant, (2) excellent chemical and physical stability, (3) rich in surface functional groups, which can be adjusted using different thermochemical conditions, (4) special porous structures and high specific surface area (SSA), which can promote the accessibility of electrolyte to the electrode and shorten the ion transport distance, (5) most biomass contain heteroatoms, leading to the formation of extra active sites in the BCMs. However, it is challenging to regulate the material features and properties containing multiscale morphology, porosity, and surface chemistry using a typical top-down approach, which limits its applications. Therefore, appropriate activation and modification are required to enhance the performances of BCMs in further applications.[5, 6]

After decades of research, advanced development has been made in BCMs production by applying novel synthesis strategies to renewable biomass. Furthermore, a diverse range of biomass feedstocks has been utilized to synthesize BCMs by pyrolysis or hydrothermal carbonization with activation and modification process, thus producing porous carbon or nanostructured carbon with rich porosity, abundant functionalities, hierarchical porous structure, and excellent conductivity. Due to the aforementioned advantages, BCMs are considered promising electrode materials for various types of electrochemical energy storage and conversion systems, including supercapacitors (SCs), lithium batteries, sodium batteries, potassium batteries, and fuel cells.

2 PRECURSORS IN BCMS

The electrode material is a key component of a supercapacitor (SC). Research findings of varying precursors are discussed in detail below and Figure 6.1a illustrates a variety of biomasses as the precursors to prepare BCMs.

DOI: 10.1201/9781003387831-6

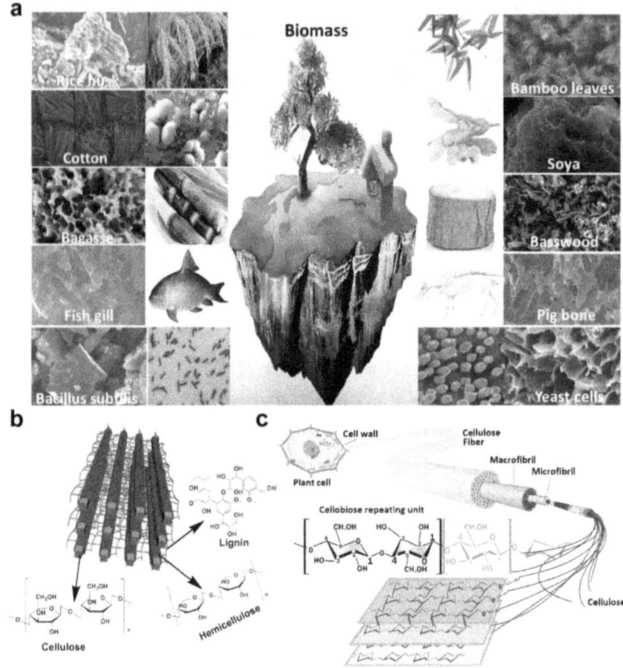

FIGURE 6.1 (a) A variety of biomass as the precursor is transformed into electrode material for supercapacitors. Reproduced with permission.[7] Copyright 2017, Elsevier. (b) Schematic illustration of the cellulose, hemicellulose, and lignin in the plant cell. Reproduced with permission.[8] Copyright 2017, RSC. (c) The structure of cellulose fiber in the plant cell wall. Reproduced with permission.[9] Copyright 2017, Elsevier.

2.1 Plant-based precursors

The components of plant biomass (trees, crops, fruits, vegetables, algae, etc.) include cellulose, hemicelluloses, lignin, extractives, lipids, proteins, etc.[7] Figure 6.1b,c illustrate the basic structure of the cellulose, hemicellulose, and lignin in the plant cell, and it is found that hemicellulose and cellulose are less stable to thermal decomposition as compared to lignin, and contribute low carbon yield, but the hemicellulose, lignin, and cellulose components contribute porosity of biochar yield.[8] Furthermore, it has been determined that high O content in precursors results in less crystalline and more defects also more volatile compounds get released, whereas the presence of higher levels of N content could result in N-doped carbon with enhanced electrochemical characteristics.[9] Thus selecting plant biomass precursors rich in N content and lignin fraction, and low in O content and cellulose fraction to obtain the BCMs with good conductivity, and controllable defects seems very important.

2.2 Animal remains-based precursors

Chitin (derived from crustaceans, mollusks, coleoptera, animal waste, etc.) is another promising bio-based precursor for the production of BCMs. Different from cellulose, chitin is capable of forming (1) intensive intermolecular hydrogen bonds, (2) cross-linking networks of chitin-glucan and chitin-catecholamine complexes, which endow chitin with substantially higher thermal stability than cellulose and higher carbon yields.[10] The cuticles of many contain relatively high contents of chitin ranging from 17% to 72%.[11] However, mechanical and chemical processes are required to extract chitin from the biomasses, which may involve mechanical grinding, chemical demineralization, and deproteinization. Different from the cuticles, the hairs, horns, claws, and hooves of the animals are majorly consisting of fibrous structural proteins, called keratin.[12] Similar to chitin, keratin also attains significant chemical and thermal stability as a result of its intensive intermolecular bonding.[13]

2.3 Microorganism-based precursors

The microorganisms (classified as bacteria, algae, and fungus) consist of proteins, lipids, carbohydrates, nucleic, and amino acids. The major components present in the microorganism-based biochar match with plant/animal-based biomasses, however, the elements and actual molecules that constitute these components are substantially different.[14] For example, the carbohydrates of the microorganism-based biomass consist of chitins, which are bio-crosslinked with glucan and can serve as a primary source of carbon during thermal carbonization, while the main carbohydrates in the plant/animal-based biomasses are sucrose and starch, which are non-crosslinked with low thermal stability.[15]

3 FUNCTIONALIZATION CARBONIZATION FOR BCMS

The carbonization methods would determine the chemical and physical properties of BCMs directly, such as morphology, surface area, porosity, chemical components, functional groups, degree of graphitization, etc. Moreover, the carbonization methods are of great importance for reserving some macro-/micro-structures of the biomass in BCMs. Table 6.1 lists different activation methods and capacitance performances of a variety of BCMs.[16–33]

3.1 Directly pyrolysis carbonization

Heating biomass in a certain atmosphere is the most popular carbonization method because the structures and properties of the prepared BCMs can be easily tailored by selecting carbonization parameters. The main controllable parameters include temperature, heating rate, residence time, and atmosphere. In comparison with reaction time, heating rate, and particle size of feedstock, temperature plays a key role in controlling the pyrolysis mechanism and biochar yield. The heating rate is the key factor that determines the pyrolysis products. Depending on their heating rate, pyrolysis is categorized into slow pyrolysis and fast pyrolysis. The slow pyrolysis process possesses longer residence time (>1 h) and lower heating rates (between 5 and 7°C min^{-1}), conducted at the temperature of 400–600°C.[34] In contrast, fast pyrolysis has a much higher heating rate (>200°C min^{-1}) and lower residence time (<10 s).[35] Therefore, a slow heating rate is more favorable in the synthesis of BCMs.[36]

TABLE 6.1
The different activation methods and capacitance performances of BCMs

Precursor	Activation	SSA (m^2 g^{-1})	C_g (F g^{-1})	Current density/scan rate	Electrolyte	Ref
Soybean	Steam	1,077.67	143	0.1 A g^{-1}	1 M Na_2SO_4	33
Fir wood	Steam	1,016	85	10 mV s^{-1}	1 M $NaNO_3$	34
Cornstalk	Air	1,588	407	1 A g^{-1}	1 M H_2SO_4	35
Clover stem	Air	1,459	451	0.5 A g^{-1}	1 M H_2SO_4	36
Cattail	CO_2	441.12	126.5	0.5 A g^{-1}	6 M KOH	37
Banana	CO_2	1,414.97	178.9	1 A g^{-1}	6 M KOH	38
Apricot shell	H_3PO_4	1,474.82	169.1	0.5 A g^{-1}	6 M KOH	39
Lacquer wood	H_3PO_4	1,419.46	354	0.2 A g^{-1}	1 M H_2SO_4	40
Cotton stalk	H_3PO_4	1,342.93	338	0.2 A g^{-1}	1 M H_2SO_4	41
Lignin fibers	H_3PO_4	2,340	240	10 mV s^{-1}	1 M H_2SO_4	42
Argy wormwood	H_3PO_4 + KOH	927	344	1 A g^{-1}	6 M KOH	43
Sakura	KOH	1,433.76	265.8	0.2 A g^{-1}	6 M KOH	44
Bamboos	KOH	2,221.1	293	0.5 A g^{-1}	3 M KOH	45
Green tea	KOH	1,057.8	162	0.5 A g^{-1}	1 M H_2SO_4	46
Rice straw	KOH	3,333	400	0.1 A g^{-1}	6 M KOH	47
Jujube fruits	NaOH	1,135	587	0.1 A g^{-1}	6 M KOH	48
Mangosteen peel	NaOH	2,623	357	1 A g^{-1}	6 M KOH	49
Tea leaves	$ZnCl_2$	1,143.9	296	0.5 A g^{-1}	2 M KOH	50
Chitin	$ZnCl_2$	2,450	294.4	1 A g^{-1}	6 M KOH	51
Mango-stone	$ZnCl_2$ + KOH	1,497.8	353.8	0.5 A g^{-1}	6 M KOH	52
Wood scraps	CO_2 + KOH	703.5	285.6	10 mA cm^{-2}	1 M Na_2SO_4	53
Paper flower	$ZnCl_2$ + CO_2	1,801	118	1 A g^{-1}	1 M H_2SO_4	54
Jute fiber	HTC + KOH	1,902	346	1 A g^{-1}	6 M KOH	55
Toboco rods	HTC + KOH	2,115	287	0.5 A g^{-1}	6 M KOH	56
Microalgae	HTC + KOH	2,876	280	1 A g^{-1}	1 M H_2SO_4	57
Amaranthus	Direct carbonization	1,172	326	0.5 A g^{-1}	6 M KOH	58
Perilla frutescens	Direct pyrolysis	655	270	0.5 A g^{-1}	6 M KOH	59
Jute sticks	Direct pyrolysis	2,000	150	1 A g^{-1}	KOH/ Glycerol-gel	60
Betel nut	HNO_3	NA	423	0.5 A g^{-1}	6 M KOH	61

The carbonizing temperature influences the element compositions and the degree of graphitization directly, and the main products are CO, CO_2, and H_2O. Besides, the overall N and S contents in the BCMs generally decrease when the annealing temperature increases. Meanwhile, temperature also influences the chemical states of heteroatoms.[37] In the carbonizing process, the annealing time is an important factor that is closely related to temperature. A long annealing time at low temperatures will result in the formation of more char and tar. The high-temperature carbonization with a long annealing time is suitable for producing BCMs with less heteroatom and higher graphitization degree. Carbonization processes are generally carried out in inert gases, such as nitrogen and argon. These inert gases will not react with the biomass during the heating process, providing a suitable environment for carbonization. Some other reactive gases can be used to modify the chemical compositions or introduce new elements during the carbonization of biomass.[38]

3.2 Activation

Activation is the most convenient method to increase the surface area of carbon materials and adjust the mesopore/micropore proportion.

3.2.1 Physical activation

Physical activation usually uses a two-step process. Biomass material is first pyrolyzed to produce biochar (400–850°C), which is then activated by regulated gasification (600–1,200°C) utilizing gases, such as CO_2, steam, air, or their mixture.[39] Subsequently, controlled gasification can facilitate the as-prepared biochar to proceed a further decomposition and obtain a fully developed, accessible, and interconnected porous structure. The development of porosity often results from carbon calcination and volatile substance generation and greatly depends on the activating gas, and the extent of carbon calcination is affected by activation temperature, time, gas flow rate, and choice of the furnace.[40] In general, the higher activation temperature can result in larger SSA and micropore volume; however, a higher temperature above 900°C leads to pore deformation.[41] Rising activation time favored the micropores and mesopores generation, but overly prolonged activation time (>60 min) resulted in the collapse and deterioration of pores[42] because the development and widening of micropores were less effective than the deterioration of high porosity.[43]

3.2.2 Chemical activation

For chemical activation, the precursor is first impregnated with certain chemicals, followed by carbonization at a temperature from 450 to 900°C. Compared with physical activations, chemical activations have several advantages: (1) the pores are well developed and pore size is more controllable; (2) carbon yield is high; (3) pyrolysis temperature is lower and the activation time is shorter; (4) carbon materials possess high SSA.[44] However, the washing process is necessary due to the corrosive properties of activating reagents. Various activating regents have been used, such as KOH, NaOH, H_3PO_4, $ZnCl_2$, $MgCl_2$, $FeCl_3$, and K_2CO_3. Acting as an oxidant and creating oxygen functional groups on biochar, KOH is the most frequently used chemical activation agent.[45] During the chemical activation process, micropore formation is initiated by adding activation agents to the raw material at the beginning. Pore widening results from enlarged existing micropores, due to higher activation agent dose, higher activation temperature, or longer reaction time. Thus the micropores are widened and even collapsed, which can contribute to micropore volume reduction and mesopore volume growth.[46] To achieve the optimal SSA and pore-size distribution, micropore formation and mesopore growth should be well balanced. For different biomass materials, suitable activation and process conditions should be carefully selected to obtain appropriate physicochemical properties and surface chemistry, thus further achieving excellent electrochemical performance in their applications.

3.2.3 Self-activation

Unlike chemical and physical activation, no additional activation reagents are needed for preparing BCMs. Self-activation process contains two activation mechanisms. One is utilizing the gases emitted during the pyrolysis process of biomass to activate the converted carbon (physical self-activation). Another one is making use of the inorganic materials (like K^+) already existing within the biomass to in-situ activate the converted carbon (chemical self-activation).[47]

3.3 Template Method

To control the porosity of BCMs for specific applications, template methods, including hard templates and soft templates have been adopted. By using hard templating/nano-casting approaches, the obtained carbon materials reversely replicate the morphology of templates and can also be adapted to introduce more mesopores in the resulting carbon matrix.[48] Up to now, various architectures have been prepared, such as ordered mesoporous carbon (OMC) matrix, hierarchically porous carbon monoliths, carbon nanosheets, carbon spheres, as well as peanut-like carbon. By the soft-templating method of block-co-polymer micelles, OMC materials have also been prepared using biomass, like fructose and dopamine, as carbon precursors.[49] This soft-templating method simplifies the production of porous carbon greatly by avoiding the step of removing the template. Template methods provide advantages in controlling morphology and porous structure, but the largest challenge is the use of the template, which involves time-consuming steps and high-cost expenses.[50]

3.4 Hydrothermal Carbonization

Hydrothermal carbonization (HTC) is a powerful method that converts biomass into BCMs in a subcritical water state. There are three kinds of reactions in the HTC process, dehydration of the biomass, polymerization of the fragments, and carbonization via intermolecular dehydration. By modifying the reaction conditions, it is easy to realize the precise control of the BCMs including structure, component, and morphology. The main controllable parameters of HTC can be summarized as biomass precursor, temperature, residence time, and doping. There were a lot of works about synthesizing globular BCMs by HTC of different saccharides, including glucose, sucrose, starch, sugar, and others.[51]

3.5 Microwave-Assisted Carbonization (MAC)

MAC is a facile approach to the carbonization of biomass. The mechanism of MAC is the conversion of electromagnetic energy into heat within the material being irradiated, which offers several advantages compared with conventional heating, such as non-contact volumetric heating, rapid heating, heating from the interior of materials, high safety level, quick start-up, and stopping.[52] However, the high temperature and the rapid heating rate will cause the rapid weight loss of the biomass precursors. Compared with other carbonization methods, the pyrolysis of biomass in MAC is easier and faster, which enables the effective doping effect to tune the structures and components of BCMs.[53]

3.6 Molten salt carbonization

The hydrothermal reaction usually ends at around 200°C because of the limitation of solvents boiling. After high-temperature pyrolysis, the final carbon structures shrink significantly due to mass loss. On the contrary, molten salt carbonization, with much higher heat capacity, can provide a much wider operating temperature window. In addition, the carbonaceous material mechanically stabilizes against shrinkage because of the molten salt, resulting in a greatly enhanced yield.[54] Molten salt carbonization is an efficient and green method to prepare carbon materials from biomass.[55]

3.7 Laser-induced carbonization (LIC)

The laser can produce an extremely high temperature within a short time and thus it can be used to induce the carbonization of biomass. As for biomass, the high carbonization temperature and rapid heating rate of laser processing will cause burning and rapid weight loss of the biomass precursors.[56]

In short, each of the above carbonization methods for biomass has advantages and drawbacks, as summarized in Table 6.2. All these aspects should be considered while choosing carbonization methods for the preparation of BCMs.

TABLE 6.2
Characteristics of the typical carbonization methods of biomass

Method	Operating temperature	Preponderances	Shortcomings
Pyrolysis carbonization	≈3,000°C	• Facile control • Wide operating temperature • Controllable atmosphere	• Uneven heating • High-energy consumption
Activation	600–1,200°C	• Convenient to regulate SSA and pores • Diverse activating reagents • Low energy consumption	• Difficult regulating the activation parameters • Activating reagents pollution
Template method	–	• Regulate the morphology and porous structure	• Time-consuming steps • Need to remove the template
Hydrothermal carbonization	≈350°C	• Homogeneous carbonization • Fitting for nanostructural control • Easy doping	• Low carbonization temperature • Low graphitization degree
Microwave carbonization	≈1,000°C	• Rapid and volumetric heating • Material-selective carbonization • Low energy consumption	• Difficult temperature control • Bad reproducibility • Lab-scale
Laser-induced carbonization	>2,500°C	• Ultrafast and local heating • Pattern-designable carbonization • Low energy consumption	• Surface carbonization • Large biomass loss • Lab-scale

4 STRUCTURE OF BIO-CARBON AND SUPER-CAPACITIVE PERFORMANCE

4.1 Pore structure and specific morphologies

During the charging and discharging of SC devices, the wide pore-size distribution of BCMs plays an important role in enhancing their electrochemical performance. Macropores (≥50 nm in size) serve as ion buffers for mesopores and micropores, while mesopores (2–50 nm in size), also known as transition pores, provide efficient ion diffusion, and micropores (<2 nm in size) are used to store charges (Figure 6.2a).[57] In the case of hierarchical porous carbon, there is an interconnection between these different pore-size structures, which facilitates the diffusion of electrolyte ions inside the channels. Meanwhile, the pore structure of carbon materials can be made to have hierarchical pores in two-dimension (2D) or three-dimension (3D) to facilitate ion transport and provide a robust interface for charge storage.[58] A porous carbon with macropores, mesopores, and micropores well-distributed in 2D or 3D with low-resistant ion transport paths is ideal for EDLCs.[59]

4.2 Heteroatom-doping

Heteroatom-doping can significantly adjust the surface properties of carbon electrode materials, such as conductivity, wettability, and pseudocapacitance.[60] The doping pattern can be divided into self-doping and external doping. For self-doping, it can be transformed into heteroatom-contained carbon by simple one-step carbonization without the addition of any other doping agents. For external doping, the doping agents normally include melamine, urea, phosphoric acid, etc. Table 6.3 illustrates some biomass-derived heteroatoms-doping carbon materials and their application in SCs.[61–75]

4.2.1 N-doping carbon

Nitrogen (N), the most common doping element, is similar to a carbon atom for radius, which can maintain the original carbon skeleton. The N atoms in the carbon structure mainly emerge with four states, including pyridine-N, pyrrole-N, quaternary-N, and oxidized pyridine-N (Figure 6.2c).[76] N atom has one more electron in the valence electron layer than the carbon atom, resulting in the increased conductivity of N-doping carbon materials due to the electron donor of the N atom. At the same time, the large electronegativity of N can improve the hydrophilicity and make the carbon electrode fully in contact with the aqueous electrolyte.[77]

4.2.2 S-doping carbon

The radius of the S atom is larger than that of the carbon atom, resulting in the deformation and destruction of the original carbon structure with the introduction of the S atom, which leads to lots of defects used as redox-active sites to increase the capacitance for supercapacitors.[78] The S atoms doped in carbon materials mainly exist in many forms of thiol, thiophene, thioether, sulfoxide, sulfonyl, etc. (Figure 6.2d).

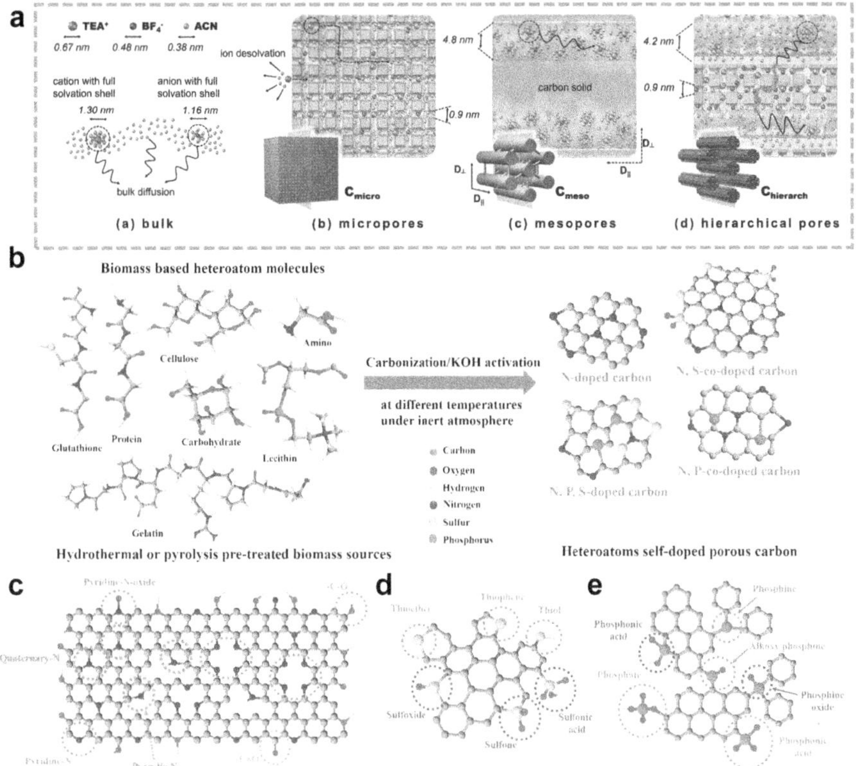

FIGURE 6.2 (a) Schematic representation of the different sizes of pore structures. Reprinted with permission.[57] Copyright 2018, Wiley. (b) Schematic illustration of heteroatoms-doped porous carbon derived from biomass sources and types of nitrogen (c), sulfur (d), and phosphorus (e) functional groups in the carbon matrix. Reprinted with permission.[76] Copyright 2020, Elsevier.

4.2.3 P-doping carbon

Phosphorous (P) elements, belonging to the same main group as N, have similar properties of N as a doping element in carbon material. However, the carbon skeleton structure doped with P is easy to deform, causing a ripple effect due to the larger atom radius of P.[79] The presence of P atoms in carbon materials improves the electronic distribution, wettability, and additional capacitance, thereby enhancing their electrochemical performance.[80] Interestingly, the P incorporation also can even broaden the operating voltage.[81] Since the reduced state of the P element is not stable, P eventually exists in a certain degree of oxidation state. The main forms are phosphine, alkoxy phosphine, phosphorus oxide, phosphonic acid, and so on (Figure 6.2e).

4.2.4 Other heteroatom-doping carbon

The doping of other heteroatoms in biomass-based carbon material, such as oxygen (O) and boron (B), has also received much attention in improving electrochemical

TABLE 6.3
The heteroatom contents and corresponding electrochemical performances of BCMs

Precursors	Heteroatoms (at %)	SSA (m^2 g^{-1})	C_g (F g^{-1})	Current density (A g^{-1})	Electrolyte	Ref.
Pomelo peel	O (7.84)	38.44	222.6	0.5	6 M KOH	61
Coffee bean	P (13.3)	742	157	1	1 M H_2SO_4	62
Soybean	N (1.43)	2,130	258	0.2	1 M H_2SO_4	63
Watermelon rind	N (2.48)	1,958	333.4	1	6 M KOH	64
Soybean milk	N (2.4)	1,208	240.7	1	6 M KOH	65
Houttuynia	N (7.22)	2,090	473.5	1	6 M KOH	66
Poplar power	N (8.7)	1,612	508	1	6 M KOH	67
Celery stem	N (7.9)	1,640	402	1	1 M H_2SO_4	68
Bamboo fiber	S (2.2)	2,561	328	1	1 M H_2SO_4	69
Wood	P (9.24)	958	384	1 mA cm^{-2}	6 M KOH	70
Elaeocarpus tectorius	P (8.1)	858	358	0.2	1 M H_2SO_4	71
Dragon fruit skin	O (22.26)	911.2	286.9	0.5	2 M KOH	72
Jujube pit	O (6.7)	2,438	398	0.5	6 M KOH	73
Pitch and sawdust	O (6.79)	2,361	324	0.5	6 M KOH	74
Dandelion fluff	N (2.2), B (4.6)	1,420	355	1	6 M KOH	75

performances (Figure 6.2b). The obtained BCMs exhibited a high degree of hydrophilicity due to abundant heteroatoms, conducive to the transport of electrolyte ions and electrons in the carbon electrode and the improved electrochemical performances. The heteroatom-doping content in BCMs is closely related to the temperature of carbonization or activation processes. It is found that the heteroatom content gradually decreases along with the enhancement of conductivity via the increased graphitization degree at higher carbonization temperatures.[82] Therefore, there is a compromise between the degree of graphitization and the heteroatom-doping content to maximize electrochemical performance.

4.3 Graphitization Degree

Improving the graphitization degree of BCMs simultaneously enhances the electric conductivity of the electrode and its surface wettability toward an aqueous electrolyte, which can facilitate ion diffusion and electron transfer, thus improving the electrochemical performance.[83] High-temperature treatment can enhance graphitization but it is of high-energy consumption. Besides, it will also decrease the SSA and pore volume of the BCMs. Therefore, there is a compromise between the degree of graphitization and the pore structure to maximize electrochemical performance.[84]

Catalytic graphitization by means of a transition metal is an effective way to obtain BCMs with a certain graphitization degree.[85] Coupling chemical activation with catalytic graphitization enables the preparation of porous carbons with a high SSA and excellent electrocapacitive performance.[86] For instance, Sun et al.[87] synthesized porous graphene-like nanosheets (PGNSs) with a large SSA via a simultaneous activation-graphitization route using coconut shells as the precursor. $FeCl_3$ and $ZnCl_2$, functioning as a graphitic catalyst and activating agent, respectively, were simultaneously introduced into the skeleton of the coconut shell through coordination of the metal precursors with the functional groups in the coconut shell, thus making simultaneous activation and graphitization. The obtained PGNSs possessed good electrical conductivity due to a high graphitization degree, a high SSA of 1874 $m^2\ g^{-1}$, a pore volume of 1.21 $cm^3\ g^{-1}$, and a specific capacitance of 268 $F\ g^{-1}$ at 1 $A\ g^{-1}$ in KOH electrolyte.

5 BCMS FOR ADVANCED SUPERCAPACITORS

5.1 Supercapacitors (SCs)

SC is the most prominent and promising energy storage device technology, stores energy the same way a conventional capacitor does, but they are more suited for rapid energy release and storage. According to the energy storage mechanism, SCs are divided into electrical double-layer capacitances (EDLCs), pseudocapacitors, and hybrid SCs (Figure 6.3).

5.1.1 EDLCs

EDLCs have two identical carbon-based materials used as electrodes, an electrolyte parted by a separator. The charges can either be electrostatically stored via forming an EDL or transferred between the electrode and electrolyte using a non-Faradaic method. As seen in Figure 6.3,[88] EDLC is illustrated with a PC electrode, with a high SSA, and reducing the distances between the electrodes, can decrease the electrolyte ion transport resistance and subsequently improve the power density.

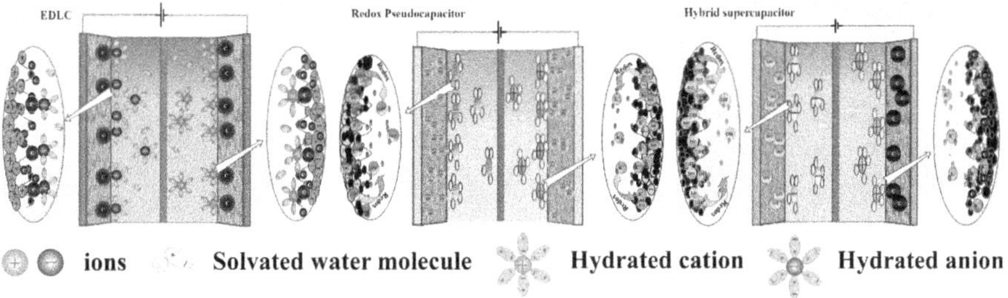

FIGURE 6.3 Schematic illustration of EDLCs (carbon electrode and KOH electrolyte), Redox pseudocapacitance (metal oxide-MO electrode and KOH electrolyte), and hybrid supercapacitor (carbon and MO electrode and KOH electrolyte). Reprinted with permission.[88] Copyright 2023, Elsevier.

5.1.2 Pseudocapacitors

Pseudocapacitor is based on the redox processes of electroactive materials like metal oxides and conducting polymers. Transferring charges between the electrode and the electrolyte faradaically stores energy. Electrochemical energy storage via pseudocapacitor involves three different mechanisms, they are adsorption/desorption (underpotential deposition), reduction/oxidation (redox pseudocapacitor), and intercalation/deintercalation (intercalation pseudocapacitor). Because of their poor mechanical stability and short cycle-life, pseudocapacitors are more prone to expand and shrink during charging and discharging cycles. Therefore, the cyclic stability of the pseudocapacitor will be slightly lower than that of EDLCs.

5.1.3 Hybrid SCs

A hybrid SC combines the benefits of high power density, specific capacitance, good cycle stability of EDLCs, and high-energy density of pseudocapacitors. Faradaic activities with EDLC charge storage improve an electrode's specific capacitance. Because the limiting feature of the EDLC is absent from the pseudocapacitor and vice versa, their combination overshadows the combining component restrictions. Hybrid SCs can be either asymmetric or symmetric, depending on the assembly of the electrodes. In asymmetric hybrid SCs, two different electrodes are employed, usually with a combination of an EDLC electrode and a pseudocapacitor electrode or otherwise a combination of a battery-type electrode as anode and a capacitor-type electrode as the cathode. Compared to the symmetrical EDLC, the hybrid SCs have greater specific capacitance values and a higher rated voltage corresponding to higher specific energy.

5.2 Calculations for the Capacitive Performance of SCs

Cyclic charge/discharge, cyclic voltammetry, and linear sweep voltammetry are the most commonly performed electrochemical tests. The calculations for the main parameters of SCs are as follows (Eqns 6.1–6.4)[5]:

$$C_g = \Delta Q/(m \times \Delta V) \text{ (3 electrode system)} \tag{6.1}$$

$$C_s = (2 \times \Delta Q)/(m \times \Delta V) \text{ (2 electrode system)} \tag{6.2}$$

where C = specific capacitance; ΔQ = difference in charge during charge and discharge cycles respectively; ΔV = difference in potential.

$$E = (C \times \Delta V^2)/(2 \times 3.6) \tag{6.3}$$

where E = energy density; C = specific capacitance; ΔV = difference in potential.

$$P = (3.6 \times E)/\Delta t \tag{6.4}$$

where P = power density; E = energy density; Δt = discharge time.

5.3 High capacitive SC

Specific capacitance is the basic parameter to evaluate the performance of SCs, including mass-specific capacitance and volume-specific capacitance. The total of specific capacitance can be defined as the sum of EDLC and pseudocapacitance (PSC). For instance, Lu et al.[89] prepared porous carbon materials with intrinsic defects and nitrogen doping using microcrystalline cellulose as the precursor (Figure 6.4a,b), the obtained BCMs displayed an exceptional specific capacitance of 426 F g^{-1}, an excellent rate capability, and superior cycling stability. Liu et al.[90] fabricated a mass of 3D O/N co-doped carbon nanosheet from woody precursor via sequential alkaline oxidation (Figure 6.4c). The doped BCMs delivered an impressive superior capacitance of 508 F g^{-1} at 1 A g^{-1} and a long lifetime of over 12,000 cycles with 95% capacitance retention.

5.4 High volumetric SC

Volumetric performance reflects how much and how fast energy can be stored in a unit the volume of material/packed device (Figure 6.5a,b).[91, 92] For instance, Tao

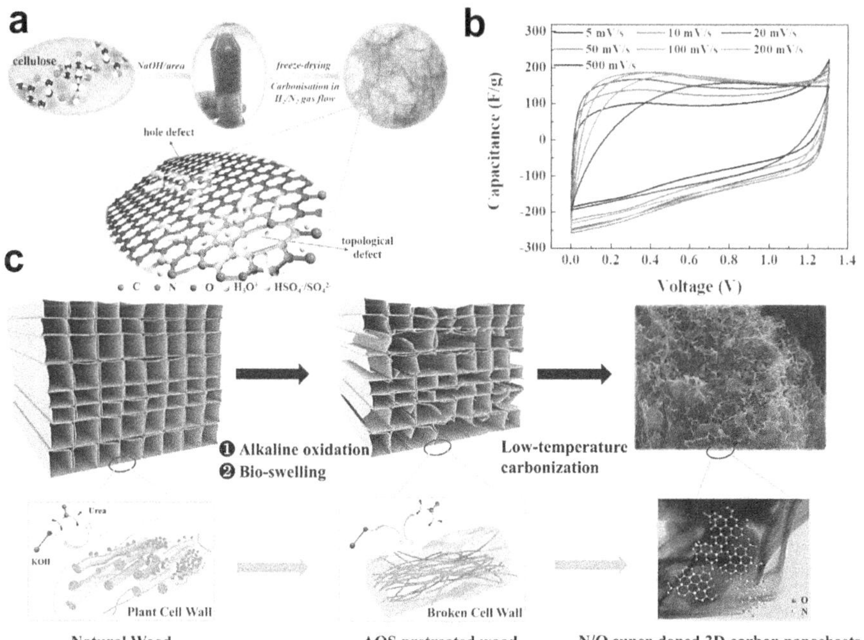

FIGURE 6.4 (a) A schematic of the preparation process of microcrystalline cellulose-derived porous carbon. (b) CV curves at scan rates 5–500 mV s^{-1} of the as-prepared sample. Reprinted with permission.[89] Copyright 2019, Royal Society of Chemistry. (c) Schematic illustration of sustainable synthesis and assembly of O/N co-doped 3D carbon nanosheet. Reprinted with permission.[90] Copyright 2019, Elsevier.

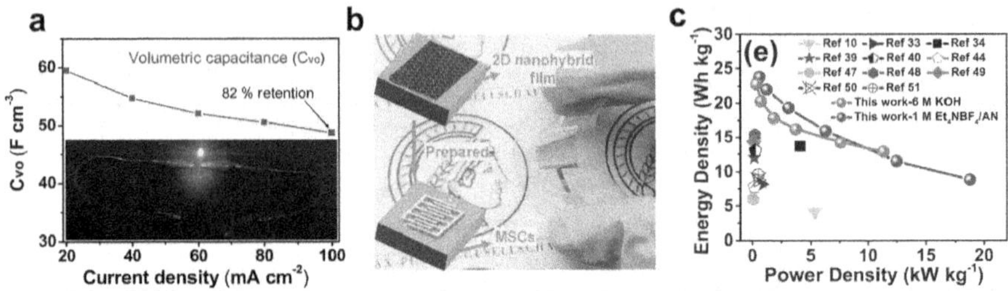

FIGURE 6.5 (a) Volumetric capacitances of the fabricated ASC (inset: blue light-emitting diode [LED] powered by the fabricated ASC). Reprinted with permission.[91] Copyright 2015, Royal Society of Chemistry. (b) Illustration of the fabrication procedure for in-plane MSCs (inset: the fabricated MSCs exhibited highly flexible character). Reprinted with permission.[92] Copyright 2015, Wiley Online Library. (c) Comparison of Ragone plots between the as-assembled cells and the reported carbon materials, reprinted with permission.[97] Copyright 2020, Elsevier.

and coworkers[93] have devoted much effort toward the synthesis of porous graphene macro-assembly (HPGM) for high-performance SCs electrode, the obtained electrode exhibited excellent volumetric capacitance up to 376 F cm^{-3} with maximum energy density as high as 13.1 Wh L^{-1} in aqueous electrolytes Gogotsi and coworkers[94] fabricated flexible and sandwich-like MXene/CNT composite paper electrode with high volumetric capacitance through alternating filtration, the sandwich-like electrode yielded a volumetric capacitance of 390 F cm^{-3} in SCs. Despite these tremendous achievements, there are still many challenges that need to be overcome on the road to the design and synthesis of BCMs with both high gravimetric and volumetric performances.

5.5 High-rate SC

High-rate SC can maintain high specific capacitance and fast charging/discharging ability even under ultra-high current/scanning speed. In general, heteroatom-doped is the main method to improve the rate performance of SCs. For instance, Zhang and cooperators[95] synthesized O-doped BCMs with stacked layer texture employing gelatin as the precursor. The as-prepared O-doped BCMs demonstrated a specific capacitance of 276.6 F g^{-1} at 1 A g^{-1} in a two-electrode system and still retained 72.3% of capacitance retention at 100 A g^{-1}, a large proportion of O (16.22 wt%) can effectively improve rate performance.

5.6 High-energy SC

In accordance with the equation $E = C_s \Delta V^2/(8 \times 3.6)$, the energy density ($E$) of SCs could be theoretically improved by either achieving higher electrode capacity (C_s) or by expanding the operating potential window of the device (V). Thus, the key for high E is to find both high-capacitance material and high-voltage electrolytes. For example, Bao Men et al.[96] reported high-performance N-doped BCMs derived from cauliflower for advanced SCs, the sample exhibits a high specific capacitance of 311 F g^{-1}. Furthermore, the energy density of the assembled symmetric SC is as high as 20.5 Wh kg^{-1} at a power density of 448.8 W kg^{-1}, meanwhile showing excellent rate capability and long-term cyclic stability. Viengkham Yang et al.[97] (Figure 6.5c) reported a new highly porous natural carbon material, that displays the specific capacitance of 587 F g^{-1} at 100 A g^{-1}. Moreover, the assembled symmetrical coinlike SCs with the wide potential window of 2.5 V in 1 M Et_4NBF_4/AN organic electrolyte offer a high-energy density of 23.7 Wh kg^{-1} at 0.629 kW kg^{-1} with remaining 94% capacitance over 10,000 cycles at 30 A g^{-1}.

5.7 Long cycle-life SC

The cycle stability directly determines the service life of SC, which can effectively reduce the cost of frequent device replacement in practical applications. Recently, Bose Nirosha et al.[98] (Figure 6.6a) demonstrated that P-doped porous carbon is prepared from new biomass (*Elaeocarpus tectorius*), the BCMs deliver high gravimetric capacitance (385 F g^{-1} at 0.2 A g^{-1}) in 1 M H_2SO_4 aqueous electrolyte. Additionally, a coin cell asymmetric SCs fabricated using this carbon works in a wide potential window from 0 to 1.5 V with 96% capacitance retention in 1 M H_2SO_4 aqueous electrolyte for 10,000 cycles and yields a high-energy density of 27 Wh kg^{-1}, showing the utility for the development of long cycle-life electronic devices.

5.8 Flexible SC

The manufacture of flexible electrodes from biomass generally does not require flexible collectors (e.g., Nickel foam), loading inactive polymer binders (e.g., PTFE, PVDF), and conducting agents. Manufacturing electrodes with low cost, high capability, and exceptional mechanical flexibility poses one of the primary challenges in the production of flexible SC. For instance, Dan Wu et al.[99] (Figure 6.6c,d) fabricated a mass of flexible B/N co-doped carbon via a hard template and subsequent annealing method. The B/N co-doped BCMs deliver a capacitance of 268 F g^{-1} at 0.1 A g^{-1}, the assembled flexible SC delivers a long lifetime over 15,000 cycles with high-capacitance retention at different bending conditions.

5.9 High/low-temperature SC

In some situations, energy storage devices need to be applied in harsh environmental conditions, especially in severe low and high temperatures. Selecting suitable BCMs plays a critical role in designing high/low-temperature SC that can work in harsh environments. Bin Yao et al.[100] (Figure 6.6b) fabricate a 3D-printed multiscale porous carbon aerogel (3D-MCA) via a unique combination of chemical methods and the direct ink writing technique. At −70°C, the symmetric SC achieves an outstanding capacitance of 148.6 F g^{-1} at 5 mV s^{-1}. Dan Wu et al.[99] (Figure 6.6e) fabricate a nacre-mimetic graphene-based film (CNT@PANI/rGO/TA), wherein electroactive biomass tannin (TA) serves as glue, the obtained all-solid-state flexible supercapacitor assembled with CNT@PANI/rGO/TA demonstrates a high capacitance of 548.6

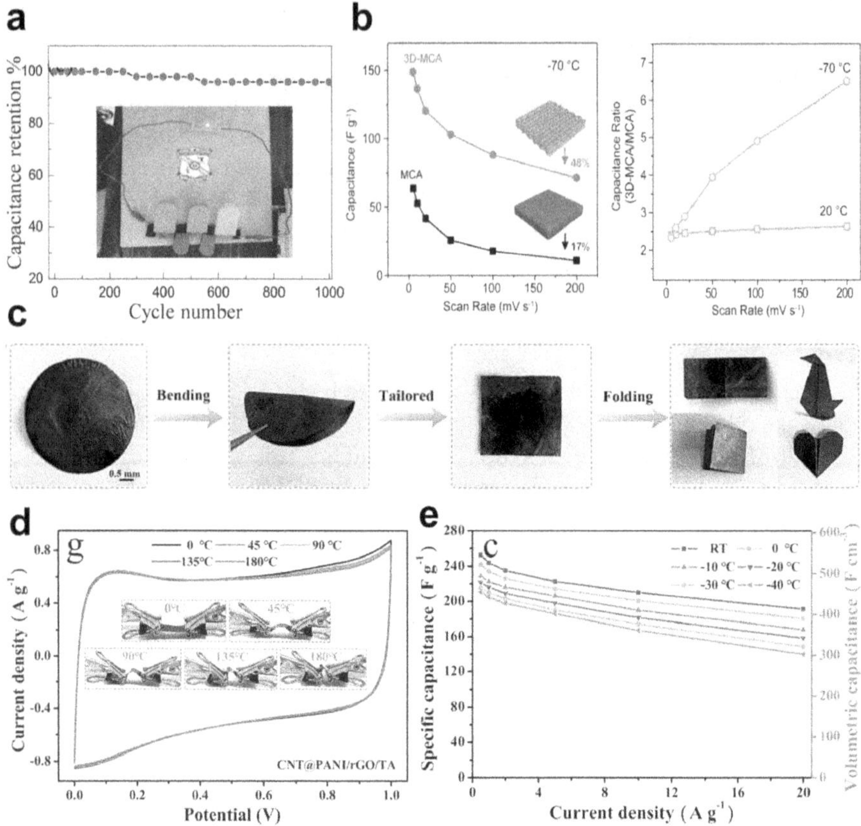

FIGURE 6.6 (a) Capacitance retention as a function of cycle number (inset: photograph of a powered red LED). Reprinted with permission.[98] Copyright 2020, Royal Society of Chemistry. (b) Gravimetric capacitances of 3D-MCA and MCA obtained at –70°C plotted as a function of scan rate and capacitance ratio of 3D-MCA to MCA obtained respectively at 20°C, and –70°C plotted as a function of scan rate. Reprinted with permission.[100] Copyright 2021, ACS. (c) Optical images of BCMs film in the flat, bent, tailored, and different folding states, respectively. (d) CV curves at 10 mV s^{-1} for different bending angles. (e) specific capacitance versus different current densities (1–20 A g^{-1}). Reprinted with permission.[99] Copyright 2021, Royal Society of Chemistry.

F cm^{-3} and outstanding rate performance of 70.5% from 1 A g^{-1} to 50 A g^{-1}. Even at –40°C, the specific capacitance of the supercapacitor is up to 454.9 F cm^{-3}, and approximately 83% of the capacitance is delivered at room temperature.

6 CONCLUSION AND OUTLOOK

In summary, the applications of plentiful biomass in fabricating BCMs with excellent electrochemical behaviors for supercapacitors (SCs) are sustainable and low cost. The electrochemical performance such as the specific capacitance, energy and power densities, rate performance, and cycling stability is strongly dependent on the chemical composition, SSA, pore structure of the biomass precursors, and experimental conditions. Therefore, it needs proper design in the selection of precursor material and appropriate pyrolysis or activation method to attain well-quality BCMs.

REFERENCES

1. L. Peng, Y.R. Liang, J.Y. Huang, L.L. Xing, H. Hu, Y. Xiao, H.W. Dong, Y.L. Liu and M.T. Zheng, Mixed-biomass wastes derived hierarchically porous carbons for high-performance electrochemical energy storage, *ACS Sustain. Chem. Eng.*, 2019, **7**, 10393–10402.
2. H.T.T. Dang, C.V. Dinh, K.M. Nguyen, N.T.H. Tran, T.T. Pham and R.M. Narbaitz, Loofah sponges as bio-carriers in a pilot-scale integrated fixed-film activated sludge system for municipal wastewater treatment, *Sustainability*, 2020, **12**, 4758.
3. S.N. Sunaina, C. Prakash and J. Aman, Biomass derived carbon for supercapacitor applications: Review, *J. Energy Storage*, 2021, **39**, 102646.
4. N. Liu, Y. Gao, N. Liu and Y. Gao, Recent progress in micro-supercapacitors with in-Plane interdigital electrode architecture, *Small*, 2017, **13**, 1701989.
5. S.Y. Yu, B. Liu, M. Yang, H.B. Chen, Y.J. Liu and H.M. Li, Flexible solid-state supercapacitor with high energy density enabled by N/B/O-codoped porous carbon nanoparticles and imidazolium-based gel polymer electrolyte, *ACS Sustain. Chem. Eng.*, 2022, **10**, 5548–5558.
6. W. Tang, Y. Zhang, Y. Zhong, T. Shen, X.L. Wang, X.H. Xia and J.P. Tu, Natural biomass-derived carbons for electrochemical energy storage, *Mater. Res. Bull*, 2017, **88**, 234–241.
7. R. Hao, H. Lan, C. Kuang, H. Wang and L. Guo, Superior potassium storage in chitin-derived natural nitrogen-doped carbon nanofibers, *Carbon*, 2018, **128**, 224–230.
8. W.-J. Liu, H. Jiang and H.-Q. Yu, Thermochemical conversion of lignin to functional materials: A review and future directions, *Green Chem.*, 2015, **17**, 4888.
9. L. Costes, F. Laoutid, S. Brohez and P. Dubois, Bio-based flame retardants: When nature meets fire protection, *Mater. Sci. Eng. A*, 2017, **117**, 1.
10. B. Duan, X. Gao, X. Yao, Y. Fang, L. Huang, J. Zhou and L. Zhang, Unique elastic N-doped carbon nanofibrous microspheres with hierarchical porosity derived from renewable chitin for high rate supercapacitors, *Nano Energy*, 2016, **27**, 482–491.
11. M. Kaya, E. Lelesius, R. Nagrockaite, I. Sargin, G. Arslan, A. Mol, T. Baran, E. Can and B. Bitim, Differentiations of chitin content and surface morphologies of chitins extracted from male and female grasshopper species, *PLoS One*, 2015, **10**, e0115531.
12. J. McKittrick, P.Y. Chen, S.G. Bodde, W. Yang, E.E. Novitskaya and M.A. Meyers, The structure, functions, and mechanical properties of keratin, *JOM*, 2012, **64**, 449–468.
13. W. Qian, F. Sun, Y. Xu, L. Qiu, C. Liu, S. Wang and F. Yan, Human hair-derived carbon flakes for electrochemical supercapacitors, *Energy Environ. Sci.*, 2014, **7**, 379.
14. J. Arroyo, V. Farkaˇs, A.B. Sanz and E. Cabib, Strengthening the fungal cell wall through chitin–glucan cross-links: Effects on morphogenesis and cell integrity, *Cell. Microbiol.*, 2016, **18**, 1239–1250.
15. J.A. Pineda-Insuasti, C.P. Soto-Arroyave and L. Beltran, Stoichiometry equation to describe the growth of the Pleurotus ostreatus ceba-gliie-po-010606 strain, *Biotecnol. Apl.*, 2014, **31**, 48–52.
16. S. Wang, W. Sun, D.-S. Yang and F. Yang, Conversion of soybean waste to sub-micron porous-hollow carbon spheres for supercapacitor via a reagent and template-free route, *Mater. Today Energy*, 2019, **13**, 50–55.
17. F.-C. Wu, R.-L. Tseng, C.-C. Hu and C.-C. Wang, Effects of pore structure and electrolyte on the capacitive characteristics of steam- and KOH-activated carbons for supercapacitors, *J. Power Sources,* 2005, **144**, 302–309.

18. C. Wang, D. Wu, H. Wang, Z. Gao, F. Xu and K. Jiang, A green and scalable route to yield porous carbon sheets from biomass for supercapacitors with high capacity, *J. Mater. Chem. A,* 2018, **6**, 1244–1254.
19. C. Wang, D. Wu, H. Wang, Z. Gao, F. Xu and K. Jiang, Biomass derived nitrogen-doped hierarchical porous carbon sheets for supercapacitors with high performance, *J. Colloid Interface Sci.*, 2018, **523**, 133–143.
20. M. Yu, Y. Han, J. Li and L. Wang, CO_2-activated porous carbon derived from cattail biomass for removal of malachite green dye and application as supercapacitors, *Chem. Eng. J.*, 2017, **317**, 493–502.
21. L. E, W. Li, C. Ma, Z. Xu and S. Liu, CO_2-activated porous self-templated N-doped carbon aerogel derived from banana for high-performance supercapacitors, *Appl. Surf. Sci.*, 2018, **457**, 477–486.
22. L. Wang, J. Wu, H. Ma, G. Han, D. Yang, Y. Chen and J. Zhou, H_3PO_4-assisted synthesis of apricot shell lignin-based activated carbon for capacitors: Understanding the pore structure/electrochemical performance relationship, *Energy Fuels,* 2021, **35**, 8303–8312.
23. S.-C. Hu, J. Cheng, W.-P. Wang, G.-T. Sun, L.-L. Hu, M.-Q. Zhu and X.-H. Huang, Structural changes and electrochemical properties of lacquer wood activated carbon prepared by phosphoric acid-chemical activation for supercapacitor applications, *Renew. Energy,* 2021, **177**, 82–94.
24. Z. Liao, H.-Y. Su, J. Cheng, G.-T. Sun, L. Zhu and M.-Q. Zhu, $CoFe_2O_4$-mesoporous carbons derived from *Eucommia ulmoides* wood for supercapacitors: Comparison of two activation method and composite carbons material synthesis method, *Ind. Crop. Prod.*, 2021, **171**, 113861.
25. J. Cheng, S.-C. Hu, G.-T. Sun, K. Kang, M.-Q. Zhu and Z.-C. Geng, Comparison of activated carbons prepared by one-step and two-step chemical activation process based on cotton stalk for supercapacitors application, *Energy,* 2021, **215**, 119144.
26. F.J. García-Mateos, R. Ruiz-Rosas, J. María Rosas, E. Morall´on, D. Cazorla-Amor´os, J. Rodríguez-Mirasol and T. Cordero, Activation of electrospun lignin-based carbon fibers and their performance as self-standing supercapacitor electrodes, *Separ. Purif. Technol.*, 2020, **241**, 116724.
27. F. Ma, S. Ding, H. Ren and Y. Liu, Sakura-based activated carbon preparation and its performance in supercapacitor applications, *RSC Adv.*, 2019, **9**, 2474–2483.
28. G. Zhang, Y. Chen, Y. Chen and H. Guo, Activated biomass carbon made from bamboo as electrode material for supercapacitors, *Mater. Res. Bull.*, 2018, **102**, 391–398.
29. S. Sankar, A.T.A. Ahmed, A.I. Inamdar, H. Im, Y.B. Im, Y. Lee, D.Y. Kim and S. Lee, Biomass-derived ultrathin mesoporous graphitic carbon nanoflakes as stable electrode material for high-performance supercapacitors, *Mater. Des.*, 2019, **169**, 107688.
30. H. Jin, J. Hu, S. Wu, X. Wang, H. Zhang, H. Xu and K. Lian, Three-dimensional interconnected porous graphitic carbon derived from rice straw for high performance supercapacitors, *J. Power Sources*, 2018, **384**, 270–277.
31. V. Yang, R. Arumugam Senthil, J. Pan, T. Rajesh Kumar, Y. Sun and X. Liu, Hierarchical porous carbon derived from jujube fruits as sustainable and ultrahigh capacitance material for advanced supercapacitors, *J. Colloid Interface Sci.*, 2020, **579**, 347–356.
32. V. Yang, R.A. Senthil, J. Pan, A. Khan, S. Osman, L. Wang, W. Jiang and Y. Sun, Highly ordered hierarchical porous carbon derived from biomass waste mangosteen peel as superior cathode material for high performance supercapacitor, *J. Electroanal. Chem.*, 2019, **855**, 113616.
33. G. Ma, J. Li, K. Sun, H. Peng, E. Feng and Z. Lei, Tea-leaves based nitrogen-doped porous carbons for high-performance supercapacitors electrode, *J. Solid State Electrochem.*, 2017, **21**, 525–535.

34. J.J. Many`a, M. Azuara and J.A. Manso, Biochar production through slow pyrolysis of different biomass materials: Seeking the best operating conditions, *Biomass Bioenergy*, 2018, **117**, 115–123.
35. K. Qian, A. Kumar, H. Zhang, D. Bellmer and R. Huhnke, Recent advances in utilization of biochar, *Renew. Sustain. Energy Rev.*, 2015, **42**, 1055–1064.
36. X. Dong, H. Jin, R.Wang, J. Zhang, X. Feng, C. Yan, S. Chen, S. Wang, J. Wang and J. Lu, High volumetric capacitance, ultralong life supercapacitors enabled by waxberry-derived hierarchical porous carbon materials, *Adv. Energy Mater.*, 2018, **8**, 1702695.
37. C. Wang, N.H. Xie, Y. Zhang, Z. Huang, K. Xia, H. Wang, S. Guo, B.Q. Xu and Y. Zhang, Silk-derived highly active oxygen electrocatalysts for flexible and rechargeable Zn-air batteries, *Chem. Mater.*, 2019, **31**, 1023.
38. S.H. Wang, Y.Y. Li, Q. Xu, Q.P. Fu, X.L. Guo, Y.M. Zheng, W.J. Zhang, Z. Cao, R.T. Lia and J.X. Ren, Facile preparation of graphene@polyaniline nanofiber network/oxidized carbon cloth composites for high-performance flexible solid-state supercapacitors, *Nanoscale*, 2022, **14**, 15908–15917.
39. Z. Gao, Y. Zhang, N. Song and X. Li, Biomass-derived renewable carbon materials for electrochemical energy storage, *Mater. Res. Lett.*, 2017, **5**, 69–88.
40. A.M. Abioye and F. Niu, Recent development in the production of activated carbon electrodes from agricultural waste biomass for supercapacitors: A review, *Renew. Sustain. Energy Rev.*, 2015, **52**, 1282–1293.
41. H.-M. Lee, H.-G. Kim, K.-H. An and B,-J. Kim, The effect of CO_2 activation on the electrochemical performance of coke-based activated carbons for supercapacitors, *J. Nanosci. Nanotechnol.*, 2015, **15**, 8797–8802.
42. C.L. Jiang, G.A. Yakaboylu, T. Yumak, J.W. Zondlo, E.M. Sabolsky and J.X. Wang, Activated carbons prepared by indirect and direct CO_2 activation of lignocellulosic biomass for supercapacitor electrodes, *Renew. Energy*, 2020, **155**, 38–52.
43. S. Rawat, T. Boobalan, M. Sathish, S. Hotha and B. Thallada, Utilization of CO_2 activated litchi seed biochar for the fabrication of supercapacitor electrodes, *Biomass Bioenergy*, 2023, **171**, 106747.
44. M. Shanmuga Priya, P. Divya and R. Rajalakshmi, A review status on characterization and electrochemical behaviour of biomass derived carbon materials for energy storage supercapacitors, *Sustain. Chem. Pharm.*, 2020, **16**, 100243.
45. W. Jiang, L. Li, J. Pan, R.A. Senthil, X. Jin, J. Cai, J. Wang and X. Liu, Hollow-tubular porous carbon derived from cotton with high productivity for enhanced performance supercapacitor, *J. Power Sources*, 2019, **438**, 226936.
46. Z. Zhu, D.J. Macquarrie, R. Simister, L.D. Gomez and S.J. McQueen-Mason, Microwave assisted chemical pretreatment of Miscanthus under different temperature regimes, *Sustain. Chem. Process.*, 2015, **3**, 15.
47. C.H. Hsiao, S. Gupta, C.Y. Lee and N.H. Tai, Effects of physical and chemical activations on the performance of biochar applied in supercapacitors, *Appl. Surf. Sci.*, 2023, **610**, 155560.
48. W. Zhang, R.R. Cheng, H.H. Bi, Y.H. Lu, L.B. Ma and X.J. He, A review of porous carbons produced by template methods for supercapacitor applications, *New Carbon Mater.*, 2021, **36**, 69–81,
49. W.X. Cao and F.Q. Yang, Supercapacitors from high fructose corn syrup-derived activated carbons, *Mater. Today Energy*, 2018, **9**, 406–415,
50. C. Wang, D. Wu, H. Wang, Z. Gao, F. Xu and K. Jiang, A green and scalable route to yield porous carbon sheets from biomass for supercapacitors with high capacity, *J. Mater. Chem. A*, 2018, **6**, 1244–1254.
51. H.Y. Ma, C. Li, M. Zhang, J.D. Hong and G.Q. Shi, Graphene oxide induced hydrothermal carbonization of egg proteins for high-performance supercapacitors, *J. Mater. Chem. A*, 2017, **5**, 17040–17047.

52. B.Q. Li, Y.F. Cheng, L.P. Dong, Y.M. Wang, J.C. Chen, C.F. Huang, D.Q. Wei, Y.J. Feng, D.C. Jia and Y. Zhou, Nitrogen doped and hierarchically porous carbons derived from chitosan hydrogel via rapid microwave carbonization for high-performance supercapacitors, *Carbon*, 2017, **122**, 592–603.
53. Y.F. Cheng, B.Q. Li, Y.J. Huang, Y.M. Wang, J.C. Chen, D.Q. Wei, Y.J. Feng, D.C. Jia and Y. Zhou, Molten salt synthesis of nitrogen and oxygen enriched hierarchically porous carbons derived from biomass via rapid microwave carbonization for high voltage supercapacitors, *Appl. Surf. Sci.*, 2018, **439**, 712–723.
54. D, Zeng, Y. Dou, M. Li, F. Yang and J.J. Peng, Wool fiber-derived nitrogen-doped porous carbon prepared from molten salt carbonization method for supercapacitor application, *J. Mater. Sci.*, 2018, **53**, 8372–8384.
55. Z.Y. Pang, G.S. Li, X.L. Xiong, L. Ji, Q. Xu, X.L. Zou and X.G. Lu, Molten salt synthesis of porous carbon and its application in supercapacitors: A review, *J. Energy Chem.*, 2021, **61**, 622–640.
56. G.F. Hawes, D. Yilman, B.S. Noremberg and M.A. Pope, Supercapacitors fabricated via laser-induced carbonization of biomass-derived poly (furfuryl alcohol)/graphene oxide composites, *ACS Appl. Nano Mater.*, 2019, **2**, 6312–6324.
57. L. Borchardt, D. Leistenschneider, J. Haase, M. Dvoyashkin, Revising the concept of pore hierarchy for ionic transport in carbon materials for supercapacitors, *Adv. Energy Mater.*, 2018, **8**, 1800892.
58. X. Zheng, J. Luo, W. Lv, D.W. Wang and Q.H. Yang, Two-dimensional porous carbon: Synthesis and ion-transport properties, *Adv. Mater.*, 2015, **27**, 5388–5395.
59. X. Fan, C. Yu, J. Yang, Z. Ling, C. Hu, M. Zhang and J. Qiu, A Layered-nanospace-confinement strategy for the synthesis of two-dimensional porous carbon nanosheets for high-rate performance supercapacitors, *Adv. Energy Mater.*, 2015, **5** (7), 1401761.
60. Z. Guo, X. Kong, X. Wu, W. Xing, J. Zhou, Y. Zhao and S. Zhuo, Heteroatom-doped hierarchical porous carbon via molten-salt method for supercapacitors, *Electrochim. Acta*, 2020, **360**, 137022.
61. J. Li, W. Liu, D. Xiao and X. Wang, Oxygen-rich hierarchical porous carbon made from pomelo peel fiber as electrode material for supercapacitor, *Appl. Surf. Sci.*, 2017, **416**, 918–924.
62. C. Huang, T. Sun and D. Hulicova-Jurcakova, Wide electrochemical window of supercapacitors from coffee bean-derived phosphorus-rich carbons, *Chem. Sus. Chem.*, 2013, **6**, 2330–2339.
63. G.A. Ferrero, A.B. Fuertes and M. Sevilla, From soybean residue to advanced supercapacitors, *Sci. Rep.*, 2015, **5**, 16618.
64. R.-J. Mo, Y. Zhao, M. Wu, H.-M. Xiao, S. Kuga, Y. Huang, J.-P. Li and S.-Y. Fu, Activated carbon from nitrogen rich watermelon rind for high-performance supercapacitors, *RSC Adv.*, 2016, **6**, 59333–59342.
65. M. Chen, D. Yu, X. Zheng and X. Dong, Biomass based N-doped hierarchical porous carbon nanosheets for all-solid-state supercapacitors, *J. Energy Storage,* 2019, **21**, 105–112.
66. Z. Shang, X. An, H. Zhang, M. Shen, F. Baker, Y. Liu, L. Liu, J. Yang, H. Cao, Q. Xu, H. Liu and Y. Ni, Houttuynia-derived nitrogen-doped hierarchically porous carbon for high-performance supercapacitor, *Carbon*, 2020, **161**, 62–70.
67. M. Liu, K. Zhang, M. Si, H. Wang, L. Chai and Y. Shi, Three-dimensional carbon nanosheets derived from micro-morphologically regulated biomass for ultrahigh-performance supercapacitors, *Carbon*, 2019, **153**, 707–716.
68. W. Du, X. Wang, X. Sun, J. Zhan, H. Zhang and X. Zhao, Nitrogen-doped hierarchical porous carbon using biomass-derived activated carbon/carbonized polyaniline composites for supercapacitor electrodes, *J. Electroanal. Chem.*, 2018, **827**, 213–220.
69. L. Ji, B. Wang, Y. Yu, N. Wang and J. Zhao, N, S co-doped biomass derived carbon with sheet-like microstructures for supercapacitors, *Electrochim. Acta*, 2020, **331**, 135348.

70. F. Wang, J.Y. Cheong, Q. He, G. Duan, S. He, L. Zhang, Y. Zhao, I.-D. Kim and S. Jiang, Phosphorus-doped thick carbon electrode for high-energy density and long-life supercapacitors, *Chem. Eng. J.*, 2021, **414**, 128767.
71. B. Nirosha, R. Selvakumar, J. Jeyanthi and S. Vairam, Elaeocarpus tectorius derived phosphorus-doped carbon as an electrode material for an asymmetric supercapacitor, *New J. Chem.*, 2020, **44**, 181–193.
72. W. Feng, P. He, S. Ding, G. Zhang, M. He, F. Dong, J. Wen, L. Du and M. Liu, Oxygen-doped activated carbons derived from three kinds of biomass: Preparation, characterization and performance as electrode materials for supercapacitors, *RSC Adv.*, 2016, **6**, 5949–5956.
73. K. Sun, S. Yu, Z. Hu, Z. Li, G. Lei, Q. Xiao and Y. Ding, Oxygen-containing hierarchically porous carbon materials derived from wild jujube pit for high-performance supercapacitor, *Electrochim. Acta*, 2017, **231**, 417–428.
74. S. Cao, J. Yang, J. Li, K. Shi and X. Li, Preparation of oxygen-rich hierarchical porous carbon for supercapacitors through the co-carbonization of pitch and biomass, *Diam. Relat. Mater.*, 2019, **96**, 118–125.
75. J. Zhao, Y. Li, G. Wang, T. Wei, Z. Liu, K. Cheng, K. Ye, K. Zhu, D. Cao and Z. Fan, Enabling high-volumetric-energy-density supercapacitors: Designing open, low-tortuosity heteroatom-doped porous carbon-tube bundle electrodes, *J. Mater. Chem. A*, 2017, **5**, 23085–23093.
76. A. Gopalakrishnan and S. Badhulika, Effect of self-doped heteroatoms on the performance of biomass-derived carbon for supercapacitor applications, *J. Power Sources*, 2020, **480**, 228830.
77. Z. Shang, X. An, H. Zhang, M. Shen, F. Baker, Y. Liu, L. Liu, J. Yang, H. Cao, Q. Xu, H. Liu and Y. Ni, Houttuynia-derived nitrogen-doped hierarchically porous carbon for high-performance supercapacitor, *Carbon*, 2020, **161**, 62–70.
78. X. Yu, S.K. Park, S.-H. Yeon and H.S. Park, Three-dimensional, sulfur-incorporated graphene aerogels for the enhanced performances of pseudocapacitive electrodes, *J. Power Sources*, 2015, **278**, 484–489.
79. P.A. Denis, Band gap opening of monolayer and bilayer graphene doped with aluminium, silicon, phosphorus, and sulfur, *Chem. Phys. Lett.*, 2010, **492**, 251–257.
80. Y. Wang, M. Zhang, Y. Dai, H.-Q. Wang, H. Zhang, Q. Wang, W. Hou, H. Yan, W. Li and J.-C. Zheng, Nitrogen and phosphorus co-doped silkworm-cocoon-based self-activated porous carbon for high performance supercapacitors, *J. Power Sources*, 2019, **438**, 227045.
81. Y. Wen, B. Wang, C. Huang, L. Wang and D. Hulicova-Jurcakova, Synthesis of phosphorus-doped graphene and its wide potential window in aqueous supercapacitors, *Chem. Eur J.*, 2015, **21**, 11538.
82. G.A. Ferrero, A.B. Fuertes and M. Sevilla, From soybean residue to advanced supercapacitors, *Sci. Rep.*, 2015, **5**, 16618.
83. L. Sun, C. Tian, M. Li, X. Meng, L. Wang, R. Wang, J. Yin and H. Fu, From coconut shell to porous graphene-like nanosheets for high-power supercapacitors, *J. Mater. Chem. A*, 2013, **1**, 6462–6470.
84. Q.G. Wang, C.L. Li, L. He, X.F. Yu, W.P. Zhang and A.H. Lu, Outside-in catalytic graphitization method for synthesis of dispersible and uniform graphitic porous carbon nanospheres, *J. Colloid Interface Sci.*, 2021, **599**, 586–594.
85. L. Jiang, J. Yan, L. Hao, R. Xue, G. Sun and B. Yi, High rate performance activated carbons prepared from ginkgo shells for electrochemical supercapacitors, *Carbon*, 2013, **56**, 146–154.
86. J. Hou, C. Cao, F. Idrees and X. Ma, Hierarchical porous nitrogen-doped carbon nanosheets derived from silk for ultrahigh-capacity battery anodes and supercapacitors, *ACS Nano*, 2015, **9**, 2556–2564.

87. L. Sun, C. Tian, M. Li, X. Meng, L. Wang, R. Wang, J. Yin and H. Fu, From coconut shell to porous graphene-like nanosheets for high-power supercapacitors, *J. Mater. Chem. A,* 2013, **1**, 6462–6470.
88. D. Siva Priya, L. John Kennedy, G. Theophil Anand, Emerging trends in biomass-derived porous carbon materials for energy storage application: A critical review, *Mater. Today Sustain.*, 2023, **21**, 100320.
89. H. Lu, L. Zhuang, R. Gaddam, X. Sun, C. Xiao, T. Duignan, Z. Zhu and X. Zhao, Microcrystalline cellulose-derived porous carbons with defective sites for electrochemical applications, *J. Mater. Chem.*, 2019, **7**, 22579–22587.
90. M. Liu, K. Zhang, M. Si, H. Wang, L. Chai and Y. Shi, Three-dimensional carbon nanosheets derived from micro-morphologically regulated biomass for ultrahigh-performance supercapacitors, *Carbon*, 2019, **153**, 707–716.
91. J.S. Lee, D.H. Shin and J. Jang, Polypyrrole-coated manganese dioxide with multiscale architectures for ultrahigh capacity energy storage, *Energy Environ. Sci.*, 2015, **8**, 3030–3039.
92. Z.S. Wu, K. Parvez, S. Li, S. Yang, Z. Liu, S. Liu, X. Feng and K. Müllen, Alternating stacked graphene-conducting polymer compact films with ultrahigh areal and volumetric capacitances for high-energy micro-supercapacitors, *Adv. Mater.*, 2015, **27**, 4054–4061.
93. Y. Tao, X. Xie, W. Lv, D.-M. Tang, D. Kong, Z. Huang, H. Nishihara, T. Ishii, B. Li, D. Golberg, F. Kang, T. Kyotani and Q.-H. Yang, Towards ultrahigh volumetric capacitance: Graphene derived highly dense but porous carbons for supercapacitors, *Sci. Rep.*, 2013, **3**, 2975.
94. M.-Q. Zhao, C.E. Ren, Z. Ling, M.R. Lukatskaya, C. Zhang, K.L. Van Aken, M.W. Barsoum and Y. Gogotsi, Flexible MXene/carbon nanotube composite paper with high volumetric capacitance, *Adv. Mater.*, 2015, **27**, 339–345.
95. L.L. Zhang, H.H. Li, Y.H. Shi, C.Y. Fan, X.L. Wu, H.F. Wang, H.Z. Sun and J.P. Zhang, A novel layered sedimentary rocks structure of the oxygen-enriched carbon for ultrahigh-rate-performance supercapacitors, *ACS Appl. Mater. Inter.*, 2016, **8**, 4233–4241.
96. B. Men, P.K. Guo, Y.Z. Sun, Y. Tang, Y.M. Chen, J.Q. Pan and P.Y. Wan, High-performance nitrogen-doped hierarchical porous carbon derived from cauliflower for advanced supercapacitors, *J. Mater. Sci.*, 2019, **54**, 2446–2457.
97. V. Yang, R.A. Senthil, J.Q. Pan, T.R. Kumar, Y.Z. Sun and X.G. Liu, Hierarchical porous carbon derived from jujube fruits as sustainable and ultrahigh capacitance material for advanced supercapacitors, *J. Colloid Inter. Sci.*, 2020, **579**, 347–356.
98. B. Nirosha, R. Selvakumar, J. Jeyanthib and S. Vairam, Elaeocarpus tectorius derived phosphorus-doped carbon as an electrode material for an asymmetric supercapacitor, *New J. Chem.*, 2020, **44**, 181–193.
99. D. Wu, C.Y. Yua and W.B. Zhong, Bioinspired strengthening and toughening of carbon nanotube@polyaniline/graphene film using electroactive biomass as glue for flexible supercapacitors with high rate performance and volumetric capacitance, and low-temperature tolerance, *J. Mater. Chem. A*, 2021, **9**, 18356–18368.
100. B. Yao, H.R. Peng, H.Z. Zhang, J.Z. Kang, C. Zhu, G. Delgado, D. Byrne, S. Faulkner, M. Freyman, X.H. Lu, M.A. Worsley, J.Q. Lu and Y. Li, Printing porous carbon aerogels for low temperature supercapacitors, *Nano Lett.*, 2021, **21**, 3731–3737.

7 Carbon-Based Fibers for Supercapacitors

Tieqi Huang

1 INTRODUCTION

Carbon-based fibers are normally carbon-based materials with high length-diameter ratio, showing linear/semi-linear shapes with nano-, micro-, or macro-sized in length. They are derived from numerous methods, typically from the precursors of graphene, carbon nanotube (CNT), or polymer and pitch, with more than 160 years of history.[1] Since last century, the carbon-based fiber industry has been greatly promoted due to the explosive consumption of advanced high-performance materials, triggering accelerated improvements of carbon-based fiber electrodes for supercapacitors in specific fields.

They often have unique intrinsic properties of satisfactory strength, considerable flexibility, outstanding conductivity, and ideal surface, making them suitable for supercapacitors. They can be assembled into yarns, fabrics, and other flexible materials for special energy-supplying systems in smart clothing, flexible electronics, and wearable devices. According to the material characteristics of the prepared carbon-based fibers, they can be used as electrode materials or current collectors for supercapacitors, or even as substrate materials to load other highly active materials. Therefore, carbon-based fibers have significant value in manufacturing portable and deformable supercapacitors.

In comparison with other carbon-based materials, unidimensional structural design usually brings about more regular carbon frameworks, basically leading to highly oriented building blocks. Generally, this indicates more conductive networks and stronger mechanical strength than non-fiber carbon assemblies, which on the other hand inevitably attenuates the specific surface area (SSA) and sacrifices the bulk electrochemical performance. To note, it is usually a trade-off consideration to balance the mechanical and electrochemical properties of carbon-based electrodes, which should depend on the specific application demands in principle. With respect to carbon-based fibers, their potential application mainly relies on the surface-related field with requirements of flexibility and durability. Typically, surface engineering projects can enable carbon-based fibers with porous and hollow inner structures for efficient surface capacitive processing. Another huge difference between fiber-shaped and non-fiber-shaped carbons is the electrode preparation process. Most carbons need the help of binders to fabricate composite materials on additional metal current collectors as supercapacitor electrodes, making them inappropriate for flexible and wearable devices. On the contrary, carbon-based fibers can be used as binder-free

DOI: 10.1201/9781003387831-7

and self-standing electrodes directly, which broadens the scope of the application scenario of supercapacitors. Thus, carbon-based fibers have their irreplaceable merits in enlarging the supercapacitor application limits for special cases.

This chapter mainly introduces the relevant preparation methods of carbon-based fibers and their corresponding application status in supercapacitors, as well as related challenges and prospects in nowadays.

2 FABRICATION OF CARBON-BASED FIBERS

Fabrication of carbon-based fibers is vital in determining the key chemical and structural properties of obtained electrodes, which further influence the supercapacitor behavior as a result. As aforementioned, introduced carbons aiming at supercapacitor electrodes can be used directly or as a substrate to anchor electrochemical active materials, which depends on the organization states of carbon and other heteroatoms. Typically, commercially produced and widely researched carbon-based fibers can be classified into three types due to the skeleton precursors, including graphene (oxide) fibers, CNT fibers, and carbonaceous fibers. Each category of preparation approach emphasizes the rational design of electronic conductivity, shape plasticity, and surface constructability of carbon-based materials, which are fully studied by numerous characterizations such as scanning electron microscope (SEM) and transmission electron microscope (TEM). This section will discuss the general manufacturing strategies of corresponding fibers in detail.

2.1 Graphene (oxide) fibers

One of the important technologies for fabricating graphene-based fibers is the wet-spinning strategy. Wet-spinning method was initially utilized to achieve highly flexible graphene oxide (GO) fibers by Gao's group (Figure 7.1a),[2] which is the subsequent work of their previous finding that GO in suitable aqueous solution shows macromolecular liquid crystals due to electrostatic repulsion.[3] Additionally, by injection of GO hydrogel into liquid nitrogen within the followed freeze-drying treatment, numerous macro- and micro-pores can be generated inner the wet-spun GO fibers with ice-induced alignment (Figure 7.1b).[4] The obtained oriented aerogel fiber can be reversibly folded and stretched, endowing it with potential application in pressable and bendable fabrics. For following treatments, annealing[5] and doping[6] can efficiently improve the conductivity of graphene fibers, and optimizing the surface chemistry is beneficial to mechanical strength.[7] Besides, this wet-spinning strategy was used to prepare GO two-dimensional films[8] and three-dimensional aerogels[9], indicating the universality and versatility of this technology in GO-based constructing. In addition, the wet-spun GO fibers can be further interfused into nonwoven fabrics when they are in liquid-crystal states (Figure 7.1c), which are composed of linked short GO ribbon-like sheets.[10] After hydrothermal reduction, the shank nonwoven fabrics displayed good flexibility and high surface area, showing outstanding performance in wearable electrodes.[11] Interestingly, the fusion and fission of GO fibers within water is reversible due to solvent evaporation and infiltration, meaning that a number of macroscopic fibers can fuse into a thicker one and can also separate into original individual fibers under the stimulation of solvents.[12] Additionally, Cheng's group used basified GO with non-liquid-crystal properties to obtain graphene-based fibers, which indicates the generality of wet-spinning skills for fabricating graphene-based fibers under rational parameters.[13]

Also, there are other methods to realize the scalable preparation of graphene or GO fibers. Gao's group used blow-spinning to extrude GO dope with selected high-level viscosity (Figure 7.1d), showing a high preparation speed of 550 m min^{-1}.[14] The fibers lapped on each other to form interconnected mats, joining into united fabrics with high flexibility on drum collector. Electrospinning had been successfully realized by Gao's group to fabricate nanofibers with linked networks.[15] Moreover, templating methods have been proposed to guide the skeleton forming of GO from fluid dispersions to steady hydrogels. Qu's group used a sealed pipeline to force the densification of GO aqueous solution under hydrothermal conditions, also achieving GO fibers with plentiful micro-nano structures and satisfactory mechanical strength.[16] As displayed in Figure 7.1e, Terrones's group proposed a preparation method by scrolling a piece of centimeter-sized sprawl GO film into a strong fiber, showing a helically twisted surface structure with fused inner connections.[17] Zhu's group invented a facile method to obtain graphene fibers by directly drawing CVD-graphene film out of ethanol solvent using a tweezer, depending on graphene's different behavior in water and ethanol and the capillary force from the surface meniscus.[18]

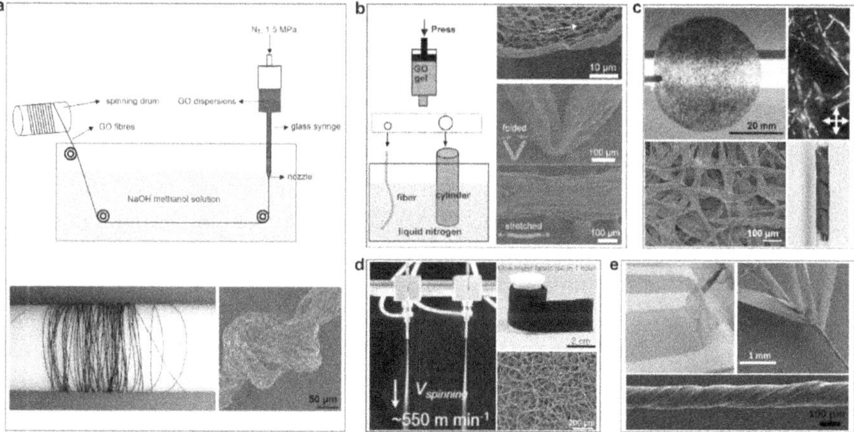

FIGURE 7.1 Some preparation methods for graphene (oxide) fibers. (a) Wet-spinning process of GO fibers from GO dispersion across coagulation bath, and the optical and SEM image of obtained GO fiber. Reprinted with permission from ref.[2]. Copyright 2011 Springer Nature. (b) GO aerogel fiber and its oriented porous structure, as well as its SEM images showing folded and stretched states. Reprinted with permission from ref.[4]. Copyright 2012 American Chemical Society. (c) GO nonwoven fabrics assembled from GO fibers, their polarization microscope image showing liquid-crystal state and SEM image after drying. Reprinted with permission from ref.[10]. Copyright 2016 Springer Nature. (d) High-speed blow-spinning process of GO fibers, and the optical and SEM images of the resulted fiber-connected mat. Reprinted with permission from ref.[14]. Copyright 2021 American Chemical Society. (e) Optical image of GO film and SEM images of film-derived scrolled helical fiber. Reprinted with permission from ref.[17]. Copyright 2014 American Chemical Society.

Graphene (oxide) fibers are mostly composed of folded graphene (oxide) sheets by strong interlayer interactions, while the edges of these sheets extend to the surface and create numerous crumples. These unique structure properties enable graphene (oxide) fibers with efficient electron pathways and abundant surface area.

2.2 CNT fibers

The development of CNT-based fibers initially benefited from research on chemical vapor deposition (CVD) technology. Initially, laser-vaporization skills,[19] floating catalytic decomposition technology,[20] and vertical floating technique[21] had been proposed to self-organize CNTs into winding microsized nanofibers, as well as the innovation of continuous spinning of CNT fibers on modified procedures was reported by Windle's group.[22] By drawing CNT aerogel at various winding rates, fiber orientation, assembly density, and mechanical properties of spun CNT fibers can be precisely controlled.[23] Li's group upgraded the spinning technology by using water densification through a water tank (Figure 7.2a), leading to a continuous CNT sock fabrication with a kilometer-level production stream achieved.[24] The prepared yarns had a unique "tube-in-tube" hollow structure, which boosted the mass flow of electrolytes for energy storage systems. Wet-spinning technology had also been carefully researched as an efficient preparation tool to get CNT fibers. As reported by Poulin's group, by the sequential processes of surfactant dispersing, nanotube recondensing, and mesh collating, indefinitely long-aligned CNT fibers can be prepared.[25] Another modification related to the introduction of chlorosulfonic acid as the solvent to disperse CNTs by Pasquali's group[26] which induced CNTs to form liquid crystals for wet spinning.

CNT arrays or forests could be directly drawn to make macroscopic fibers when they were superaligned and the interstitial interaction of nanotubes was strong enough. In 2002, Fan's group reported the fabrication of CNT fibers by simply drawing the CNT arrays into controllable bundles with strong orientation (Figure 7.2b),[27] which was mainly attributed to the van der Waals interactions during aggregation and Joule heating. Afterward, they found that condensing the obtained CNT fibers by solvent can efficiently improve the regularity and tightness of assembled bundles, which can be realized by pulling bundles through an organic solvent membrane (Figure 7.2c).[28] Besides, twisting the yarns from CNT forests had been demonstrated as another simple method to obtain macro fibers. Baughman's group used a rotating spindle to attach a drawn yarn to obtain twisted CNT fibers due to the numerous contacting surface area of CNTs.[29] Also, twisting CNT-assembled film into fibers is feasible, which had been researched by Baughman's group initially.[30] Biscrolling process had been developed to fabricate spirals by twisting different components, making multifunctional hybrid yarns that can load varying guest materials.

CNT fibers are usually composed of bundles of CNTs by strong intertube forces with moderate interstitial structures, ensuring the CNT fibers with outstanding toughness, considerable strength and abundant surface area.

2.3 Carbonaceous fibers

Carbonization of precursor fibers has been largely promoted to be an important and mature strategy to prepare carbonaceous fibers. The mechanical properties such as

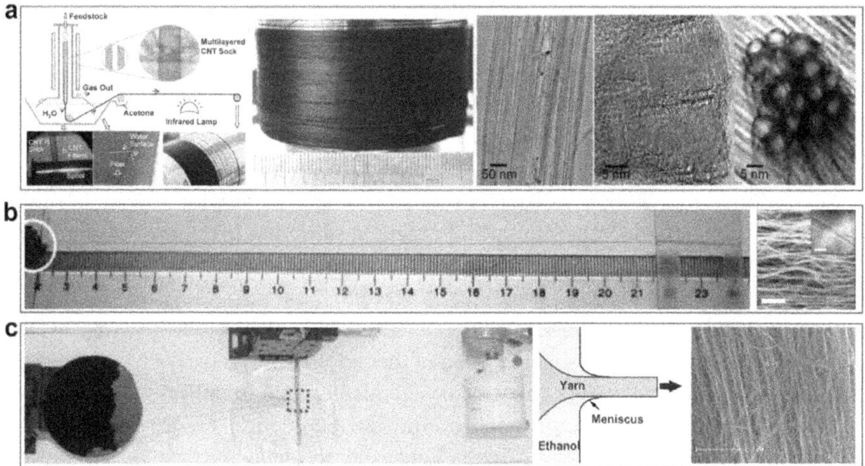

FIGURE 7.2 Some preparation methods for CNT fibers. (a) CVD synthesis and following spinning set-up for the preparation of continuous CNT yarn by water tank and acetone bath, and the optical and TEM images of obtained oriented CNT yarn. Reprinted with permission from ref.[24]. Copyright 2010 John Wiley and Sons. (b) Directly drawing CNTs from a two-dimensional superaligned array into a CNT nanoribbon-like yarn, and the SEM image of the aligned inner fiber structure. Reprinted with permission from ref.[27]. Copyright 2002 Springer Nature. (c) Ethanol-induced shrinking CNT ribbon yarn for a tight yarn, and the illustrated shrinking mechanism and the SEM image showing a superaligned inner structure. Reprinted with permission from ref.[28]. Copyright 2006 John Wiley and Sons.

tensile strength and electric conductivity of obtained carbon fibers extremely depend on the chemical and structural properties of precursors, as well as the detailed carbonization processes. There are mainly two kinds of carbonaceous fiber precursors, biomass materials and spinnable molecules.

As early as the 1880s, biomass extracted from plants had been carbonized through gas furnaces in absence of air, creating identical replicas of carbon derivates.[31] Initially, cellulose nanofibers were first prepared by homogenizing wood fiber in 1983.[32] In nature, fiber-shaped building blocks are the important components of some biomass, which can be thermally treated to obtain carbonaceous fibers directly.[33] The degradation of hemicellulose, cellulose, and lignin occurs at 220–320, 320–400, and 200–600°C, respectively.[34] The superiority of biomass-originated fibers lies in the fact that the sources are cheap, abundantly available, sustainable, and versatile. On the other hand, this method still suffers from unsatisfactory performance of low mechanical strength and conductivity, which is due to the intrinsic short fiber precursors.

Spinnable molecules (especially specific polymers) have attracted wide attention due to their outstanding ability to assemble continuous fibers due to controllable regularity, shape, roughness, diameter, and length. In a typical process of spinning, the precursors are stretched by extra forces before and after solidification into fibers if necessary. Some materials can be directly melted into suitable fluid with ideal viscosity for following spinning[35], while others should be dispersed in a solvent to be spun. As an important carbon source, the pitch can be melted for spinning directly due to the stable chemistry under high temperatures.[36] Up till now, numerous sub-technologies of spinning have been developed to deal with giving systems of precursors, including stretch-spinning, blow-spinning, electrospinning, and centrifugal-spinning (also called rotary-jet spinning). Electrospinning is one of the most suitable technologies to prepare insulated polymer nanofibers, which applies high voltage between the syringe and the collector to stretch the polymer viscous dispersion into one-dimensional material (Figure 7.3a).[37] Polyacrylonitrile (PAN), due to its instability at high temperatures and excellent solubility in common organic solvents, is more suitable to be spun under low temperatures rather than melted-spun,[38] except for the introduction of plasticizers.[39] Recently, blow-spinning technology has been largely advanced to achieve nanofibers.[40] In a typical blow-spinning system (Figure 7.3b), spinning dope that is placed in the inner nozzle is blown by air pressure from the outer nozzle, leading to a linear viscous dope on to collector for fiber collection.[41] As an important supplement to spinning techniques (Figure 7.3c), polymer emulsions and suspensions can also be centrifugally spun into fibers without the need for high-voltage powering.[42] Furthermore, some monomers such as aniline[43] and pyrrole[44] can also be polymerized by designed chemistry surroundings to spin into fibrous polymers, which can be carbonized into carbonaceous fibers for further applications. These spinning technologies aim at manufacturing different kinds of fibers in corresponding applications, e. g. microsized fibers for textiles and nanosized fibers for additives.

The carbonization process for organic precursors fibers into highly connected graphitic networks is realized by reorganizing organic chains, during which time polymers lost most heteroatoms and main carbon skeleton forms. To note, the pre-oxidization of polymers has been proven a necessary process to stabilize polymers for following high-temperature treatments, which involves complicated chemical conversion and structural arrangement.[45] Two or more processes with step-up temperatures from 1,000 to 3,000°C under an inert atmosphere have been added into the preparation of final carbonaceous fibers, resulting in increasing density, modulus, strength, and conductivity.

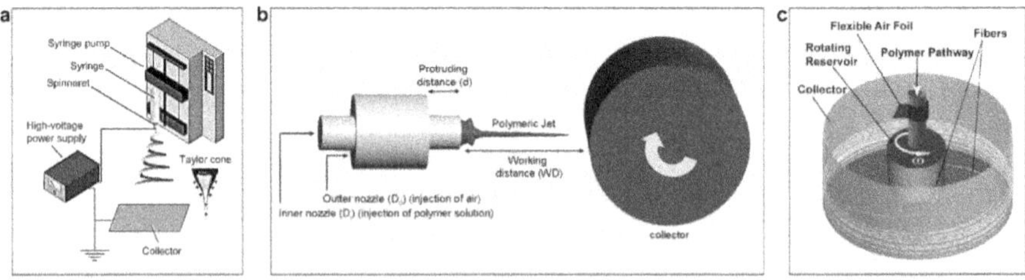

FIGURE 7.3 Some preparation technologies for carbonaceous fiber precursors. (a) Schematic illustration of the electrospinning process. Reprinted with permission from ref.[37]. Copyright 2017 American Chemical Society. (b) Schematic illustration of the blow-spinning process. Reprinted with permission from ref.[41]. Copyright 2011 John Wiley and Sons. (c) Schematic illustration of centrifugal-spinning process. Reprinted with permission from ref.[42]. Copyright 2010 American Chemical Society.

Carbonization for carbon-based fiber preparations is always energy-costing and time-wasting due to various thermal annealing processes and a series of pre-/pro treatments. On the other hand, the obtained carbonaceous fibers have uniform chemical and structural properties with smooth surfaces, making them strong yet brittle.

3 APPLICATIONS IN SUPERCAPACITORS

Supercapacitors, as the aforementioned chapters discussed, show irreplaceable contributions in some application fields that need high power density, long-working lifespan, and durable environmental adaptability. In brief, there are two branches of supercapacitors that involve carbon-based materials, including electrochemical double-layer capacitor (EDLC) and pseudocapacitors.[46] Commercially, activated carbon has been used for supercapacitors of EDLC type for the last two centuries, mainly due to the rich raw sources, high electronic conductivity, and satisfactory surface area.[47] For ordinary supercapacitor electrodes, active materials and binder should be mixed and spread on the current collector into bulky and rigid electrodes. The introduction of pseudocapacitive materials such as conductive polymers and metal oxides largely improves the specific capacitance (Cs).

In comparison with traditional materials that are normally in powder states for three-dimensional electrodes of supercapacitors, carbon-based fibers have unique advantages in size, continuity, flexibility, and weavability, meaning that they can directly assemble into self-standing macro materials without using binders. These outstanding merits lead to explosive research and findings in the field of flexible, bendable, and portable supercapacitors. Furthermore, these mini-sized fibers facilitate the assembling of microdevices, which are used in specific applications that need energy-powering systems in tiny size. This part will introduce in detail the research advances of carbon-based fibers in supercapacitors.

3.1 Device configuration and construction

For a typical energy storage system, the power-supplying device is usually constructed from positive electrodes, negative electrodes, electrolytes, separators, current collectors, and device packages. A supercapacitor normally contains all the above units when assembled into devices, while their electrodes can be the same if needed.[48] For a conventional device, a viscous and uniform dispersion containing active materials (such as activated carbons), binders, and conductive additives is spread onto the collectors with a uniform thickness of hundred micrometers. After drying, the two electrodes are separated by a separator and then are imbedded into a package filled with electrolytes in button cells or spiral wounds.[49] Besides conventional electrode preparation patterns, carbon-based fibers, as important members of smart fibers, can also be used for fiber-shaped electronics, especially meeting the needs for fiber-shaped energy-powering system.[50] Carbon-based fibers can be used as both self-standing substrates and active materials for EDLC's type energy storage, or as only conductive substrates to anchor redox materials for pseudocapacitive performance. To note, the fabrication of such electrodes without polymer binders, conductive additives, or even current collectors can improve the proportion of active materials, which as a result will increase the energy density of the whole device.

At present, four main types of supercapacitor devices have been designed to make use of the unique properties of carbon-based fibers:

1. Planar-structured supercapacitors (shown in Figure 7.4a). Similar to conventional supercapacitors, carbon-based fibers can be assembled into interconnected woven or nonwoven mats with numerous pores generated, and two mats can be packed into one supercapacitor device. Han's group has used wet-spinning technology to fabricate two-dimensional mats that are composed of GO fibers, showing a Cs of 188 F g^{-1} at a scan rate of 5 mV s^{-1} when packed in a planar supercapacitor device.[51]
2. Parallelly-constructed fiber supercapacitors (shown in Figure 7.4b). Two carbon-based fibers or bundles can be placed parallelly with a small gap on a substrate, followed by coating suitable gel electrolytes and assembling into supercapacitor devices. Gao's group used reduced wet-spun GO fibers from liquid-crystal GO aqueous solution to make parallelly constructed fiber supercapacitors, with 3.3 mF cm^{-2} at 10 mV s^{-1} obtained for flexible supercapacitors.[52] Chen's group mixed CNTs and GO sheets to make porous fibers by treatments of hydrothermal and doping reactions, showing high SSA yet good electronic conductivity with Cs of 300 F cm^{-3}.[53] Cheng's group used basified GO dispersions for continuous wet-spinning of GO fibers, which after reducing had been assembled into parallel-arranged linear supercapacitors and showed 279 F g^{-1} at a current density of 0.2 A g^{-1}.[13]
3. Twisted-constructed linear supercapacitors (shown in Figure 7.4c). A carbon-based fiber or a bundle of carbon-based fibers can be coated with thin gel electrolytes, two of which can be further assembled into linear supercapacitors by twisting each other. Peng's group used in situ polymer-anchored CNTs fiber to twist into linear supercapacitors, resulting in outstanding electrochemical performance of this flexible device (274 F g^{-1} at current density of 2 A g^{-1}).[54] Similarly, Peng's group introduced pseudocapacitive materials of manganese dioxide into the carbon nanotube bundle, also obtaining high-performance flexible supercapaictors.[55]
4. Core-sheath linear supercapacitors (shown in Figure 7.4d). A carbon-based fiber or a bundle of carbon-based fibers functions as the core, while a layer of active covering functions as the sheath. The core and the sheath are separated by a thin layer of gel electrolyte. Peng's group used aligned carbon nanotube fiber as the core and aligned carbon nanotube sheet as the sheath to fabricate a coaxial linear supercapacitor, which can be further woven in common textiles.[56] Similarly, Gao's group used wet-spun GO fiber as the core to coat gel electrolyte as the separator, followed by coating another layer of GO as the sheath and reducing into a flexible supercapacitor.[57]

The listed four kinds of supercapacitor configurations have respective application requirements for corresponding carbon-based fibers. As different carbon-based fibers display different properties such as electronic conductivity, surface area, and

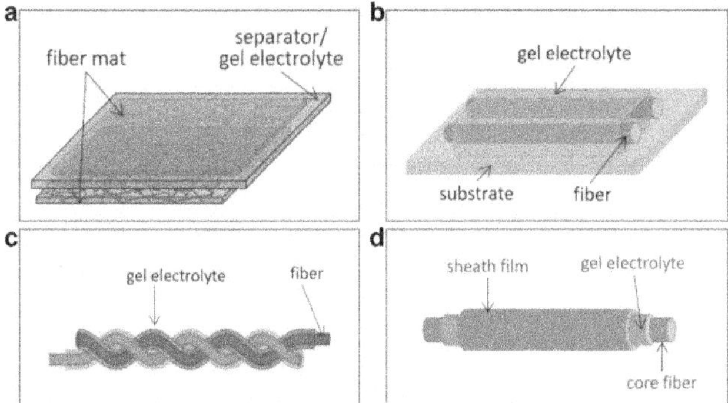

FIGURE 7.4 Schematic illustrations of four typical configurations. (a) Planar-structured supercapacitors. (b) Parallel-constructed fiber supercapacitors. (c) Twisted-constructed linear supercapacitors. (d) Core-sheath linear supercapacitors. Reprinted with permission from ref.[1]. Copyright 2020 American Chemical Society.

especially mechanical behavior, supercapacitor configuration should be carefully investigated and evaluated for the most suitable condition for assembling supercapacitor devices based on carbon-based fibers.

3.2 Performance of Carbon-Fiber Electrodes

In this part, carbon-fiber electrode (CFE) is defined as an electrode that is composed of carbon-based fibers without the introduction of additional gest pseudocapacitive materials. In common, carbons without structural or chemical modifications display EDLC behavior[58] which possess high power density but low energy density due to the intrinsic physical adsorption/desorption of ions.[59] Another important issue is that the real processes of electrochemical reactions for carbon-based fibers are far from the theoretic behavior of a single unit, which is due to the assembling mode of carbon-based fibers. For graphene-based fibers, most graphene-based fibers have an SSA of only 100–500 m^2g^{-1}, while the theoretical surface area of a single perfect graphene sheet reaches as high as 2,630 m^2g^{-1}.[60] For a single CNT of CNT-based fibers, the inner structure is normally dense due to compact carbon linkage, and the surface is too smooth to provide sufficient reaction sites. This structural property hinders the ion transmission into the nanotube bulk while the number of attached ions is basically unsatisfactory, commonly showing unideal electrochemical performance.[61] Therefore, the structural and chemical modifications on carbon-based fibers have been greatly innovated and studied these years, in order to meet the requirements of new-generation high-performance supercapacitors.[62] To realize the goal of boosting real active surfaces, researchers have proposed numerous strategies including post-treatment activation, fiber-shaping modification, and heteroatom-doping construction.

The precise process design of fiber shaping has an intrinsic influence on the properties of obtained carbon-based fibers. Pore size, pore distribution, electrolyte wettability, electronic conductivity, surface roughness, and macroscopic density can be determined during this key process. A graphene-based fiber usually has a compact inner structure due to strong interlayer interactions, which is unfavorable to ion penetration into fiber bulk. Tuning the wet-spinning processes can effectively ameliorate the limited surface-active sites, which has been realized by delicate introducing pores into the fiber during the fiber-shaping process. Konstantinov's group used a slightly acidified GO aqueous solution as the dope, which still showed liquid-crystal behavior and can be wet-spun into fibers.[63] As shown in Figure 7.5a, after the removal of the solvent of acetone and the following mild thermal reduction of obtained GO fibers, a porous inner structure can be achieved and SSA has been greatly enlarged up to 2,605 $m^2 g^{-1}$. Due to the rational porosity of this fiber, the resulting CFE has ultrahigh Cs of 394 F g^{-1} at 10 mV s^{-1}, which is the result of compromising the mechanical strength to be only ~48 MPa. Zhang's group used a mixture of integral and holey GO sheets as the spinning dope, obtaining flexible graphene-based fibers with hierarchical inner porous structure after thermal reduction and thus showing robust volumetric Cs (220.1 F cm^{-3} at 0.1 A cm^{-3}) and excellent rate capability (81% at 2.0 A cm^{-3}).[64] Other carbon-based fibers, especially carbonaceous fibers, should draw support from an extra pore-forming reagents for the fiber-shaping process to modify the surface and inner structure. Activating gas, for example, can be added into the fiber-shaping process to in situ react with pyrolyzed carbons, leaving pores on the surface of carbonaceous fibers. Zhao's group had pumped water vapor into the carbonization furnace which contained pre-treated waste wood to obtain activated carbonaceous fibers with numerous pores (SSA of 3,223 $m^2 g^{-1}$) containing extra pseudocapacitive sites of introduced oxygen-containing groups (Cs of 280 F g^{-1} at 0.5 A g^{-1}).[65] Shi's group had introduced pore-forming reagents of inorganic ($Mg(NO_3)_2$) and organic (F127 and PVP) chemicals into the phenol-formaldehyde resin to make electrospinning dope, resulting in oriented inner pores parallel to the axial direction of fiber and thus high performance (186 F g^{-1} at 0.1 A g^{-1} and 160 F g^{-1} at 20 A g^{-1}).[66]

Post treatments on obtained original carbon-based fibers are critical to optimize related properties for activation of CFE. One of the strategies is to mix activation reagents with original carbon-based fibers for thermal activation. Carbon skeleton (e. g. carbon rings of graphene) can be etched by extra chemicals at high temperatures to create ion-penetrable nanopores, even showing higher SSA (~3,100 $m^2 g^{-1}$) than the theoretical value of a single graphene sheet.[67] Based on the fact that water steam can etch the carbons under high temperatures by generating a large number of pores,[68] Kim's group used water steam to activate carbonaceous nanofibers under selected temperatures.[69] KOH, as the most used thermally activating reagent, can react with C by proceeding as $6KOH + C \leftrightarrow 2K + 3H_2 + 2K_2CO_3$, followed by decomposition of K_2CO_3 and/or reaction of $K/K_2CO_3/CO_2$ with carbon.[70] Lim's group followed the structural change of carbon nanofibers under different KOH etching degrees.[71] As shown in Figure 7.5b, they found that the specific surface had a positive correlation with the activation time, while it reached the limit if the time is enough. By SEM images, mild etching led to ladder-like CNT and the tube-like morphology can be maintained, while severe activation resulted in total deformation of nanotubes and the initial nanotube collapsed into graphene-like sheets. This result displayed that it is hard to endow CNTs with both high-degree structural uniformity and satisfactory surface area, indicating that balancing the performance and the strength should be considered carefully for the practical CFE of supercapacitors. In addition, Bismarck's group also proved this trade-off design discipline when producing CFE, finding that moderate thermal activation by KOH rather than CO_2, acid, or water can obtain high strength yet high-capacitance CFE.[72] By wet-chemistry reactions, Gao's group used wet-spun GO fibers to self-assemble nonwoven fabrics, followed by hydrothermal reactions to reduce GO and enlarge fiber interfaces for the ultra-high areal capacitance of 1,060 mF cm^{-2}.[11] Chen's group had used strong oxidants of $KMnO_4$–$NaNO_3$–H_2SO_4 to generate pores on commercial carbon-fiber tows with SSA increased from 6 to 92 $m^2 g^{-1}$.[73] By electrochemical modification, Peng's group introduced controllable pores on the carbonaceous microfibers in a strong acidic solution, showing an increased surface area and thus boosted Cs (from 133 to 239 F g^{-1}).[74] Similar to chemical modifications, some physical techniques have also been used to conduct modifications.[75]

The heteroatom-doping strategy has also been developed to boost the limited performance from neat EDLC of carbon-based materials. As evidenced by Huang's group, doping with a heteroatom such as nitrogen can improve the hydrophilicity and thus supercapacitor performance due to the residual lone-pair electrons.[76] With respect to macroscopic carbon-based self-standing electrodes, Gao's group used an N-doping strategy through facile hydrothermal reaction with ammonia solution to modify GO-based bulk electrode, showing enlarged interlayer distance, and extra Faradaic reaction sites.[77] Nonmetallic elements such as O, N, S, P, and B had been employed to replace some C atoms of carbon-based fibers, resulting in different electron distribution on the carbon skeleton within changed lattice and thus higher activity for aqueous reactions. For inner doping, Béguin's group used N-containing polymer (polyacrylonitrile) as the carbonaceous fiber precursor to fabricate nanoporous carbon nanofiber under condition of CO_2 and high temperature of 900°C, showing high Cs of 240 F g^{-1} at 0.05 A g^{-1}.[78] For surface doping, Yu's group had used pyrrole-covered carbonaceous mat for following polymerization and annealing, which displayed enhanced Cs of 202 F g^{-1} at 1 A g^{-1}.[79] For O-doping, Liu's group had used optimized mixture of CNTs and GO as the spinning dope to wet-assemble carbon-based fibers, showing Cs of 38.8 F cm^{-3} when packed into flexible devices.[80] Multi-element co-doping modifications can further change the electronic properties of the carbon skeleton, and the synergetic function of these heteroatoms can be purposely tuned to optimize the supercapacitive performance of assembled CFE. For example, Yu's group immersed bacterial cellulose mats in the aqueous solutions of $NH_4H_2PO_4$, which was followed by pyrolyzing in nitrogen gas for high-temperature carbonization (Figure 7.5c).[81] The obtained carbonaceous fibers were uniformly doped with N and P, showing higher Cs of 205 F g^{-1} and energy density of 7.76 Wh kg^{-1} than reported mono-heteroatom doped CFE and non-doped CFE.

3.3 Performance of Carbon Composite-Fiber Electrodes

Numerous active materials that can trigger reversible redox reactions on or near the surface of electrodes have been introduced to provide extra capacitance, which are normally classified as pseudocapacitive materials.[82] Conducting polymers possess conjugated structures that share delocalized electrons, which can be easily oxidized

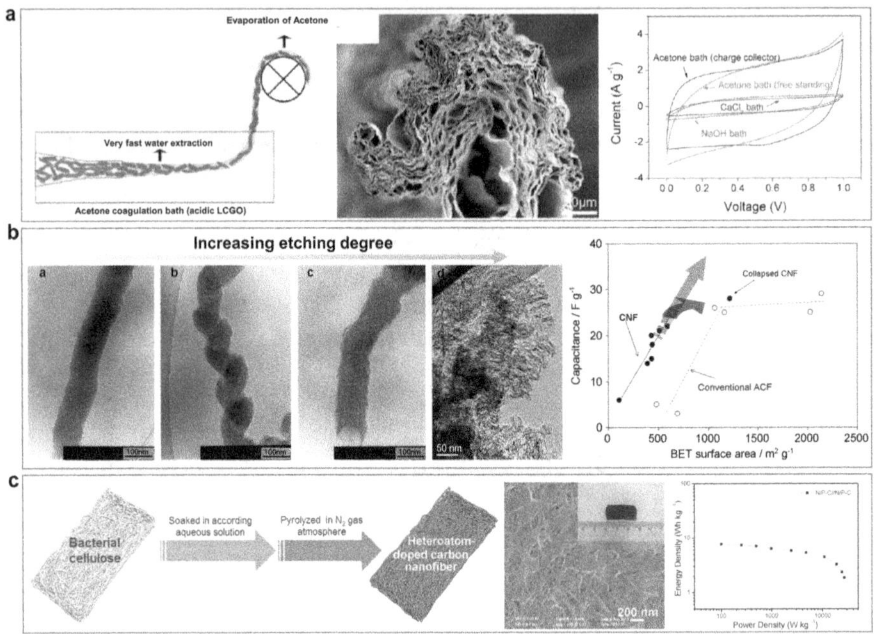

FIGURE 7.5 (a) Preparation of porous graphene fibers from modified spinning dope and coagulation bath, and the corresponding cross-section SEM image and cyclic voltammetry curves. Reprinted with permission from ref.[63]. Copyright 2014 American Chemical Society. (b) Carbon nanofiber morphology changes by TEM images depending on etching degree, and the corresponding change in SSA and Cs. Reprinted with permission from ref.[71]. Copyright 2004 Elsevier. (c) Preparation of N, P-doped carbon nanofibers, and the corresponding top-view SEM image and Ragone plots. Reprinted with permission from ref.[81]. Copyright 2014 John Wiley and Sons.

or reduced by obtaining extra electrons or losing intrinsic electrons.[83] By rational design with delicate preparation, conducting polymers can be synthesized into shapable and soft materials.[84] Metal oxides usually contain valence-changeable metal elements, which can conduct Faradaic reactions reversibly and quickly in the vicinity of the surface.[85] Metal oxides recently had also been used in flexible supercapacitor devices, for their unique structural preparation approaches.[86] Typically, the pseudocapacitive materials display high performance, while they suffer from low cycling stability and unsatisfactory mechanical strength. Many carbon-based fibers from conventional preparations hold dense inner structures, which endow them with high strength and conductivity but low SSA. Therefore, these materials can be used as a robust framework to support pseudocapacitive materials for supercapacitor electrodes. Carbon composite-fiber electrode (CCFE) is mainly composed of the host substrate of carbon-based fibers and the best decoration of pseudocapacitive materials, which takes advantage of both carbon-based fibers (e. g. flexibility, conductivity, and surface) and pseudocapacitive materials (e. g. activity and reversibility).

Conducting polymers, such as polyaniline (PANi), polypyrrole (PPy), and polythiophene (PTh), are composed of conjugated structures. Monomers of these polymers can be directly in situ polymerized on the surface or in the pores of prepared carbon-based fibers, which not only introduce pseudocapacitance but also enhance roughness. As shown in Figure 7.6a, Wei's group used CNT yarns as the substrate to deposit nanosized coniform-shaped PANi by dilute polymerization for flexible and wovenable CCFE, leading to high-performance (Cs of 12 mF cm^{-2} at 1 mA cm^{-2}) twisted CCFEs.[87] Also, it is feasible to use the composite of conducting polymer and carbon precursors to fabricate CCFE by the aforementioned approaches. Wei's group used CNT film as the substrate to electrochemical deposit PPy to form twisted CCFE, showing high Cs of ~69 F g^{-1} at a wide temperature window of –27 to 60°C.[88] Kim's group used vapor phase polymerization technology to make PTh-coated CNT nanomembrane for biscrolled CCFE, showing ultrahigh Cs (180 F cm^{-3} at 0.01 V s^{-1}, 97 F cm^{-3} at 10 V s^{-1}).[89]

Metal oxides can be reversibly reduced and oxidized during the electrochemical cycling when under a suitable potential window, which advances the progress of performance improvement of carbon-based fibers as the active gest materials. MnO_x, which holds the unique metal of Mn with numerous changeable valences, has been greatly explored to make composite with carbon-based fibers to realize high-performance CCFE. As shown in Figure 7.6b, Zhu's group used synthesized MnO_2 nanowires to combine with GO aqueous solution for wet-spinning preparation of CCFE.[90] The obtained CCFE displayed a crumpled surface and porous bulk, which were beneficial to electrolyte penetration and redox kinetics, displaying comparable performance with state-of-the-art commercial energy storage systems. Huang's group used prepared CNT sheets for in situ synthesizing amorphous MnO_2 coating, and the twisted well-aligned and porously-stacked CCFE showed satisfactory Cs (10.9 F cm^{-3} at 0.1 A cm^{-3}) with enhanced rate capability (6.9 F cm^{-3} at 5 A cm^{-3}).[91] In addition, other metal oxides had been applied in CCFE as pseudocapacitive gest materials, such as Bi_2O_3[92] and MoO_3,[93] which are similar to the preparation methods and function mechanisms of MnO_2.

In convention, the positive and negative electrode materials are set the same with constant mass loading. According to the theory, the energy density of the device is proportional to the square of working voltage.[94] Therefore, high device voltage between electrodes can largely improve the delivered energy density of the whole device. In fact, aqueous electrolyte has its low working potential due to the water-splitting threat, as well as the restriction of active materials for safe and reversible cycling.[95] However, due to the different working conditions of positive and negative materials, the supercapacitor device can be asymmetrically manufactured. To note, the mass ratio of two electrodes should be recalculated as the capacitance of each side is usually different, which should depend on the charge conservation principle to realize maximum capacitance.[96] Commonly, carbon nanomaterials are usually used as negative electrode materials due to the intrinsic electrochemical hydrogen storage ability and the stability at low potential.[97] For example, Gao's group used wet-spun reduced GO fiber as a substrate to coat the MnO_2 layer on the surface.[98] By using the MnO_2-deposited fiber as the positive electrode and the original fiber as the anode, the assembled twisted fiber supercapacitor displayed high flexibility and good electrochemical performance of 11.9 mW h cm^{-3} at a voltage of 1.6 V. Also, different CCFEs can be applied as both positive and negative electrodes according

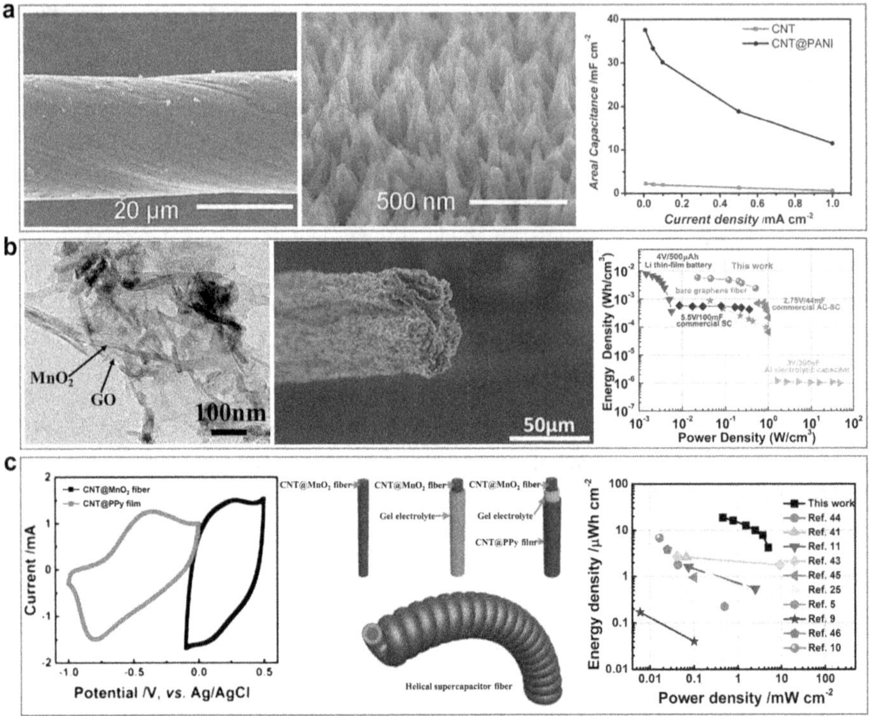

FIGURE 7.6 (a) SEM images of PANi-grafted CNT yarn surface and the corresponding rate performance of assembled twisted supercapacitor. Reprinted with permission from ref.[87]. Copyright 2013 John Wiley and Sons. (b) TEM and SEM images of the MnO_2-GO (reduced) porous fiber, and the Ragone plots of assembled flexible fiber-shaped supercapacitor. Reprinted with permission from ref.[90]. Copyright 2004 Elsevier. (c) Three-electrode cyclic voltammetry curves of PPy-deposited CNT fiber and MnO_2-deposited CNT film, the consequent asymmetric fiber-shaped supercapacitor, and the corresponding Ragone plots of this flexible supercapacitor. Reprinted with permission from ref 99. Copyright 2017 John Wiley and Sons.

to the different stable potential windows from the three-electrode test, which can further improve the energy density of the whole device. For example, Huang's group prepared MnO_2-deposited CNT fibers and PPy-deposited CNT films (Figure 7.6c), which displayed stable cyclic voltammetry curves at high and low potential windows (vs. Ag/AgCl) respectively.[99] Thus, a helical fiber supercapacitor can be assembled by a positive electrode of one MnO_2-deposited CNT fiber and a negative electrode of one PPy-deposited CNT film, with the former as the core and the latter as the sheath. The assembled stretchable fiber supercapacitor displayed a high working voltage of 1.5V.

4 CHALLENGES AND PROSPECTS

Now, the explosive research on carbon-based fibers has made them excellent candidates for next-generation supercapacitors. Some of them are of great importance in smart electronics and wearable devices, based on the reported performance, flexibility, and stability. However, there remain some key challenges that hinder their practical applications.

First, the fabrication technologies of carbon-based fibers should be further explored and promoted, aiming at finding time-saving, cost-effective, scalable, and continuous preparation methods for industrial applications. Though the origin of carbon-based fibers can be traced back to two centuries ago, their application on supercapacitors started a few decades before and is still in its infancy. Traditional technologies mainly center on the mechanical strength of the obtained carbon-based fibers, while paying little attention to the requirements of energy storage. Thus, some mature technologies should be precisely modified and optimized to realize the targeted properties of high SSA, good conductivity, suitable porosity, and satisfactory strength.

Second, the structural and chemical information of ideal carbon-based fibers should be further researched and studied by complete theoretical guidance, since the practical application relies on the overall interdisciplinary of chemistry, electrochemistry, mechanics, ergonomics, and so on. The related theories on all aspects have to keep up with the experimental developments, even making prior instructions for superior CFEs and CCFEs. As artificial intelligence is exploding, advanced theories will make great contributions to the super-performance carbon-based fibers by finding out the right way to the right material.

Third, the precise and fastidious evaluation standards on supercapacitors with different configurations should be established and executed for all the CFEs and CCFEs. Fiber-, mat- and bulk-shaped supercapacitors have their own unique constructions, leading to different quantitative criteria on electrochemical performance. Besides, Cs calculated by mass, volume, area, and length may lead to different results and conclusions, which makes researchers confused and difficult to compare different devices. In more detail, the parameters used for calculations (e.g. volume, area, and length) have always been treated with obfuscation for convenience because they are hard to be accurately measured, which undoubtedly results in great error. Therefore, they all should be definitely confirmed to develop a universal and fair assessment system.

In the future, there should be more and more advanced carbon materials discovered and innovated, some of which will be applied in carbon-based fibers for supercapacitors. Moreover, the boundary between supercapacitors and batteries is increasingly blurry, since some materials can be used as electrodes of both supercapacitors and batteries. Maybe someday, the hybrid energy systems from carbon-based fibers can deliver both ultrahigh energy density and ultrahigh power density, as well as a long-working lifespan and all-climate durability. We believe that supercapacitors from carbon-based fibers will play an important role in clean energy supplying for our daily life.

REFERENCES

1 S. Chen, L. Qiu and H.-M. Cheng, Carbon-based fibers for advanced electrochemical energy storage devices, *Chem. Rev.*, 2020, **120**, 2811–2878.
2 Z. Xu and C. Gao, Graphene chiral liquid crystals and macroscopic assembled fibres, *Nat. Commun.*, 2011, **2**, 571.
3 Z. Xu and C. Gao, Aqueous liquid crystals of GO, *ACS Nano*, 2011, **5**, 2908–2915.

4 Z. Xu, Y. Zhang, P. Li and C. Gao, Strong, conductive, lightweight, neat graphene aerogel fibers with aligned pores, *ACS Nano*, 2012, **6**, 7103–7113.
5 Z. Xu, Y. Liu, X. Zhao, L. Peng, H. Sun, Y. Xu, X. Ren, C. Jin, P. Xu, M. Wang and C. Gao, Ultrastiff and strong graphene fibers via full-scale synergetic defect engineering, *Adv. Mater.*, 2016, **28**, 6449–6456.
6 Y. Liu, Z. Xu, J. Zhan, P. Li and C. Gao, Superb electrically conductive graphene fibers via doping strategy, *Adv. Mater.*, 2016, **28**, 7941–7947.
7 P. Tang, Z. Deng, Y. Zhang, L.-X. Liu, Z. Wang, Z.-Z. Yu and H.-B. Zhang, Tough, strong, and conductive graphene fibers by optimizing surface chemistry of GO precursor, *Adv. Funct. Mater.*, 2022, **32**, 2112156.
8 Z. Liu, Z. Li, Z. Xu, Z. Xia, X. Hu, L. Kou, L. Peng, Y. Wei and C. Gao, Wet-spun continuous graphene films, *Chem. Mater.*, 2014, **26**, 6786–6795.
9 H. Sun, Z. Xu and C. Gao, Multifunctional, ultra-flyweight, synergistically assembled carbon aerogels, *Adv. Mater.*, 2013, **25**, 2554–2560.
10 Z. Li, Z. Xu, Y. Liu, R. Wang and C. Gao, Multifunctional non-woven fabrics of interfused graphene fibres, *Nat. Commun.*, 2016, **7**, 13684.
11 Z. Li, T. Huang, W. Gao, Z. Xu, D. Chang, C. Zhang and C. Gao, Hydrothermally activated graphene fiber fabrics for textile electrodes of supercapacitors, *ACS Nano*, 2017, **11**, 11056–11065.
12 D. Chang, J. Liu, B. Fang, Z. Xu, Z. Li, Y. Liu, L. Brassart, F. Guo, W. Gao and C. Gao, Reversible fusion and fission of GO-based fibers, *Science*, 2021, **372**, 614–617.
13 S. Chen, W. Ma, Y. Cheng, Z. Weng, B. Sun, L. Wang, W. Chen, F. Li, M. Zhu and H.-M. Cheng, Scalable non-liquid-crystal spinning of locally aligned graphene fibers for high-performance wearable supercapacitors, *Nano Energy*, 2015, **15**, 642–653.
14 S. Liu, Y. Wang, X. Ming, Z. Xu, Y. Liu and C. Gao, High-speed blow spinning of neat graphene fibrous materials, *Nano Lett.*, 2021, **21**, 5116–5125.
15 Z. Han, J. Wang, S. Liu, Q. Zhang, Y. Liu, Y. Tan, S. Luo, F. Guo, J. Ma, P. Li, X. Ming, C. Gao and Z. Xu, Electrospinning of neat graphene nanofibers, *Adv. Fiber Mater.*, 2022, **4**, 268–279.
16 Z. Dong, C. Jiang, H. Cheng, Y. Zhao, G. Shi, L. Jiang and L. Qu, Facile fabrication of light, flexible and multifunctional graphene fibers, *Adv. Mater.*, 2012, **24**, 1856–1861.
17 R. Cruz-Silva, A. Morelos-Gomez, H.-i. Kim, H.-k. Jang, F. Tristan, S. Vega-Diaz, L.P. Rajukumar, A.L. Elías, N. Perea-Lopez, J. Suhr, M. Endo and M. Terrones, Super-stretchable GO macroscopic fibers with outstanding knotability fabricated by dry film scrolling, *ACS Nano*, 2014, **8**, 5959–5967.
18 X. Li, T. Zhao, K. Wang, Y. Yang, J. Wei, F. Kang, D. Wu and H. Zhu, Directly drawing self-assembled, porous, and monolithic graphene fiber from chemical vapor deposition grown graphene film and its electrochemical properties, *Langmuir*, 2011, **27**, 12164–12171.
19 A. Thess, R. Lee, P. Nikolaev, H. Dai, P. Petit, J. Robert, C. Xu, Y.H. Lee, S.G. Kim, A.G. Rinzler, D.T. Colbert, G.E. Scuseria, D. Tománek, J.E. Fischer and R.E. Smalley, Crystalline ropes of metallic carbon nanotubes, *Science*, 1996, **273**, 483–487.
20 H.M. Cheng, F. Li, G. Su, H.Y. Pan, L.L. He, X. Sun and M.S. Dresselhaus, Large-scale and low-cost synthesis of single-walled carbon nanotubes by the catalytic pyrolysis of hydrocarbons, *Appl. Phys. Lett.*, 1998, **72**, 3282–3284.
21 H.W. Zhu, C.L. Xu, D.H. Wu, B.Q. Wei, R. Vajtai and P.M. Ajayan, Direct synthesis of long single-walled carbon nanotube strands, *Science*, 2002, **296**, 884–886.
22 Y.-L. Li, I.A. Kinloch and A.H. Windle, Direct spinning of carbon nanotube fibers from chemical vapor deposition synthesis, *Science*, 2004, **304**, 276–278.
23 K. Koziol, J. Vilatela, A. Moisala, M. Motta, P. Cunniff, M. Sennett and A. Windle, High-performance carbon nanotube fiber, *Science*, 2007, **318**, 1892–1895.
24 X.-H. Zhong, Y.-L. Li, Y.-K. Liu, X.-H. Qiao, Y. Feng, J. Liang, J. Jin, L. Zhu, F. Hou and J.-Y. Li, Continuous multilayered carbon nanotube yarns, *Adv. Mater.*, 2010, **22**, 692–696.

25 B. Vigolo, A. Pénicaud, C. Coulon, C. Sauder, R. Pailler, C. Journet, P. Bernier and P. Poulin, Macroscopic fibers and ribbons of oriented carbon nanotubes, *Science*, 2000, **290**, 1331–1334.
26 N. Behabtu, C.C. Young, D.E. Tsentalovich, O. Kleinerman, X. Wang, A.W.K. Ma, E.A. Bengio, R.F. ter Waarbeek, J.J. de Jong, R.E. Hoogerwerf, S.B. Fairchild, J.B. Ferguson, B. Maruyama, J. Kono, Y. Talmon, Y. Cohen, M.J. Otto and M. Pasquali, Strong, light, multifunctional fibers of carbon nanotubes with ultrahigh conductivity, *Science*, 2013, **339**, 182–186.
27 K. Jiang, Q. Li and S. Fan, Spinning continuous carbon nanotube yarns, *Nature*, 2002, **419**, 801–801.
28 X. Zhang, K. Jiang, C. Feng, P. Liu, L. Zhang, J. Kong, T. Zhang, Q. Li and S. Fan, Spinning and processing continuous yarns from 4-inch wafer scale super-aligned carbon nanotube arrays, *Adv. Mater.*, 2006, **18**, 1505–1510.
29 M. Zhang, K.R. Atkinson and R.H. Baughman, Multifunctional carbon nanotube yarns by downsizing an ancient technology, *Science*, 2004, **306**, 1358–1361.
30 M.D. Lima, S.L. Fang, X. Lepro, C. Lewis, R. Ovalle-Robles, J. Carretero-Gonzalez, E. Castillo-Martinez, M.E. Kozlov, J.Y. Oh, N. Rawat, C.S. Haines, M.H. Haque, V. Aare, S. Stoughton, A.A. Zakhidov and R.H. Baughman, Biscrolling nanotube sheets and functional guests into yarns, *Science*, 2011, **331**, 51–55.
31 M. Minus and S. Kumar, The processing, properties, and structure of carbon fibers, *JOM*, 2005, **57**, 52–58.
32 M. Jonoobi, R. Oladi, Y. Davoudpour, K. Oksman, A. Dufresne, Y. Hamzeh and R. Davoodi, Different preparation methods and properties of nanostructured cellulose from various natural resources and residues: A review, *Cellulose*, 2015, **22**, 935–969.
33 P. Luan, X. Zhao, K. Copenhaver, S. Ozcan and H. Zhu, Turning natural herbaceous fibers into advanced materials for sustainability, *Adv. Fiber Mater.*, 2022, **4**, 736–757.
34 H. Kawamoto, Lignin pyrolysis reactions, *J. Wood Sci.*, 2017, **63**, 117–132.
35 R. Hufenus, Y. Yan, M. Dauner and T. Kikutani, Melt-spun fibers for textile applications, *Materials*, 2020, **13**, 4298.
36 A.H. Wazir and L. Kakakhel, Preparation and characterization of pitch-based carbon fibers, *New Carbon Mater.*, 2009, **24**, 83–88.
37 J. Xue, J. Xie, W. Liu and Y. Xia, Electrospun nanofibers: New concepts, materials, and applications, *Acc. Chem. Res.*, 2017, **50**, 1976–1987.
38 J.C. Chen and I.R. Harrison, Modification of polyacrylonitrile (PAN) carbon fiber precursor via post-spinning plasticization and stretching in dimethyl formamide (DMF), *Carbon*, 2002, **40**, 25–45.
39 E.V. Chernikova, N.I. Osipova, A.V. Plutalova, R.V. Toms, A.Y. Gervald, N.I. Prokopov and V.G. Kulichikhin, Melt-spinnable polyacrylonitrile—An alternative carbon fiber precursor, *Polymers*, 2022, **14**, 5222.
40 Y. Gao, J. Zhang, Y. Su, H. Wang, X.-X. Wang, L.-P. Huang, M. Yu, S. Ramakrishna and Y.-Z. Long, Recent progress and challenges in solution blow spinning, *Mater. Horiz.*, 2021, **8**, 426–446.
41 J.E. Oliveira, E.A. Moraes, R.G.F. Costa, A.S. Afonso, L.H.C. Mattoso, W.J. Orts and E.S. Medeiros, Nano and submicrometric fibers of poly(D,L-lactide) obtained by solution blow spinning: Process and solution variables, *J. Appl. Polym. Sci.*, 2011, **122**, 3396–3405.
42 M.R. Badrossamay, H.A. McIlwee, J.A. Goss and K.K. Parker, Nanofiber assembly by rotary jet-spinning, *Nano Lett.*, 2010, **10**, 2257–2261.
43 J. Huang and R.B. Kaner, A general chemical route to polyaniline nanofibers, *J. Am. Chem. Soc.*, 2004, **126**, 851–855.
44 A. Wu, H. Kolla and S.K. Manohar, Chemical synthesis of highly conducting polypyrrole nanofiber film, *Macromolecules*, 2005, **38**, 7873–7875.

45 M.S.A. Rahaman, A.F. Ismail and A. Mustafa, A review of heat treatment on polyacrylonitrile fiber, *Polym. Degrad. Stab.*, 2007, **92**, 1421–1432.
46 M. Zhong, M. Zhang and X. Li, Carbon nanomaterials and their composites for supercapacitors, *Carbon Energy*, 2022, **4**, 950–985.
47 M. Sevilla and R. Mokaya, Energy storage applications of activated carbons: Super capacitors and hydrogen storage, *Energy Environ. Sci.*, 2014, **7**, 1250–1280.
48 A. González, E. Goikolea, J.A. Barrena and R. Mysyk, Review on supercapacitors: Technologies and materials, *Renew. Sustain. Energy Rev.*, 2016, **58**, 1189–1206.
49 L. Liu, Z. Niu and J. Chen, Unconventional supercapacitors from nanocarbon-based electrode materials to device configurations, *Chem. Soc. Rev.*, 2016, **45**, 4340–4363.
50 W. Ma, Y. Zhang, S. Pan, Y. Cheng, Z. Shao, H. Xiang, G. Chen, L. Zhu, W. Weng, H. Bai and M. Zhu, Smart fibers for energy conversion and storage, *Chem. Soc. Rev.*, 2021, **50**, 7009–7061.
51 Y. Chang, G. Han, D. Fu, F. Liu, M. Li and Y. Li, Larger-scale fabrication of N-doped graphene-fiber mats used in high-performance energy storage, *J. Power Sources*, 2014, **252**, 113–121.
52 T. Huang, B. Zheng, L. Kou, K. Gopalsamy, Z. Xu, C. Gao, Y. Meng and Z. Wei, Flexible high performance wet-spun graphene fiber supercapacitors, *RSC Adv.*, 2013, **3**, 23957–23962.
53 D. Yu, K. Goh, H. Wang, L. Wei, W. Jiang, Q. Zhang, L. Dai and Y. Chen, Scalable synthesis of hierarchically structured carbon nanotube–graphene fibres for capacitive energy storage, *Nat. Nanotechnol.*, 2014, **9**, 555–562.
54 Z. Cai, L. Li, J. Ren, L. Qiu, H. Lin and H. Peng, Flexible, weavable and efficient microsupercapacitor wires based on polyaniline composite fibers incorporated with aligned carbon nanotubes, *J. Mater. Chem. A*, 2013, **1**, 258–261.
55 J. Ren, L. Li, C. Chen, X. Chen, Z. Cai, L. Qiu, Y. Wang, X. Zhu and H. Peng, Twisting carbon nanotube fibers for both wire-shaped micro-supercapacitor and micro-battery, *Adv. Mater.*, 2013, **25**, 1155–1159.
56 X. Chen, L. Qiu, J. Ren, G. Guan, H. Lin, Z. Zhang, P. Chen, Y. Wang and H. Peng, Novel electric double-layer capacitor with a coaxial fiber structure, *Adv. Mater.*, 2013, **25**, 6436–6441.
57 X. Zhao, B. Zheng, T. Huang and C. Gao, Graphene-based single fiber supercapacitor with a coaxial structure, *Nanoscale*, 2015, **7**, 9399–9404.
58 J.R. Miller and P. Simon, Electrochemical capacitors for energy management, *Science*, 2008, **321**, 651–652.
59 A. Borenstein, O. Hanna, R. Attias, S. Luski, T. Brousse and D. Aurbach, Carbon-based composite materials for supercapacitor electrodes: A review, *J. Mater. Chem. A*, 2017, **5**, 12653–12672.
60 M.D. Stoller, S. Park, Y. Zhu, J. An and R.S. Ruoff, Graphene-based ultracapacitors, *Nano Lett.*, 2008, **8**, 3498–3502.
61 E. Frackowiak, K. Metenier, V. Bertagna and F. Beguin, Supercapacitor electrodes from multiwalled carbon nanotubes, *Appl. Phys. Lett.*, 2000, **77**, 2421–2423.
62 P. Simon and Y. Gogotsi, Perspectives for electrochemical capacitors and related devices, *Nat. Mater.*, 2020, **19**, 1151–1163.
63 S.H. Aboutalebi, R. Jalili, D. Esrafilzadeh, M. Salari, Z. Gholamvand, S. Aminorroaya Yamini, K. Konstantinov, R.L. Shepherd, J. Chen, S.E. Moulton, P.C. Innis, A.I. Minett, J.M. Razal and G.G. Wallace, High-performance multifunctional graphene yarns: Toward wearable all-carbon energy storage textiles, *ACS Nano*, 2014, **8**, 2456–2466.
64 C. Lu, J. Meng, J. Zhang, X. Chen, M. Du, Y. Chen, C. Hou, J. Wang, A. Ju, X. Wang, Y. Qiu, S. Wang and K. Zhang, Three-dimensional hierarchically porous graphene fiber-shaped supercapacitors with high specific capacitance and rate capability, *ACS Appl. Mater. Interfaces*, 2019, **11**, 25205–25217.

65 Z. Jin, X. Yan, Y. Yu and G. Zhao, Sustainable activated carbon fibers from liquefied wood with controllable porosity for high-performance supercapacitors, *J. Mater. Chem. A*, 2014, **2**, 11706–11715.
66 C. Ma, J. Sheng, C. Ma, R. Wang, J. Liu, Z. Xie and J. Shi, High-performanced supercapacitor based mesoporous carbon nanofibers with oriented mesopores parallel to axial direction, *Chem. Eng. J.*, 2016, **304**, 587–593.
67 Y. Zhu, S. Murali, M.D. Stoller, K.J. Ganesh, W. Cai, P.J. Ferreira, A. Pirkle, R.M. Wallace, K.A. Cychosz, M. Thommes, D. Su, E.A. Stach and R.S. Ruoff, Carbon-based supercapacitors produced by activation of graphene, *Science*, 2011, **332**, 1537–1541.
68 F. Salvador, M.J. Sánchez-Montero, J. Montero and C. Izquierdo, Activated carbon fibers prepared from a phenolic fiber by supercritical water and steam activation, *J. Phys. Chem. C*, 2008, **112**, 20057–20064.
69 C. Kim, Electrochemical characterization of electrospun activated carbon nanofibres as an electrode in supercapacitors, *J. Power Sources*, 2005, **142**, 382–388.
70 M.A. Lillo-Ródenas, D. Cazorla-Amorós and A. Linares-Solano, Understanding chemical reactions between carbons and NaOH and KOH: An insight into the chemical activation mechanism, *Carbon*, 2003, **41**, 267–275.
71 S.-H. Yoon, S. Lim, Y. Song, Y. Ota, W. Qiao, A. Tanaka and I. Mochida, KOH activation of carbon nanofibers, *Carbon*, 2004, **42**, 1723–1729.
72 H. Qian, H. Diao, N. Shirshova, E.S. Greenhalgh, J.G.H. Steinke, M.S.P. Shaffer and A. Bismarck, Activation of structural carbon fibres for potential applications in multifunctional structural supercapacitors, *J. Colloid Interface Sci.*, 2013, **395**, 241–248.
73 D. Yu, S. Zhai, W. Jiang, K. Goh, L. Wei, X. Chen, R. Jiang and Y. Chen, Transforming pristine carbon fiber tows into high performance solid-state fiber supercapacitors, *Adv. Mater.*, 2015, **27**, 4895–4901.
74 T. Qin, S. Peng, J. Hao, Y. Wen, Z. Wang, X. Wang, D. He, J. Zhang, J. Hou and G. Cao, Flexible and wearable all-solid-state supercapacitors with ultrahigh energy density based on a carbon fiber fabric electrode, *Adv. Energy Mater.*, 2017, **7**, 1700409.
75 F. Su, M. Miao, H. Niu and Z. Wei, Gamma-irradiated carbon nanotube yarn as substrate for high-performance fiber supercapacitors, *ACS Appl. Mater. Interfaces*, 2014, **6**, 2553–2560.
76 T. Lin, I.-W. Chen, F. Liu, C. Yang, H. Bi, F. Xu and F. Huang, Nitrogen-doped mesoporous carbon of extraordinary capacitance for electrochemical energy storage, *Science*, 2015, **350**, 1508–1513.
77 T. Huang, X. Chu, S. Cai, Q. Yang, H. Chen, Y. Liu, K. Gopalsamy, Z. Xu, W. Gao and C. Gao, Tri-high designed graphene electrodes for long cycle-life supercapacitors with high mass loading, *Energy Storage Mater.*, 2019, **17**, 349–357.
78 E.J. Ra, E. Raymundo-Piñero, Y.H. Lee and F. Béguin, High power supercapacitors using polyacrylonitrile-based carbon nanofiber paper, *Carbon*, 2009, **47**, 2984–2992.
79 L.-F. Chen, X.-D. Zhang, H.-W. Liang, M. Kong, Q.-F. Guan, P. Chen, Z.-Y. Wu and S.-H. Yu, Synthesis of nitrogen-doped porous carbon nanofibers as an efficient electrode material for supercapacitors, *ACS Nano*, 2012, **6**, 7092–7102.
80 Y. Ma, P. Li, I. Sedloff, X. Zhang, H. Zhang and J. Liu, Conductive graphene fibers for wire-shaped supercapacitors strengthened by unfunctionalized few-walled carbon nanotubes, *ACS Nano*, 2015, **9**, 1352–1359.
81 L.-F. Chen, Z.-H. Huang, H.-W. Liang, H.-L. Gao and S.-H. Yu, Three-dimensional heteroatom-doped carbon nanofiber networks derived from bacterial cellulose for supercapacitors, *Adv. Funct. Mater.*, 2014, **24**, 5104–5111.
82 S. Fleischmann, J.B. Mitchell, R. Wang, C. Zhan, D.-e. Jiang, V. Presser and V. Augustyn, Pseudocapacitance: From fundamental understanding to high power energy storage materials, *Chem. Rev.*, 2020, **120**, 6738–6782.
83 Q. Meng, K. Cai, Y. Chen and L. Chen, Research progress on conducting polymer based supercapacitor electrode materials, *Nano Energy*, 2017, **36**, 268–285.

84 Z. Zhao, K. Xia, Y. Hou, Q. Zhang, Z. Ye and J. Lu, Designing flexible, smart and self-sustainable supercapacitors for portable/wearable electronics: From conductive polymers, *Chem. Soc. Rev.*, 2021, **50**, 12702–12743.

85 V. Augustyn, P. Simon and B. Dunn, Pseudocapacitive oxide materials for high-rate electrochemical energy storage, *Energy Environ. Sci.*, 2014, **7**, 1597–1614.

86 S.A. Delbari, L.S. Ghadimi, R. Hadi, S. Farhoudian, M. Nedaei, A. Babapoor, A. Sabahi Namini, Q.V. Le, M. Shokouhimehr, M. Shahedi Asl and M. Mohammadi, Transition metal oxide-based electrode materials for flexible supercapacitors: A review, *J. Alloys Compd.*, 2021, **857**, 158281.

87 K. Wang, Q. Meng, Y. Zhang, Z. Wei and M. Miao, High-performance two-ply yarn supercapacitors based on carbon nanotubes and polyaniline nanowire arrays, *Adv. Mater.*, 2013, **25**, 1494–1498.

88 F.M. Guo, R.Q. Xu, X. Cui, L. Zhang, K.L. Wang, Y.W. Yao and J.Q. Wei, High performance of stretchable carbon nanotube–polypyrrole fiber supercapacitors under dynamic deformation and temperature variation, *J. Mater. Chem. A*, 2016, **4**, 9311–9318.

89 J.A. Lee, M.K. Shin, S.H. Kim, H.U. Cho, G.M. Spinks, G.G. Wallace, M.D. Lima, X. Lepró, M.E. Kozlov, R.H. Baughman and S.J. Kim, Ultrafast charge and discharge biscrolled yarn supercapacitors for textiles and microdevices, *Nat. Commun.*, 2013, **4**, 1970.

90 W. Ma, S. Chen, S. Yang, W. Chen, Y. Cheng, Y. Guo, S. Peng, S. Ramakrishna and M. Zhu, Hierarchical MnO_2 nanowire/graphene hybrid fibers with excellent electrochemical performance for flexible solid-state supercapacitors, *J. Power Sources*, 2016, **306**, 481–488.

91 P. Shi, L. Li, L. Hua, Q. Qian, P. Wang, J. Zhou, G. Sun and W. Huang, Design of amorphous manganese oxide@multiwalled carbon nanotube fiber for robust solid-state supercapacitor, *ACS Nano*, 2017, **11**, 444–452.

92 K. Gopalsamy, Z. Xu, B. Zheng, T. Huang, L. Kou, X. Zhao and C. Gao, Bismuth oxide nanotubes–graphene fiber-based flexible supercapacitors, *Nanoscale*, 2014, **6**, 8595–8600.

93 W. Ma, S. Chen, S. Yang, W. Chen, W. Weng, Y. Cheng and M. Zhu, Flexible all-solid-state asymmetric supercapacitor based on transition metal oxide nanorods/reduced GO hybrid fibers with high energy density, *Carbon*, 2017, **113**, 151–158.

94 N. Choudhary, C. Li, J. Moore, N. Nagaiah, L. Zhai, Y. Jung and J. Thomas, Asymmetric supercapacitor electrodes and devices, *Adv. Mater.*, 2017, **29**, 1605336.

95 Z. Yang, J. Tian, Z. Yin, C. Cui, W. Qian and F. Wei, Carbon nanotube- and graphene-based nanomaterials and applications in high-voltage supercapacitor: A review, *Carbon*, 2019, **141**, 467–480.

96 V. Khomenko, E. Frackowiak and F. Béguin, Determination of the specific capacitance of conducting polymer/nanotubes composite electrodes using different cell configurations, *Electrochim. Acta*, 2005, **50**, 2499–2506.

97 K. Jurewicz, E. Frackowiak and F. Béguin, Towards the mechanism of electrochemical hydrogen storage in nanostructured carbon materials, *Appl. Phys. A*, 2004, **78**, 981–987.

98 B. Zheng, T. Huang, L. Kou, X. Zhao, K. Gopalsamy and C. Gao, Graphene fiber-based asymmetric micro-supercapacitors, *J. Mater. Chem. A*, 2014, **2**, 9736–9743.

99 J. Yu, W. Lu, J.P. Smith, K.S. Booksh, L. Meng, Y. Huang, Q. Li, J.-H. Byun, Y. Oh, Y. Yan and T.-W. Chou, A high performance stretchable asymmetric fiber-shaped supercapacitor with a core-sheath helical structure, *Adv. Energy Mater.*, 2017, **7**, 1600976.

8 Supercapacitors for Energy Storage

Fundamentals, Electrode Materials, and Application

Ruchun Li and Dingsheng Yuan

1 INTRODUCTION

The development of sustainable and clean energy sources, such as solar, wind, and tidal energy, is urgent due to the continuous consumption of fossil fuels and increasing environmental pollution.[1–3] However, their intermittent, unstable, and regional features lead to a considerable challenge in the efficient usage of these sustainable energy sources. The development of advanced energy storage and conversion systems is of vital significance for efficiently promoting the utilization efficiency of these sustainable energy sources. Currently, capacitors and rechargeable batteries represent two important energy storage and conversion devices. On the one hand, rechargeable batteries (Li/Na/Al/Zn-ion batteries) can provide high-energy density, but slow diffusion kinetics usually lead to low power density and thus are used in applications that require long-term energy supply but low power output.[4–6] While capacitors are more widely used in high-power output applications owing to their large power density. Recently, supercapacitors, as ideal electrochemical energy storage technologies, have the nature of the simple principle, long cyclic life, high coulombic efficiency, wide operated temperature, and fast charging-discharging rates, and received a great deal of attention.[2, 3, 7–11]

Generally, supercapacitors are composed of electrochemical double-layer capacitors (EDLCs) and pseudo-capacitors that store energy by rapidly accumulating charge at the surface of the electrode materials or by implementing a Faradic charge transfer process.[13] Figure 8.1 shows the relationship between power–energy density for the typical electrochemical energy storage technologies. It can be clearly seen that supercapacitors can act as a bridge between batteries and conventional capacitors. In terms of the history of supercapacitors, the earliest capacitor evolved from a Leiden bottle, as shown in Figure 8.2. The original concept of the EDLCs was thus established in the middle of the 18th century. Thereafter, in 1853, the electrical charge storage mechanism in capacitors was studied and first proposed by Helmholz: the electric double-layer model.[13] Meanwhile, some innovative interfacial electrochemists, including Gouy, Chapman, Stern, and so on, have successively

DOI: 10.1201/9781003387831-8

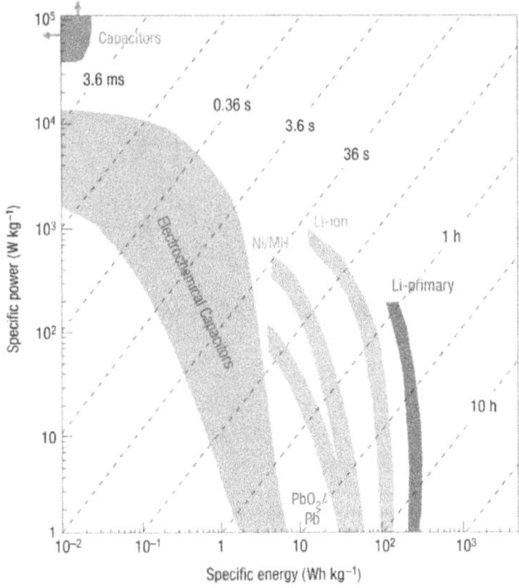

FIGURE 8.1 The Ragone plot for the representative electrical energy storage devices.[12]

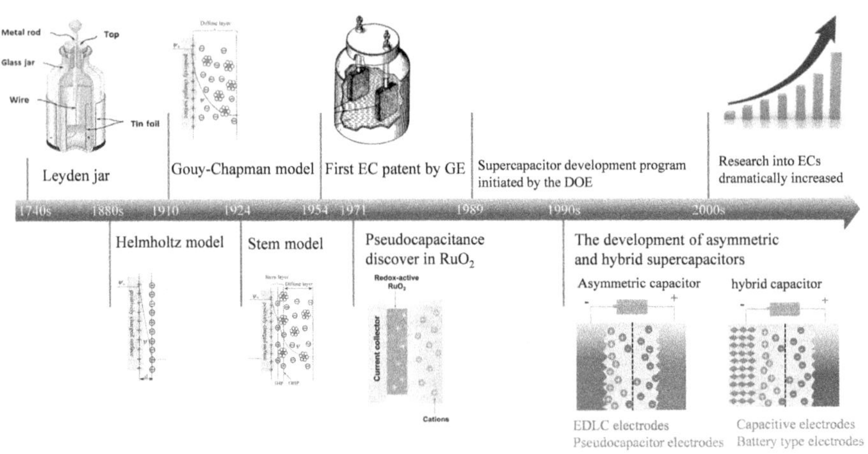

FIGURE 8.2 Historical timeline for the development of supercapacitors.[13]

proposed the modern theories of the EDLCs,[13] which also have gained wide recognition. Although significant progress has been made in the general mechanism of EDLCs, the first patent for an electrochemical capacitor was applied by Becker et al. in 1954 which first discovered the phenomenon of the electrical double layer at the electrode-electrolyte interface. Subsequently, in 1971, a new electrochemical mechanism, named pseudo-capacitor, was discovered on ruthenium dioxides which involve Faradaic reaction processes. Since then, a great deal of research related to EDLCs and pseudo-capacitors has continuously been reported.[1, 13, 14] Currently, supercapacitors have attracted significant attention with a focus on improving their energy density, cycling stability, and cost-effectiveness. Various supercapacitors (cylindrical, flexible, fibrous, self-breathing, etc.) have already been used in a wide range of applications, including transportation, industry, communication, medical devices and life.

2 THE FUNDAMENTALS OF SUPERCAPACITORS

In this section, the fundamentals and charge storage mechanisms of supercapacitors (both EDLCs and pseudo-capacitors) are extensively elaborated. Furthermore, the differences among supercapacitors, batteries, and intercalative pseudocapacitive behavior also are summarized and discussed.

2.1 The EDLCs

Generally, the charge storage, the electrode material, and the electrochemical response in EDLCs have apparent differences with that of pseudo-capacitors. The charge storage of EDLCs is derived from the physical adsorption of electrolyte ions and a double layer of electric charge at the electrode-electrolyte interface. The EDLCs model was initially proposed by Hermann von Helmholtz in 1853. Helmholtz

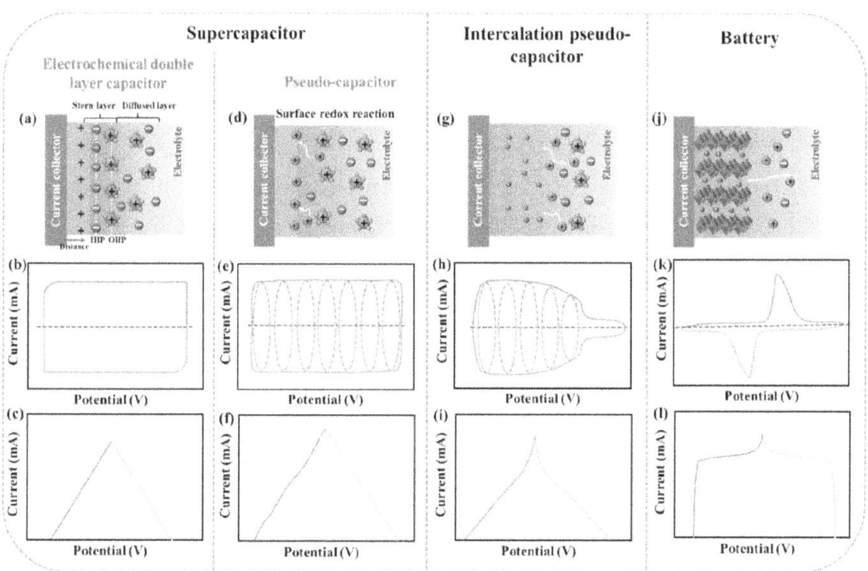

FIGURE 8.3 (a) The mechanism schematic, (b) cyclic voltammetry curves, and (c) the charge/discharge curves for EDLCs; (d) the mechanism schematic, (e) cyclic voltammetry curves, and (f) the charge/discharge curves for pseudo-capacitance; (g) the mechanism schematic, (h) cyclic voltammetry curves, and (i) the charge/discharge curves for intercalation pseudo-capacitance; (j) the mechanism schematic, (k) cyclic voltammetry curves, and (l) the charge/discharge curves for batteries.

claimed that the electrochemical double layer is formed by an electrified electrode surface accompanied by an adsorbed oppositely charged ion layer from an electrolyte which follows the fundamentals of a classical parallel plate capacitor. And then, Gouy and Chapman demonstrated the existence of a diffused layer of oppositely charged ions in the early 1900s, and the ion concentration in the diffused layer changes exponentially. In 1920, inspired by the above two models, Stern put forward that the electrochemical double layer contained two different surface charge layers, where the inner is an adsorbed de-solvated ions layer (Stern Layer, SL) and the outer is a solvated ions layer (Diffused layer, DL). This avoids the disadvantages of the model proposed by Gouy and Chapman, which failed in a highly charged surface or high ion concentration. In 1957, a complete and redefined model of EDLCs was presented by Grahame in two sub-regions: (i) inner Helmholtz plane (IHP) and (ii) outer Helmholtz plane (OHP) in 1957, as shown in Figure 8.3a. The capacitance value in EDLCs can be efficiently calculated by the following equation (Eqn 8.1)[1, 13]:

$$C = \frac{\varepsilon_r \varepsilon_0}{d} A \tag{8.1}$$

where ε_r and ε_0 are the relative permittivity and the permittivity of vacuum, respectively. A is the electrode area, and d is the distance of charge separation. The electrochemical double layers can be formed in a very short time ($\sim 10^{-8}$s) and can respond immediately to the change of voltages. Thus, EDLCs usually have a potential-independent capacitance behavior and thus display rectangular cyclic voltammogram (CV) curves (Figure 8.3b) with a triangular-shaped charging/discharging curve (Figure 8.3c).

2.2 Pseudo-capacitance

In the case of pseudo-capacitors, the charge storage mainly stems from rapid Faradaic charge transfer or/and reversible redox reaction occurring at/near the surface of the material (Figure 8.3d).[15] The capacitance in a pseudo-capacitor is determined by the linear relationship between the amount of charge (ΔQ) and the voltage change (ΔU). Thus, a pseudocapacitive electrode also displays rectangular CV curves (Figure 8.3e,f) in which the electron-transfers with the change of oxidation state of metal center ions are responsible for the charge storage rather than the accumulation of ions in EDLCs. Meanwhile, pseudo-capacitors possess enhanced energy storage than EDLCs. The RuO_2, MnO_2, polypyrrole, polyaniline, etc., represent the typical pseudo-capacitor materials. For instance, Buzzanca and co-authors first reported that RuO_2 film as electrode material exhibited a rectangular CV curve, which is a distinctive pseudocapacitive feature in 1971.[16] Since then, the pseudo-capacitance became a new term to distinguish capacitive signatures between charge transfer Faradaic reactions and the electrochemical double layer. Michael Gratzel demonstrated that polyaniline as a redox-active material could afford high pseudo-capacitor value. Afterward, other conducting polymers, such as polypyrrole and polythiophene, also were explored as electrode materials for pseudo-capacitors.

Recently, the battery-type capacitors as the new supercapacitors have been proposed.[8, 9, 11] The electrochemical behaviors showed a clear redox peak and an apparent plateau in the CV curve and a charge-discharge curve which is similar to that of the battery. Although Faradic reactions occur in both pseudocapacitive and battery-type capacitors, the charge storage mechanism is different. The Faradic reaction in pseudocapacitive materials occurs at the surface/near surface of the electrode material with shorter ion diffusion paths, faster embedding and disembedding, and no phase change during the reaction. Moreover, the battery-type materials are mainly involved in bulk phase reactions accompanied by a phase change, and their energy density and capacity are higher than those of pseudocapacitive materials. The nickel- or cobalt-based oxides, hydroxides and sulfides possess the battery-type behaviors.

2.3 Their Difference between Battery and Intercalation Pseudo-Capacitance

Rechargeable batteries have become one of the best energy storage devices due to their high storage efficiency and satisfactory energy.[4–6, 17–19] Different from supercapacitor, the charge storage mechanism for the rechargeable battery originates from the intercalation and de-intercalation of cations (such as Li^+, Na^+, or Mg^{2+}) with the redox reactions within the crystalline structure of the electrode in the particular voltage, generally accompanying with an apparent phase change (Figure 8.3j). There are distinct redox peaks in CV curves with the typical plateaus in the charge/discharge process (Figure 8.3k,l). These are different from supercapacitor that generally exhibits a rectangular CV shape with a linear voltage versus time plot and has no phase changes during the charge/discharge process. On the other hand, the performance of the battery was expressed in terms of specific capacity (mAh g^{-1}) instead of the capacitance (F g^{-1}) in the supercapacitor. Compared with batteries, supercapacitors deliver much faster power density, and more excellent cyclic stability but lower energy density.

Recently, intercalation pseudo-capacitance has emerged as a new type of energy conversion (Figure 8.3g). The charge storage mechanism of intercalation pseudo-capacitance arises from the ion intercalation (e.g., Li^+, Na^+, O^{2-} and NH_4^+) into the tunnels or layers of the electrode materials with a rapid charge transfer.[20–23] The intercalation process occurs in the bulk but instead on the surface of the material, but no new phase is generated during the intercalation process. Meanwhile, the charge/discharge curves display the linear-like profiles with a rectangular-like CV curve similar to the surface-redox pseudo-capacitance feature (Figure 8.3h,i). These indicate that the intercalation process is fast, which enables the electrode materials to offer high-power performance and rate capability like supercapacitors. Thus, the intercalation pseudo-capacitance can bridge the gap between battery and supercapacitor from the aspects of both energy density and power density.

2.4 The Performance Evaluation for Supercapacitors

2.4.1 Capacitance

The capacitance of supercapacitors can be calculated by the CV and galvanostatic charge/discharge curves which are typical and essential characterization in the energy-related fields. Firstly, the CV curves can reflect the charge storage mechanism and assess the mass-specific capacitance (C_m, F g^{-1}) according to the following equation (Eqn 8.2):

$$C_m = \frac{Q_{\text{total}}}{2\Delta Vm} \tag{8.2}$$

where Q_{total} is the total charge in coulombs based on the integral area from the CV curve, I is the current (A), m (g) represents the mass of the active materials and ΔV is the potential window from low voltage to high voltage. When the potential changes at a constant sweep rate (v, V/s), Eqn 8.3 can be described as:

$$C_m = \frac{Q_{\text{total}}}{2\Delta Vm} = \frac{1}{2mv\Delta V}\int I \, \mathrm{d}V \tag{8.3}$$

It can be seen from Eqn (8.3) that the sweep rate plays a pivotal role in the capacitance. Usually, the capacitance from the different sweep rates can reflect the rate property of the electrode materials. In addition, the specific capacitance (C_s) can also be determined by the galvanostatic charge/discharging curves according to Eqn 8.4:

$$C_{\text{m}} = \frac{I\Delta t}{m\Delta V} \tag{8.4}$$

where I (A) is the current during the discharging process, t (s) is the discharging time, m (g) is the mass of the active materials, and V is the potential window during the discharging process. Moreover, the area/volume-specific capacitance also is calculated similar to the mass-specific capacitance (Eqn 8.5):

$$C_{\text{A}} = \frac{I\Delta t}{A\Delta V} \tag{8.5}$$

where A (cm^2) is the area/volume of electrodes/capacitors. In many self-supported electrode materials, it is difficult to calculate the mass of the active materials accurately. Thus, the area/volume-specific capacitance is more appropriate to estimate the property of electrode materials relative to mass-specific capacitance.

2.4.2 Energy density and power density

The energy density (E, Wh kg^{-1}) and power density (P, W kg^{-1}) are crucial parameters in the supercapacitor devices for practical application, which can be expressed as the following formulas (Eqns 8.6 and 8.7):

$$E = \int Q \, \mathrm{d}V = \frac{CV^2}{2} \tag{8.6}$$

$$P = \frac{E}{t} \tag{8.7}$$

where C denotes the capacitance of the supercapacitor device from the discharging process, and t and V are the discharging time and the voltage window of the supercapacitor device, respectively. However, in many actual supercapacitor devices, a nonlinear discharging curve usually is acquired, and the energy density should be assessed by a modified equation (Eqn 8.8):

$$E = \int Q \, \mathrm{d}V = \int_{t_1}^{t_2} IV_{(t)} \mathrm{d}t \tag{8.8}$$

In Eqn 8.8, t_1 is the time after the initial IR drop, t_2 is that when the discharging process is finished, and I is the constant current at the discharging measurement.

2.4.3 The reaction kinetics for electrode materials

The CV measurement is an efficient instrument for investigating the reaction kinetics of electrode materials. In theory, the current response of the active materials with different scan rates can be described by the following formula (Eqns 8.9 and 8.10):

$$I = av^b \; I = av^b \tag{8.9}$$

$$\log I = \log a + b \log v \tag{8.10}$$

where I and v are the current and the sweep-speed, and both a and b are the tunable parameters. The characteristic parameter b can be calculated from the slope of Eqn 8.10 by the relationship between the CV peak current and the sweep-speed, which provides feedback for electrochemical kinetic reactions. As shown in Figure 8.4, when the value of b is 1, it implies the capacitive behavior with fast surface-controlled processes, including the pseudo-capacitor and the EDLCs in the supercapacitor. While the value of b is 0.5, it represents the low diffusion-controlled processes with the Faradaic redox reactions, which usually proceed in the bulk of electrode materials, such as rechargeable batteries. In addition, the particular transitional region ($0.5 < b < 1$) exists in the experiments. Generally, the smaller b value indicates the dominant diffusion-controlled processes during the charge storage reactions, and conversely, s is the capacitive contribution with fast surface-controlled processes dominating the charge storage mechanism.

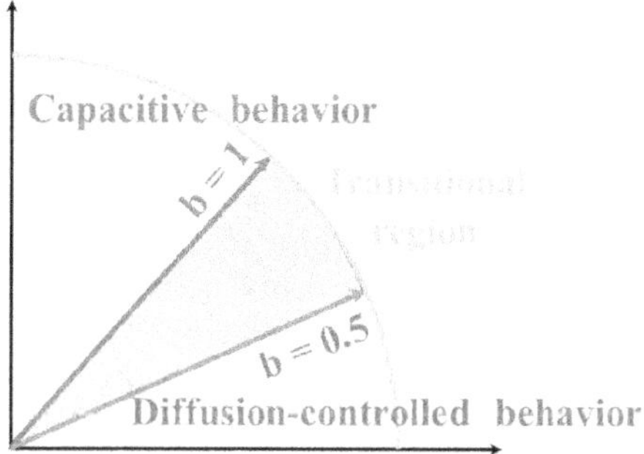

FIGURE 8.4 The reaction kinetics of electrode materials for capacitive behavior ($b = 1.0$) and diffusion-controlled behavior ($b = 0.5$). The transition region represents the existence of both capacitive and diffusion-controlled behavior ($0.5 < b < 1$).

2.4.4 Cyclic stability

Long-term stability is a crucial indicator of the supercapacitor for application in various fields. Generally, the EDLCs show high long-term stability, but the stability of the pseudo-capacitor is lowered. This is because the surface-redox reaction in a pseudo-capacitor could result in irreversibility. However, the storage mechanism of EDLCs involves the physical adsorption of electrolyte ions which is highly reversible. In reality, the long-term stability of the supercapacitor relates to many factors, such as the feature of electrode material, the type of electrolyte, the charge/discharge rate, and the operated temperature.

3 ELECTRODE MATERIALS FOR SUPERCAPACITORS

Electrode materials play a critical role in the property of supercapacitors. The features of electrode materials, such as surface area, structure, morphology, and conductivity, influence the specific capacitance, power/energy density, and cycle life of supercapacitors, as well as the energy storage mechanisms. The relationship between the properties of electrode materials and the performance of supercapacitors has been extensively explored. In this section, recent progress in electrode materials will be systematically reviewed.

3.1 Carbon Materials

Carbon materials, including carbon nanotubes (CNTs), graphene (GR), activated carbon, biomass-derived carbon and others, have been widely studied as electrode materials for EDLCs.[24–26] The capacitance performance strongly relies on the specific surface area and conductivity of carbon materials. At present, great efforts have been devoted to the design and development of carbon materials with high capacitance properties.

3.1.1 Carbon nanotubes (CNTs)

CNTs are composed of single-walled carbon nanotubes (SWNTs) and multiwalled carbon nanotubes (MWNTs). Due to their unique structure and attractive stability, both of them have attracted significant attention from EDLCs.[27–36] Moreover, their high mechanical stability, excellent conductivity, and network structure make them a good support to load active materials. The capacitive property of the CNTs is in the range from 2 F g^{-1} to 200 F g^{-1} according to the reported literatures. For instance, Ma's group[27] fabricated the electrochemical capacitors using the CNTs materials that deliver the volume-specific capacitance of 15–25 F cm^{-3} in H_2SO_4. Tennent et al.[28] reported carbon nanotube sheet electrodes with a narrow diameter distribution of around 80 Å which showed a maximum capacitance value of 102 F g^{-1} in 38 wt%H_2SO_4, accompanied by high power and excellent cyclic stability. In addition, Lee and co-authors[30] investigated the main factors affecting the capacitive properties of the single-walled CNT electrodes. The results found that the BET surface area of the CNT directly relates to theoretical specific capacitance and the low contact resistance plays a decisive role in the power density. By optimizing the annealing temperature of the CNTs, a maximum specific capacitance of 180 F g^{-1} was obtained.

Furthermore, the correlation among electrochemical characteristics, the microstructure, and the elemental composition of the multiwalled carbon nanotubes also was built.[31] The capacitance values of the multiwalled CNTs varied from 4 to 135 F g^{-1} due to the different types of nanotubes and/or different post-treatments. The mesopore structure provides easy accessibility of the ions to the surface of the electrode. The surface functionality of the CNTs with element/functional groups modification could lead to not only defect structure and enhanced conductivity but also unique pseudo-Faradaic reactions, and thus improve the capacitance performance.[32–37] First, many studies have shown that surface oxidation can significantly improve capacitive performance.[36–38] The functional groups of C–O–C, –COOH, C=O, C–OH, and other bonds existed in the carbon skeleton. Among them, C–OH, C=O, and –COOH have been proposed to increase capacitance by the following pseudo-Faradaic reactions (Eqns 8.11 and 8.12):

$$-\text{C–OH} \Leftrightarrow \text{C=O} + \text{H}^+ + e^- \tag{8.11}$$

$$-\text{COOH} + e^- \Leftrightarrow -\text{COO}^- + \text{H}^+ + e^- \tag{8.12}$$

Simultaneously, the self-discharge can also be reduced by the surface oxygen functionality and the moderate oxygen functional group would inhibit the self-discharge process.[24]

Besides, the N element has higher electron negativity (3.04) than the C element (2.55) and similar atomic radii (N: 0.74 Å; C: 0.77 Å), and thus N introduction can tune the asymmetric valence charge and electronic structure of carbon materials. The N species is usually composed of four different types (Figure 8.5): pyridinic-N (N-6), pyrrolic/pyridone-N (N-5), quaternary-N (N-Q) and pyridine-N-oxide (N-X).[38–42] Among them, the positively charged N-Q/N-X and the negatively charged N-5/N-6 could promote electron transfer. Especially the edge pyridinic-N and pyrrolic-N can provide pseudo-capacitance through Faradaic reactions (Eqns 8.13 and 8.14).[42, 43] Hence, the carbon materials with N-doping usually show much more excellent supercapacitors than pure carbon materials.[38–41] Zhuo et al. reported.[39] a N-CNTs with a hierarchical porosity and high specific surface area of 3,253 $m^2 g^{-1}$. The obtained N-CNTs exhibited high capacitive properties with a capacitive value of 365.9 F g^{-1} at 0.1 A g^{-1}. The excellent capacitive performance could be attributed to

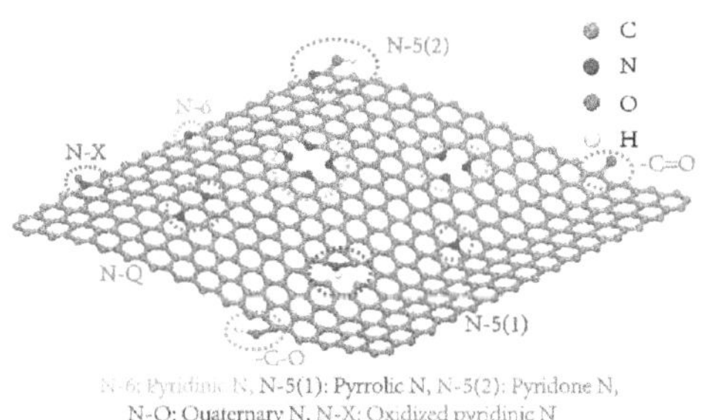

FIGURE 8.5 Various N-containing functional groups in carbon materials.[24]

the electrochemical double layer and Faradaic capacitance contributions caused by high specific surface area and N-doping (Eqns 14.13 and 14.14).

$$+ H_3O^+ \longleftrightarrow + H_2O \qquad (8.13)$$

$$+ H_3O^+ \longleftrightarrow + H_2O \qquad (8.14)$$

In addition to N and O, P and S elements are also the ordinary doping atom.[43–46] Generally, P (~1.07 Å) and S (~1.04 Å) elements have much larger atomic radii than C elements, and P and S-doping could lead to rich defects and distortion in the carbon skeleton. Meanwhile, P and S-doping often exist in the form of oxidized species (such as PO_x and SO_x) in the edge region of the carbon skeleton, which could contribute to Faradic capacitance. Moreover, the electron negativities of P and S elements are 2.19 and 2.58 respectively, which are lower or close to the C atom (2.55). Therefore, P and S-doping could not only bring structural distortion and improved surface wettability but also has the ability to offer electrons, accordingly enhancing electrochemical capacitive property. Yu's group[45] developed a new strategy to synthesize P-doped carbon nanotubes with unstable oxygen species. The P-doping could effectively enhance the electrons donating ability, and produce the abundant P-O-H and P=O active sites and defective surface for the adsorption of ions, thus giving rise to a fast Faradaic reaction. The obtained P-doped carbon nanotubes possessed high specific capacitance and good stability. Additionally, Kim et al.[34] demonstrated S-CNTs as a conducting agent showing high specific capacitance (120.2 F g^{-1}) at a large current density of 100 mA cm^{-2}. The charge transfer resistance of the S-CNT electrode was lowered by S-doping.

3.1.2 Graphene

As a typical two-dimensional material, Graphene was first discovered in 2004 by Andre Geim and Konstantin Novoselov, which is made up of a single layer of carbon atoms with a hexagonal lattice. Graphene has sparked a tremendous amount of interest in the scientific community due to its unique properties, including high conductivity, high mechanical strength, and high surface area. Especially, Graphene has shown great potential as an electrode material for supercapacitors. The large surface area (theoretical surface area up to 2,630 $m^2 g^{-1}$) of graphene allows for more electrochemical reactions to occur at the electrode surface, leading to high capacitance (theoretical capacitance value 540 F g^{-1})[47–54] much higher than that of other carbon materials such as activated carbon, CNTs, and fullerenes. The high electrical conductivity of graphene also allows for efficient charge transfer between the electrode and electrolyte, which is essential for increased power and fast charging and discharging. For instance, Rao's group[50] reported that the exfoliated graphene, as a good electrode material, displayed a large capacitive value of 117 F g^{-1} in an H_2SO_4 solution. Ruoff's group[51] developed a chemically modified graphene with 1-atom thick sheets of carbon, which delivered the specific capacitances of 135 and 99 F g^{-1} in aqueous and organic electrolytes, respectively. Afterward, to further improve the performance, Ruoff et al.[52] reported a chemical activation of graphene with an ultra-high surface area of 3,100 $m^2 g^{-1}$ which is even larger than the theoretical value of graphene. The high surface area could be attributed to primarily 0.6-–5-nm-width pores which led to a maximum specific capacitance of 200 F g^{-1}. In addition, Ma and co-authors[53] synthesized mesoporous graphene by scalable self-propagating high-temperature synthesis methodology, as shown in Figure 8.6a,b. The obtained graphene (Figure 8.6c–f) not only has monodispersed mesoporous structure but also possesses high electrical conductivity, thus leading to the high specific capacitance of 244 F g^{-1} at 2 A g^{-1}. This work opens up a new strategy for scalable synthesis of graphene materials.

Although 2D graphene has acted as the ideal electrode material for supercapacitors, the restacking of the graphene sheets will lower the electrochemically available specific surface area and in turn affect the improvement of capacitive performance. To address this, doping or surface functionalization has been proposed to reduce the restacking of the graphene and enhance supercapacitor performance.[54] Specially, doping graphene with heteroatoms (such as N, B, S, and P)[55–65] reveals unique physical and electrochemical properties. Liu et al.[55] reported that N-doped graphene was constructed by heating graphene oxide and NH_4HCO_3 at low temperatures (150°C). The capacitance value of the N-doped graphene is 170 F g^{-1} at 0.5 A g^{-1}, which is 3.6 times larger than that of RGO (47 F g^{-1}). Gryglewicz and co-authors[56] prepared N-doped reduced graphene oxides with a high N content of 13.4 at.% by a facile hydrothermal reaction. Results confirmed that introducing N can induce pseudocapacitive behavior and promote charge propagation and ion diffusion. The superiority of the N-doped reduced graphene oxides revealed a very high capacitance value of 250 F g^{-1} at 1 mV s^{-1} and good stability. The pseudocapacitive effect of the N element was also explored. As shown in Figure 8.7a,b, the CV curves showed the prominent humps in reduced graphene oxides (at around −0.55V) and N-doped reduced graphene oxides (at around −0.6 V) shows reversible humps (at approximately −0.65 V),

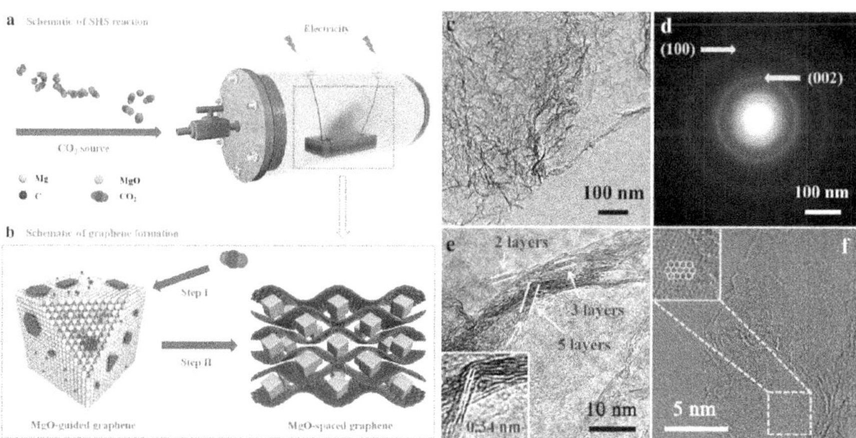

FIGURE 8.6 (a) A schematic illustration of the reaction chamber and the self-propagating high-temperature synthesis process. (b) The dual role of MgO in forming few-layered graphene. (c) TEM, (d) SAED image, and (e) HR-TEM with an inset demonstrating its few-layer feature. (f) Spherical aberration-corrected HR-TEM image.[53]

implying the existence of pseudocapacitive contribution. The reduced graphene oxides contained rich surface oxygen groups in which the reversible redox reactions of quinone and hydroxyl groups gave rise to pseudocapacitive (Figure 8.7c). In the case of N-doped reduced graphene oxides, the electrochemically active nitrogen groups (Figure 8.7d) were responsible for the reversible redox reactions.

Like N-doping, B-doping also shows an enhanced electrochemical capacitive property.[58–60] Manthiram et al.[58] reported an effective strategy inspired by the 'Fried-Ice' concept to synthesize B-doped graphene. The content of B could be well controlled by adjusting the reaction temperature. A large gravimetric capacitance of 281 F g^{-1} was delivered on the B-doped graphene electrode. Kim's group[59] also reported a B-doped graphene nanoplatelet with a tunable boron-doping content. A high boron concentration of ~6.04 at.% in B-doped graphene nanoplatelets was obtained, which showed excellent capacitive performance with the highest specific capacitance value of 448 F g^{-1}. This confirmed that B-doping has tremendous potential for capacitive applications. In addition, S-doping has been evidenced as an efficient method for the regulation of electronic properties.[61–64] S-doped reduced graphene oxides were successfully synthesized using a modified Hummers' method.[61] The electrochemical performance of S-doped reduced graphene oxides had been obviously enhanced than reduced graphene oxides. Ning et al.[62] developed S-doped nanomesh graphene with a low-temperature treatment. The S-doped nanomesh graphene possessed a high specific surface area, excellent conductivity, and good hydrophilicity, and suitable S content of 5 wt.%. Thus, a specific capacitance of 257 F g^{-1} could be achieved at 0.25 A g^{-1}, which is 23.6% larger than that of the pristine graphene. Recent research reported by Bandosz and co-authors[63] confirmed that the attractive performance was due to the presence of S element: (1) S-doping gives a positive charge to adsorb anions; (2) the sulfones, sulfoxides and quinones functional group can produce Faradaic reactions; (3) S-doping brings good wettability. Furthermore, the P element has been doped into graphene edges to improve the supercapacitor performance. Sahin's group[65]

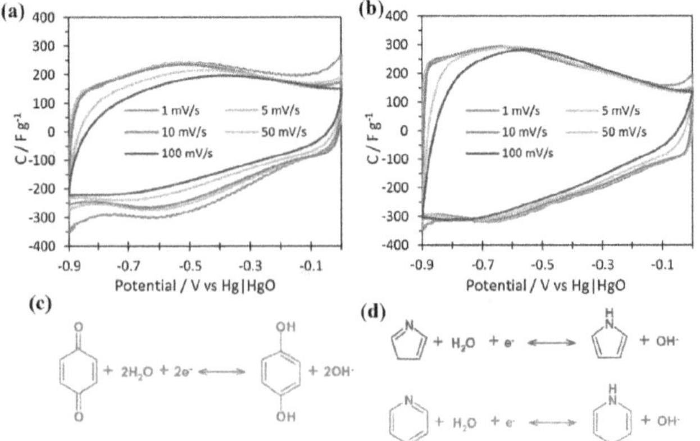

FIGURE 8.7 Cyclic voltammograms of reduced graphene oxides (a) and N-doped reduced graphene oxides (b) in 6 mol L^{-1} KOH. Redox reactions of electrochemically active oxygen (c) and nitrogen groups (d) in an alkaline solution.[56]

synthesized P-doped graphene which contained rich surface functional groups such as $-[(-P_2O_7)]^{4-}$ and $-(PO_3^-)$. The areal capacitance of 301.3 mF cm^{-2} at 10 mA cm^{-2} was realized on P-doped graphene. Hulicova-Jurcakova et al. prepared P-doped graphene with the P concentration of 1.30 at.% by the heat treatment of the mixture of graphene and phosphoric acid. The effect of P introduction had been extensively explored and the results exhibited that the P-doping in graphene showed a great enhancement in specific capacitance, IR drop, and cycling stability.

More importantly, dual or multi-element co-doping leads to unexpected electrochemical performance.[66–69] Mullen et al.[67] fabricated a B and N co-doped graphene with a remarkable capacitive behavior. The micro-supercapacitors assembled by B and Nco-doped graphene showed a superior volumetric capacitance of ~488 F cm^{-3} accompanied by an excellent rate capability. Qiu's group[68] prepared N/P co-doped thermally reduced graphene oxide by using the $(NH_4)_3PO_4$ as P and N sources with high electrochemical capacitive performance. They demonstrated that P-doping could increase the content of the surface O element due to three possible P-related O groups (P=O, P−O−Caliphatic, and P−O−Caromatic). They also proposed the basal plane configuration as shown in Figure 8.8, in which the formation of the out-of-plane P−O−Caliphatic σ bonds probably broke the π conjugation of the graphitic network. Thus, the N and P co-doping contributed to the good capacitive performance. Moreover, Jia and co-authors[69] reported the N/S co-doped graphene and explored the interactions between N/S co-doping and the presence of oxygen. Compared with N-doped graphene or S-doped graphene, the N/S co-doped graphene contains much higher N and S concentrations, especially pyrrolic-N groups, which can enhance pseudocapacitance. Simultaneously, the surface chemistry of carbon could be tuned by S and N co-doping with the presence of oxygen. The synergetic effect between N and S occurred through neighboring carbon atoms which induced a large capacitive value of 566 F g^{-1} at 0.5 A g^{-1}.

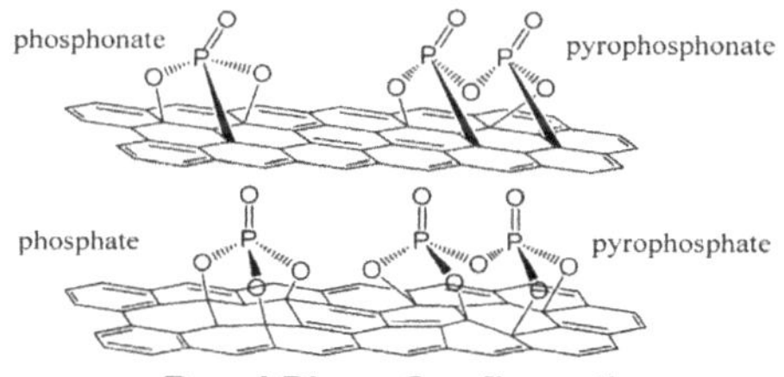

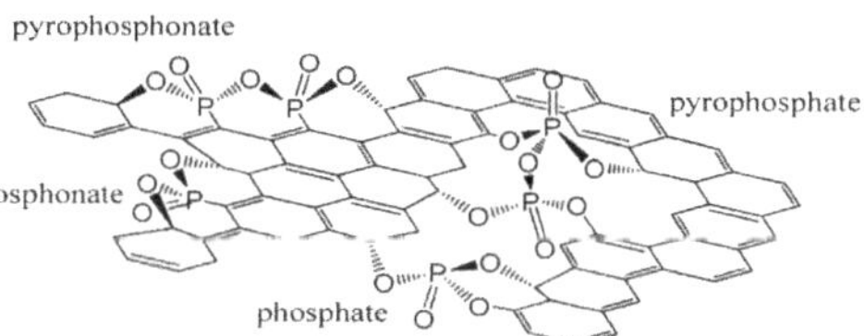

FIGURE 8.8 Proposed structural model for P-containing groups.[68]

In the case of graphene, doping can enhance conductivity, surface area, and stability. Meanwhile, doped graphene has a higher density of states near the Fermi level, which means that it has more available energy levels for storing and transferring charges. It can be concluded that the doped advance in graphene for supercapacitors shows excellent potential for developing high-performance and energy-efficient energy storage devices.

3.1.3 Activated carbon

Activated carbons have become vital capacitive materials for supercapacitors due to their good conductivity, high surface area and rich porosity. Activated carbon has been commercialized in supercapacitors, which was the first to be commercialized among carbon materials. The activated carbon can be synthesized by chemical activation, physical activation, and template synthesis of carbon precursor materials for supercapacitor applications. These methods can tune the pore size, surface area, and pore distribution of the carbon material, which leads to the enhanced energy density and power density of the supercapacitor. Generally, the chemical activation involves the mixture of carbon precursor material (coconut shells or wood) with a chemical reagent (H_3PO_4 or KOH, etc.) to calcine them in the N_2/Ar atmosphere for the synthesis of the activated carbon.[70–76] Lu et al. reported the activated carbon nano-onion by KOH activation with a large capacitive value of 122 F g^{-1}, which is five-fold as high as the pristine carbon nano-onion (25 F g^{-1}).[71] During the activation process, the following reaction usually takes place (Eqn 8.15):

$$KOH + C \rightarrow K + H2 + K_2CO_3 \tag{8.15}$$

Chong's group developed an activated carbon with a high specific surface area derived from the NaOH activation of rice husks as the electrode material. The specific surface areas can be controlled by the different annealing temperatures and a specific surface area of 2,696 $m^2 g^{-1}$ can be reached. The obtained activated carbon exhibited the maximum specific capacitance of 147 F g^{-1}.[71] Geng et al. reported a one-step H_3PO_4 activation process to treat the cotton stalk for generating pf activated carbon with a large specific capacitance of 338 F g^{-1}.[76] In addition, Elmouwahidi prepared a series of activated carbons by KOH and H_3PO_4-activation of olive residues.[77] The results pointed out that the KOH activation provided a high surface area and H_3PO_4 activation created P-containing functional groups and mesopores. The activated carbon with H_3PO_4 activation showed more excellent capacitive properties than that with KOH activation, indicating the presence of P-containing functional groups contributes to a large capacitance value.

Physical activation is usually to activate the precursor materials (coal or coconut shells) in the presence of the oxidizing gas (steam, carbon dioxide, or nitrogen) for the synthesis of the activated carbon.[78–84] The steam and CO_2 are the most common oxidizing gases for physical activation. The CO_2 activation can increase the surface oxygen group and porosity of the activated carbon but no apparent change can be detected during the steam activation process.[83] For instance, Wang and co-authors prepared the microporous activated carbons derived from lignocellulosic biomass by using CO_2 activation, which had the maximum specific capacitance of 92.7 F g^{-1}.[78] Yang's group prepared apricot shell-derived activated carbons by water vapor and CO_2 activation.[80] The multistep activation strategy can produce a hierarchical porous structure relative to the typical single physical activation. Thus, the as-obtained apricot shell-derived activated carbons displayed superior specific capacitance due to high specific surface area and pore volume.

The template synthesis method mainly utilizes a sacrificial template material (such as SiO_2, MgO, ZnO, Al_2O_3, and TiO_2) as the template for synthesizing activated carbon.[85–90] A well-defined pore size distribution can be obtained accordingly. Simon's group[87] reported a mesoporous carbon monolith which was prepared using the self-assembled Na_2SiO_3 salt particles as a template. After the removal of the salt templates, the MCM sample with a high specific surface area and high mesoporous volume showed the capacitance property as high as 75 F g^{-1} and outstanding rate performance. Among these templates, SiO_2 is one of the most common templates for synthesizing activated carbon which has been commercially produced. As shown in Figure 8.9a, a meso-structure surfactant/silicate template was used as a direct template for the large-scale preparation of mesoporous activated carbon [90]. Moreover, the porous MgO templates can be used to prepare the 3D pillared-porous carbon nanosheets (Figure 8.9b–d) in which the unique structure endowed the excellent transportation of electrolyte ions, resulting in a distinguished electrochemical performance (Figure 8.9e).[91]

Overall, activated carbon has been commercialized as a capacitor material. These recent strategies for the production of activated carbon aim to optimize the properties of the resulting material, including surface area, porosity, and pore size distribution,

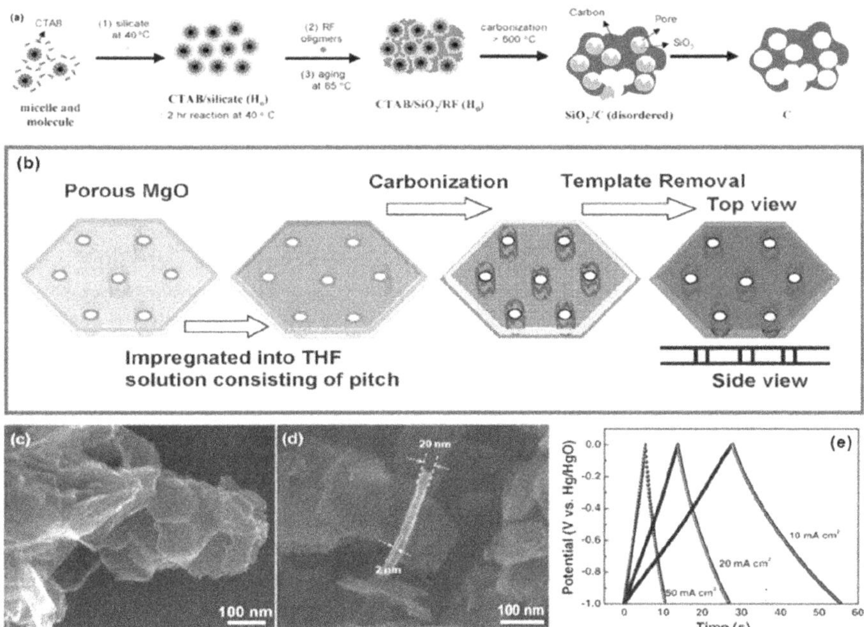

FIGURE 8.9 (a) Schematic explanation of synthesis using a meso-structure silicate template.[90] (b) Illustration of the formation of 3D pillared-porous carbon nanosheet architectures; (c) SEM images, (d) nitrogen adsorption/desorption isotherm, and (e) Galvanostatic charge-discharge curves.[91]

to improve the energy density and power density of supercapacitors for practical applications.

3.1.4 Biomass-derived carbon

Biomass is known as an excellent precursor to prepare carbon-based materials because of its renewability, environmental efficiency, and diversity. Thus, biomass-derived carbon materials have been widely studied as electrode materials in supercapacitors. Generally, Biomass-derived carbon is composed of agricultural/forestry residues, industrial biomass residues, sewage sludge, and algal biomass.[92–97] Yan's group[96] reported that the N-doped carbon microspheres (NCSs) were prepared by hydrothermal carbonization of oatmeal which showed high capacity and excellent rate performance. In addition, multistep activated carbonization can further improve the specific surface area of biomass-derived carbon materials. For example, Ma's group[98] reported on the fabrication of extremely porous carbon materials using hydrothermal carbonization and KOH activation of egg protein with GO sheets as templates. The as-prepared porous carbon materials have a substantial surface area of 1,365 $m^2 g^{-1}$. As an electrode material for supercapacitors, it exhibits exceptional cycle stability and a high specific capacitance of 482 F g^{-1} at 0.1 A g^{-1}. Combining hydrothermal carbonization and KOH activation, Long's group[99] produced the porous graphene-like carbon compounds from Auricularia (Figure 8.10a,b). Auricularia-derived carbon possessed a multi-space framework with high specific surface area, high packing density, and layered interconnections that supply electrolytic ions and abundant storage sites as well as restricted transport routes (Figure 8.10c–g).

Moreover, many plants or vegetables can also be used as precursors for the synthesis of porous carbon.[101–105] F. Kurosaki et al.[101] reported that amorphous porous carbon materials using wood chips as carbon source were prepared by flash heating (5°C min^{-1}) and low-temperature heat treatment (380°C), which exhibited significant porosity, controllable three-dimensional structure, and large specific surface area. In particular, Li et al.[103] prepared nitrogen-doped porous carbon sheets from eggplant by freezing and then carbonization at a high temperature in one step (Figure 8.11). The eggplant-derived carbon displayed a complex multilayer pore framework with 950 $m^2 g^{-1}$ of specific surface area and 100 F g^{-1} of specific capacity at 0.5 A g^{-1}. Biswal et al.[104] reported that the carbon material obtained from carbonizing dead leaves directly by pyrolysis displays a significant specific surface area as well as a specific capacitance of 400 F g^{-1}at 0.5 A g^{-1}. Furthermore, the capacitive performance can be improved by introducing transition metal compounds based on the direct carbonization of biomass-derived carbon materials.

Activation carbonization (chemical or physical activation) is an effective method to increase the porosity of biomass-derived carbon materials.[106–118] Liu's group[107] reported that a 3D porous carbon material using jujube as a carbon source was prepared by KOH active carbonization with a higher particle density (1.06 g cm^{-3}), while the particle density of the blank sample was only 0.18 g cm^3. Zou's group[106] reported that nitrogen-doped carbon prepared by using sugarcane bagasse as a carbon precursor and KOH as an activation agent has a hierarchical porous structure and high pore volume of 2.05 mL g^{-1}. When assembled as a symmetric supercapacitor, it achieves an ultra-high-energy density of around 40 Wh kg^{-1} in 2 M Li_2SO_4 electrolyte. Yang

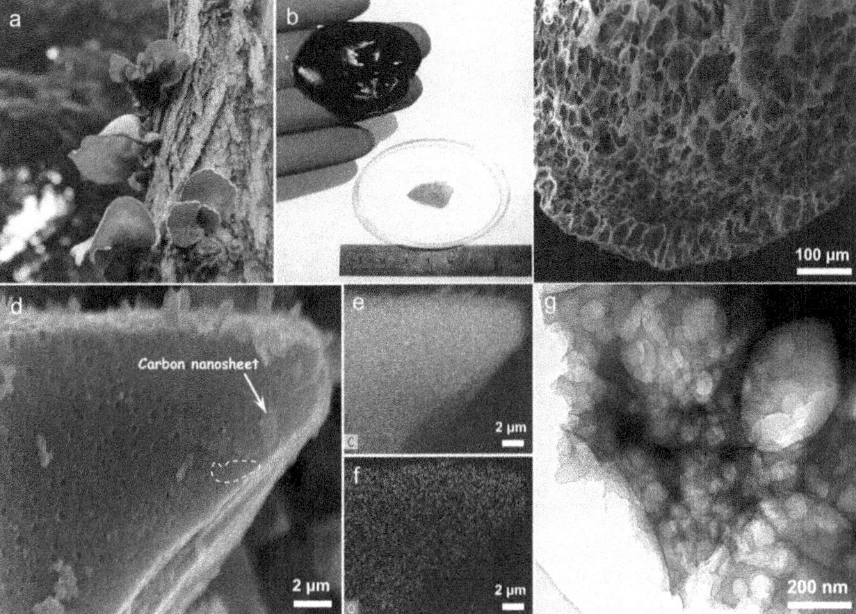

FIGURE 8.10 (a) Photograph of naturally abundant fungus; (b) photograph of dried fungus; (c) SEM image of freeze-dried fungus; (d) SEM image of the obtained porous graphene-like carbon materials; (e, f) corresponding elemental mapping images of C and O atom in (d); (g) HR-TEM image of the porous graphene-like carbon materials.[99]

et al.[108] reported that porous carbon using natural macroporous corn leaf (CL) as carbon precursor and H_3PO_4 as activation reagents, displayed an ultra-high surface area (2,507 $m^2 g^{-1}$) and a high specific capacitance of 230 F g^{-1} at 0.1 A g^{-1}. At the same time, it has excellent high-rate performance and good cycle stability. Generally, H_3PO_4-activated biomass-derived carbon materials display a smaller specific surface area and porosity compared to KOH. In addition, water vapor, CO_2, and NH_3 usually react with carbon atoms or other atoms to produce CO or H_2, which are conducive to pore formation and are often considered as physical activation. Yang's group[109] tried to obtain activated carbon from coconut shells using a one-step CO_2 activation. The ideal circumstances produced activated carbon with a micropore volume of 0.8949 $cm^3 g^{-1}$ and a BET surface area of 1,667 $m^2 g^{-1}$. Guo's group[111] reported the preparation of porous carbon by the CO_2 activation method, in which the spore precursor is pretreated, and carbonized for two hours in an Ar gas flow at 700°C, and then in a CO_2 gas flow at 900°C. The obtained activated carbon possesses a high specific surface area of 3,053 $m^2 g^{-1}$ and a sizable pore volume of about 1.43 $cm^3 g^{-1}$.

Overall, biomass-derived carbons are an effective strategy to handle difficult waste disposal issues and investigate outstanding performance energy storage materials. Developing biomass-derived porous carbon materials with high performance, high efficiency, and long lifetime through one or more carbonization processes has become the next research focus.

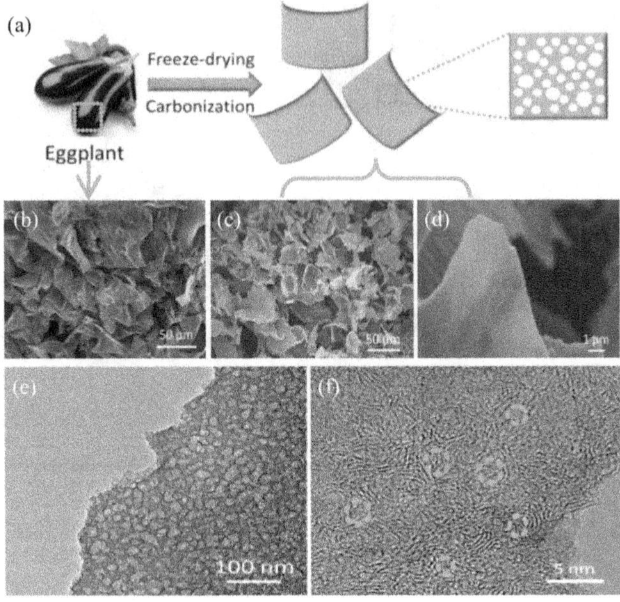

FIGURE 8.11 (a) Schematic of the preparation of nitrogen-doped porous carbon sheets from eggplant; (b) SEM of the dried eggplant; SEM images (c and d), and TEM images (e and f).[103]

3.1.5 Others carbon materials

Other carbon materials including carbon spheres,[119–122] carbon aerogels,[123–126] mesoporous carbon,[127, 128] and carbon nanofibers[129–134] have also been studied for supercapacitor applications. Carbon spheres/microspheres with unique structural and chemical properties possess a high surface area-to-volume ratio and also have excellent electrical conductivity.[119–122] Currently, carbon spheres as electrode materials have been studied extensively which can be synthesized by carbonization of organic precursors, template method, and hydrothermal synthesis. For example, Kim et al.[120] reported an easy approach to preparing pure carbon microspheres by a Friedel-Crafts reaction. The carbon spheres with hierarchical porous carbon can deliver good electric double-layer capacitors due to faster ion transport/diffusion and increased surface area. Qiao's group[119] fabricated the double-shelled hollow carbon spheres with a unique "in situ replicating" strategy (Figure 8.12a,b). After KOH activation, rich micropores and homogeneously N/O-doping could be introduced. Especially, the hollow macroporous cavity and inter-shell mesopores (Figure 8.12c,d) can act as "ion-buffering reservoirs" to minimize the diffusion distance of ions, providing a preeminent high-rate performance.[119] Finally, the hollow carbon spheres revealed a high capacitance value of 270 F g^{-1} at 90 A g^{-1}. The method presented by this work can be propagable for the synthesis of other electrode materials for high-performance supercapacitors.

Carbon aerogels with unique macro-, meso- and micropores are also promising electrode material[123–125] which were first synthesized by R. W. Pekala[124] via a sol-gel process using resorcinol and formaldehyde.[126] Various carbon aerogels including

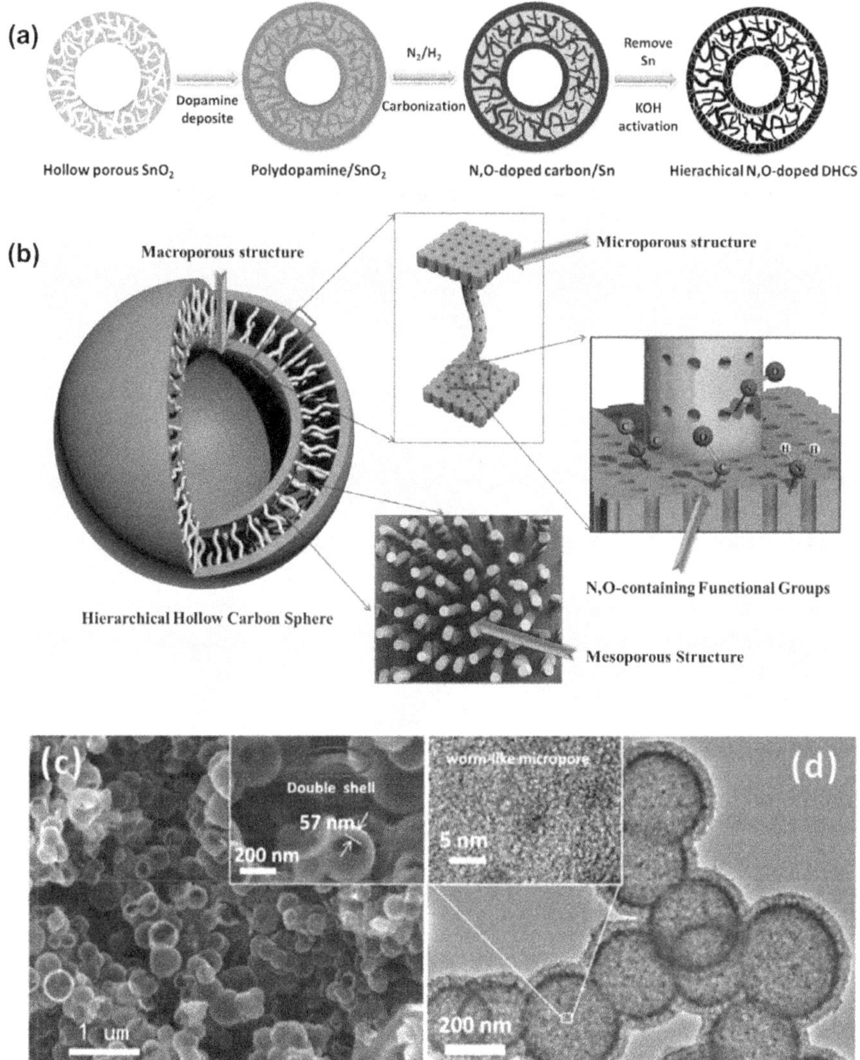

FIGURE 8.12 (a) The preparation schematic, (b) model diagram, (c) SEM, and (d) TEM of the hollow carbon spheres.[119]

CNTs carbon aerogels,[123] graphene carbon aerogels,[125] etc., have been applied to supercapacitor fields and usually provide a superior rate performance. Gogotsi's group[123] developed single-walled carbon nanotube aerogels with tunable shapes and sizes, as shown in Figure 8.13a,b. Electrochemical results (Figure 8.13c,d) indicated stable capacitive performance with a high-rate capability in common room temperature ionic liquid electrolyte. Qiao's group[125] prepared the graphene aerogels by reducing self-assembled from graphene oxide. These obtained graphene aerogels also contributed to the preeminent high-rate property for supercapacitors. In addition, carbon nanofibers are one-dimensional nanostructures derived from the pyrolysis of

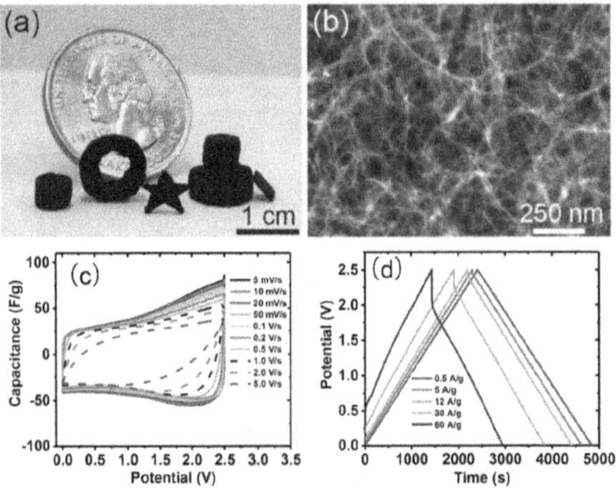

FIGURE 8.13 (a) Images of single-walled carbon nanotube aerogels in various sizes and shapes; (b) SEM image of single-walled carbon nanotube aerogels; (c) cyclic voltammograms and (d) galvanostatic charge/discharge of the single-walled carbon nanotube aerogels cells.[123]

organic materials[127, 128] and porous carbons showed an interconnected pore structure with high surface area and tunable pore size,[129–134] which also have been widely used as electrodes in supercapacitors.

3.2 Transition metal compounds

Transition metal compounds have been widely employed as essential electrode material[135–139] for the application of supercapacitors. The earliest transition metal oxide for supercapacitors is RuO_2. However, the scarcity and expensive price of Ru limited the practical application. Nowadays, a lot of research has been put toward developing alternative, low-cost electroactive materials based on transition metal hydroxides, oxides, sulfides, selenides, and phosphates.

3.2.1 Transition metal hydroxide

Transition metal hydroxide is a typical electrode material with distinctive layered structures, however, it presents poor conductivity and durability. Up to date, chemical doping, and structural adjustment have been verified as efficient methods for enhancing electrical conductivity and stability. For example, Gao's group[136] found that the single-layer β-$Co(OH)_2$ nanosheets had a high surface area (Figure 8.14) which displayed a large capacitance of 2,028 $F\,g^{-1}$. Additionally, the addition of metal cations can enhance the cyclic life of the electrode material and stabilize its crystal phase. Xie et al.[137] reported that Mg-$Ni(OH)_2$ with a stable structure was grown on nickel foam via $Mg(OH)_2$ sacrificial substrate and in situ ion exchange. After 10,000 cycles at 10 A g^{-1}, the Mg-$Ni(OH)_2$ preserves 95% of the original specific capacitance, whereas pure $Ni(OH)_2$ retains just 51% of the initial specific capacitance.

Compared to other metal hydroxide materials, the layered double hydroxide (LDH) can enhance supercapacitor performance due to the following

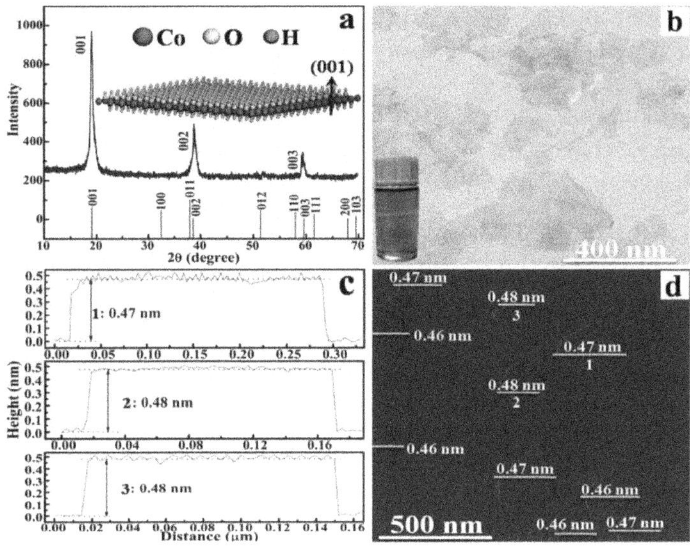

FIGURE 8.14 (a) XRD pattern, (b) TEM image, (c) height profiles derived from the atomic force microscopy image in (d) of the single-layer β-$Co(OH)_2$.[136]

reasons[140–145]: (1) Bimetals effectively stabilize the layered structure to obtain better cycle durability as well as rate performance; (2) Bimetals have a unique synergistic effect in improving the performance of supercapacitors. Chen's group[140] reported a simple method to grow Ni-Co LDH with ultra-thin nanosheets and porous nanostructures on nickel foam. The obtained supercapacitor showed an extremely high specific capacitance of 2,682 F g^{-1} at 3 A g^{-1} and an ultra-high-energy density of 77.3 Wh kg^{-1} at 623 W kg^{-1}. The ratio of nickel/cobalt can affect the thickness and properties of the nanosheet structure. Moreover, a hierarchically hollow sphere-shaped NiMn-LDH material was developed by Jai and his colleagues.[141] The optimized NiMn-LDH-12 with large surface area demonstrated an exceptional specific capacitance of 1,010.4 F g^{-1} at 0.2 A g^{-1} and a capacitance loss of only 30% at 5 A g^{-1} after 5,000 cycles.

3.2.2 Transition metal oxide

Transition metal oxides including NiO, CoO, Co_3O_4, and Fe_2O_3 have also been widely employed as electrode materials. Many endeavors have revealed that transition metal oxides possess desirable long-term charge-discharge cycles than metal hydroxides. For example, Ding's group[146] confirmed that the mesoporous hexagonal nanosheet-based flower-like Co_3O_4 material showed an excellent stability with capacitance retention of 96.07% after a 10,000 cyclic test at 5 A g^{-1}. Guan's group[147] also reported an excellent cycle stability of 2D MOF-derived Co_3O_4 with only 14.5% capacitance loss even after 20,000 cycles. However, the capacitive activity of transition metal oxides is reduced compared with hydroxides due to the feature of semiconductors or insulators. Interestingly, bimetallic or polymetallic oxides can exhibit higher electrical conductivity and capacitive activity than single-metal oxides.[148–151] For example, Wu's group[148] synthesized $NiCo_2O_4$@CNTs composite material network, which exhibited a high specific capacitance of 1,590 F g^{-1} at 0.5 A g^{-1}, excellent rate performance and favorable cycle stability with approximately 137% initial capacitance after 7,500 cycles. Puratchimani's group[152] reported $NiMn_2O_4$ nanostructures were prepared by hydrothermal method. The nanostructure $NiMn_2O_4$ reveals a higher specific capacitance (854.7 F g^{-1} at 5mV s^{-1} in 2M KOH aqueous electrolyte) and cycle stability (94.6% capacitance retention after 1,000 cycles at 4 A g^{-1}).

3.2.3 Transition metal sulfides

Transition metal sulfides are an important material for supercapacitors[153–159] which display a higher electrical conductivity than transition metal oxides. Huo's group[156] synthesized a Ni_3S_2 nanosheet array grown on nickel foam with a specific capacitance of 1,370.4 F g^{-1} at 2 A g^{-1}, which shows a good rate performance and cycle stability. CoS_2 hollow structures with adjustable interiors were easily synthesized by Peng's group using a simple solution-based approach.[157] The CoS_2 hollow spheres were obtained and exhibited an outstanding supercapacitor performance thanks to their high surface area, strong charge/discharge stability, and extended cycle life. Similar to bimetallic oxides and hydroxides, the abundance of active redox sites and increased electrical conductivity of bimetallic sulfides have drawn more interest as a viable electrode material. A S-vacancy-containing reduced $CoNi_2S_4$(r-$CoNi_2S_4$) ultra-thin nanosheet was effectively synthesized on nickel foam by Liu et al.[159] using a one-step hydrothermal process (Figure 8.15a–d). The specific capacitance (Figure 8.15e) of the r-$CoNi_2S_4$ composites was higher than that of the $CoNi_2S_4$ electrode (1,226 F g^{-1}) at a current density of 1 A g^{-1} (1,918.9 F g^{-1}). It also exhibits excellent rate performance and excellent cycling stability (Figure 8.15f), which is attributed to the formation of S-vacancies and ultra-thin nanostructures, which increase the active site on the surface of the nanostructure, and reduce the Gibbs free energy of the surface reaction.

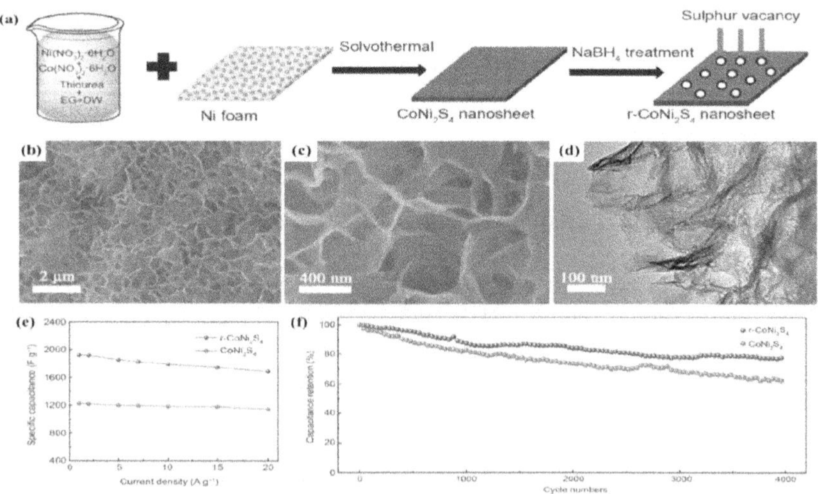

FIGURE 8.15 (a) The synthetic schematic illustration, (b,c) SEM images and (d) TEM image of the r-$CoNi_2S_4$ nanosheets. (e) specific capacity and (f) long-term cycling performance.[159]

3.2.4 Transition metal phosphates and selenides

Transition metal phosphates also store energy through redox reactions and exhibit high theoretical specific capacitance.[160–164] Gu et al.[161] reported that Ni-doped CoP (NiCoP) that has been encased in an amorphous carbon skeleton and is then adhered to CNT (Ni-CoP@C@CNT), which shows a considerable specific surface area, a significant amount of internal porosity and a continuous conducting network. Therefore, the NiCoP@C@CNT electrode showed a high specific capacitance of 708.1 F g^{-1} at 1 A g^{-1} and excellent rate performance (444.1 F g^{-1} at 20 A g^{-1}). Lan's group[162] prepared nickel foam (NF) with porous binary nickel-cobalt phosphide (NiCoP) nanosheet arrays. Compared with Ni-Co LDH@NF and $NiCoO_2$@NF, the NiCoP nanosheet arrays exhibited more outstanding electrochemical characteristics, due to the combination of composition characteristics and the advantages of array architecture, which have exhibited an ultra-high specific capacitance of 2,143 F g^{-1} at 1 A g^{-1} and retains 1,615 F g^{-1} at 20 A g^{-1}.

The electron negative of selenium is lower than that of oxygen and sulfur and thus transition metals selenides have better conductivity and are prone to redox reactions for excellent energy storage performance.[163–167] $Co_{0.85}Se$ nanowires were created on a carbon cloth substrate by Banerjee's group[166] using a "two-step" hydrothermal process, which produced an area-specific capacitance of 929.5 mF cm^{-2} at a current density of 1 mA cm^{-2}. Guo et al.[167] successfully synthesized Ni-Co-Se nanowire which showed excellent performance with a maximum energy density of 17 Wh kg^{-1} obtained at 1,526 W kg^{-1}.

3.3 Conductive Polymers

Conductive polymers,[168–172] as the typical class of pseudocapacitive materials, possess high electrical conductivity, which contribute to the fast charge transfer. The energy storage mechanism of conductive polymers can be described by the following formulas (Eqns 8.16 and 8.17):

$$\text{p-doping: conductive polymer} \rightarrow \text{conductive polymer}^{n+}*(A^-)_n + ne^- \quad (8.16)$$

$$\text{n-doping: conductive polymer} + ne^- \rightarrow \text{-do}^+)_n*\text{conductive polymer}^{n-} \quad (8.17)$$

where A^- and C^+ is the anion and cation in the electrolyte, respectively. Currently, many conductive polymers, including polyaniline (PANI),[168] Polypyrrole (PPy),[169] Polythiophene (PTh),[170] and others, have exhibited attractive electrochemical capacitance performances.

3.3.1 Polyaniline

Polyaniline (PANI) exhibits remarkable conductivity, environmental stability, and cost-effectiveness and has been considered a highly promising electrode material for pseudocapacitive supercapacitors. Theoretical specific capacitances of PANI are as high as 2,000 F g^{-1}.[173–175] However, the limited charge transfer rate of PANI leads to relatively lower specific capacitances, and the specific capacitances of PANI in the reported literature range from 200 to 1,000 F g^{-1}. [176–183] Currently, the preparation of polyaniline can be achieved through chemical oxidation, electrochemical synthesis, and biosynthesis.

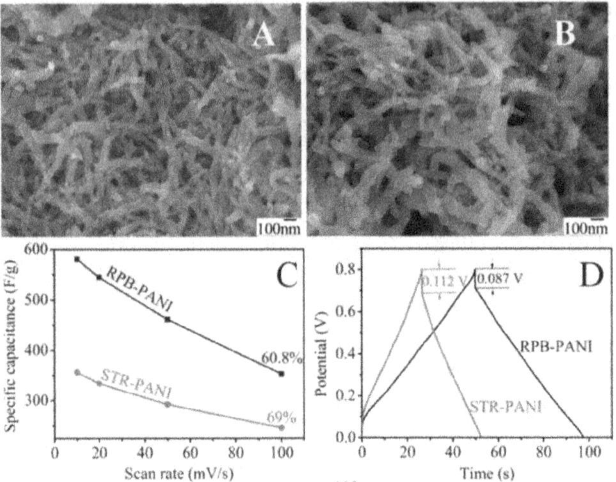

FIGURE 8.16 SEM images of (a) RPB-PANI and (b) STR-PANI; (c) specific capacitances of the as-prepared PANI nanofibers at different scan rates; (d) GCD curves of RPB-PANI and STR-PANI at 10 A g^{-1}.[185]

Chemical oxidation remains one of the widely adopted techniques for obtaining PANI.[184–188] For example, Wen et al.[184] successfully achieved the synthesis of clip-like PANI nanofibers through the incorporation of a mixed surfactant system within a wet chemical approach. The resultant nanofibers exhibit exceptional electro rheological properties, owing to their distinctive structural characteristics. Zhao et al.[185] synthesized PANI nanofibers through high gravity chemical oxidative polymerization (HGCOP) in a rotating packed bed (RPB). The resulting nanofibers exhibited superior specific capacitance of 667.6 F g^{-1} (Figure 8.16a,b) compared to 375.9 F g^{-1} of those synthesized in a stirred tank reactor (STR) at 10 A g^{-1} (Figure 8.16c,d). Additionally, the PANI nanofibers demonstrated excellent capacitance retention over repeated cycling. Electrochemical synthesis offers several advantages, including rapid reaction speed and high product purity. Bhandari et al.[187] synthesized polyaniline from aniline sulfate (AS) by electrochemical synthesis in an acidic medium. The electrochemical and supercapacitive behavior of the synthesized PANI showed significant improvement. Additionally, Navale et al.[189] prepared nodule-like PANI electrodes using the electrodeposition method, which exhibited excellent specific capacitance of 508 Fg^{-1} in acidic electrolyte and high specific power and specific energy values of 13.39 kW kg^{-1} and 32.12 Wh kg^{-1}, respectively. In addition, biosynthesis represents a novel preparation method involving the use of microorganisms or enzymes.[190–192] For instance, Cholli et al.[191] successfully synthesized conductive PANI nanoparticles using a biocatalytic method with a simple two-step enzymatic approach. The obtained PANI has electroactive properties and a great potential for electronic applications.

3.3.2 Polypyrrole

Polypyrrole (PPy) has attracted extensive attention due to its low cost, excellent flexibility, high conductivity, and favorable redox properties.[193–198] Yuan et al.[193]

successfully prepared spherical silver-doped conductive PPy composites by oxidation polymerization of pyrrole in aqueous solution containing Ag^+. The citrate produced stable and charged micelles by adjusting the surface polymerization of pyrrole. Grijalva-Bustamante et al.[194] reported the use of the bile salt sodium taurine cholate as a surfactant for the synthesis of electroactive PPy nanoparticles, in which hydrogen peroxide was used as a green oxidizer. In addition, Zhang et al.[198] synthesized a horn-shaped micro-nano polypyrrole film (GA-hPPy) with high electrical activity by pulsed potentiostatic method. At a scanning rate of 10 mV s^{-1}, the specific capacitance can reach 360 F g^{-1}. It has been proved that after 10,000 cycles, its capacitive value can be maintained at 88.2%, implying excellent cyclic stability.

3.3.3 Polythiophene

Among them, Polythiophene (PTh), as a new type of organic polymer, has high electrical conductivity and good optical properties, and is suitable for electronic devices and optoelectronic applications.[199–204] In addition, polythiophene can be prepared into flexible films for the application of flexible electronics and wearable devices. However, the complex synthetic process of polythiophene usually requires harsh techniques and conditions. Recently, a template method with the non-oxidizing process has been developed which possesses high molecular weight and good conductivity. For instance, Kitao et al.[200] used a porous coordination polymer (PCP) as a template to achieve a highly ordered chain arrangement of non-substituted PTh. Due to the extended conjugation system, the PTh particles have higher conductivity than the PTh obtained by solution polymerization. Moreover, Yoshino et al.[201] synthesized a conducting polymer using a chemical oxidation method with $FeCl_3$ as a catalyst, resulting in a conductivity of 10^{-9} S cm^{-1}. Gök et al.[202] prepared polythiophene in a $CHCl_3$ system using $FeCl_3$ as an oxidant, achieving a conductivity of 1.7×10^{-5} S cm^{-1}. The addition of different surfactants can alter the morphology and enhance the conductivity. Vijeth et al.[205] introduced a novel polythiophene/Al_2O_3 (PTCA) composite through the process of camphorsulfonic acid-assisted chemical polymerization of thiophene. The electrochemical investigations revealed that the PTCA composite with a 50% Al_2O_3 content (PTCA5) exhibited a remarkable specific capacitance of 780.40 F g^{-1} at 0.5 A g^{-1}, signifying its superior performance as an electrode material for energy storage applications.

3.4 Carbon-based composites

Due to their excellent features, carbon materials including nanocarbon materials, activated carbon, and graphene were commonly used as substrates to synthesize carbon-based composites. Many researchers have attempted to explore synergistic advantages by combining different materials to enhance the overall electrochemical performance of electrode materials. Here, the CNT-based composites, CNFs-based composites and graphene-based composites were discussed in detail.

3.4.1 CNTs and their composite materials

CNTs exhibit excellent one-dimensional electrical conductivity, high mechanical stability, and remarkable catalytic activity, which make them suitable as substrates for the preparation of electrode composites for supercapacitors. Recently, CNTs-based metal oxides or metal sulfides have shown enhanced electrochemical properties.[206–209] Wang et al.[206] successfully developed the CNT-MnO_2 as the electrode material to construct the stretchable linear supercapacitor, which exhibited remarkable electrochemical performance. It exhibited a high specific capacitance of 77.83 F g^{-1} (9.03 F cm^{-3}) at a current density of 0.689 A g^{-1} (0.08 A cm^{-3}) and retained good capacity even after 8,000 galvanostatic charge-discharge cycles. Furthermore, this asymmetrical supercapacitor demonstrated a stable output voltage of 1.7 V, a power density of 571.3 W kg^{-1}, and an energy density of 29.84 Wh kg^{-1}. Lv et al.[207] prepared CNT/MoS_2 materials by chemically adsorbing MoS_2 solution onto CNT films using chemical vapor deposition. When used as the electrode for supercapacitors, the CNT/MoS_2 material exhibited a high specific capacitance of 13.16 F cm^{-3}. Lai et al.[209] utilized a network structure of conducting CNTs as a support structure and nickel oxide as the active material of the supercapacitor to prepare CNTs@NiO composites. It resulted in a synergistic effect and yielded a high specific capacitance of up to 713.9 F/g. Li et al.[210] prepared $Co_2(CO_3)(OH)_2$@CNTs composite material through a hydrothermal method, which was then converted into $CoS_{1.29}$@CNTs-6.1% through sulfidation treatment. The specific capacitance of the composite carbon nanotube material reached 99.7 mAh g^{-1}, benefiting from the enhanced ion diffusion rate due to the increased specific surface area of carbon nanotubes. The supercapacitors assembled from this material retained 91.3% of their initial specific capacitance after 2,000 cycles. Hou et al.[211] created a hologram-like γ-MnS@CNT nanocomposite by coating γ-MnS nanoparticles on a carbon nanotube (CNT) skeleton (Figure 8.17a,b). The resulting γ-MnS@CNT electrode exhibited excellent electrochemical properties, showcasing a high specific capacitance of 641.9 F g^{-1} (Figure 8.17c,d) and remarkable cycle stability.

3.4.2 Carbon nanofibers and their composite materials

Carbon Nanofibers (CNFs) also exhibit superior electrochemical performance compared to many other carbon materials.[212–219] So far, CNFs have been explored

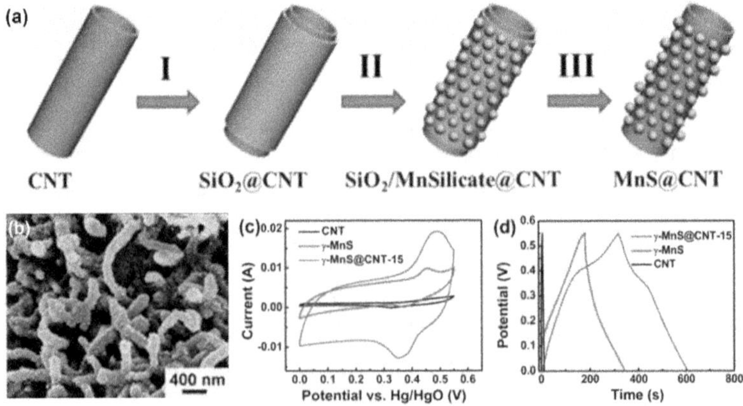

FIGURE 8.17 (a) Schematic fabrication mechanism and (b) SEM image of the MnS@CNT; (c) CV curves at 50 mV s^{-1} and (d) GCD curves at 1 A g^{-1}.[211]

as substrates for the preparation of electrode composites. Levitt et al.[214] prepared binder-free supercapacitor electrodes by incorporating $Ti_3C_2T_x$ MXene into carbon nanofibers using electrospinning. This composite achieved a high areal capacitance of up to 244 mF cm^{-2}, which is three times higher than pure carbonized PAN nanofibers (70 mF cm^{-2}). Liu et al.[217] successfully synthesized flexible and porous Co_3O_4-carbon nanofibers (Co_3O_4-CNFs) using electrospinning, followed by carbonization. These optimized Co_3O_4-CNFs, utilized as binder-free electrodes, demonstrated a specific capacitance of 369 F g^{-1} at a current density of 0.1 A g^{-1}. Even at a high current density of 2 A g^{-1}, the specific capacitance retained 181 F g^{-1}, suggesting an impressive rate property. Xu et al.[218] employed electrospinning technology to prepare $LiMn_2O_4$-modified polyacrylonitrile nanofibers (Figure 8.18a). Subsequently, the MnO/CNF composites were synthesized by annealing treatment of $LiMn_2O_4$-modified polyacrylonitrile nanofibers. The SEM images in Figure 8.18b–e showed the uniform CNF structure with a rough surface. The MnO/CNF composites delivered a high capacitance of 409.7 F g^{-1} at 1 A g^{-1}. Meanwhile, the MnO/CNF composites were used as the negative electrode and the positive electrode to assemble the supercapacitor cell (Figure 8.18f). The supercapacitor achieved a working voltage of 1.8 V and exhibited excellent stability with a capacitance retention of 94.2% after 10,000 cycles (Figure 8.18g) and could power an LED light (Figure 8.18h).

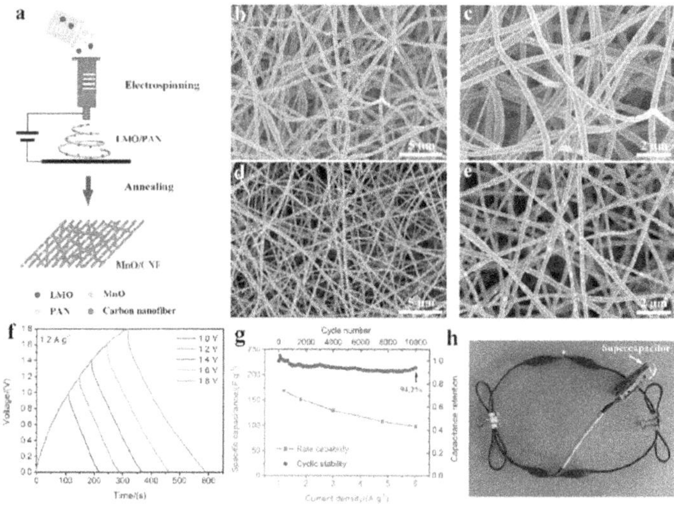

FIGURE 8.18 (a) Preparation scheme for MnO/CNF composite. SEM images of CNFs with (b) low and (c) high magnifications. SEM images of MnO/CNF composite with (d) low and (e) high magnifications. (f) GCD curves charged to different voltages of an asymmetric supercapacitor with MnO/CNF composite as both negative and positive electrodes. (g) Rate capability and cyclic stability of the asymmetric supercapacitor. (h) Photo of an LED lightened by the supercapacitor.[218]

3.4.3 Graphene and their composite materials

Graphene is a 2D hexagonal carbon nanomaterial with a honeycomb lattice structure comprising sp^2 hybridized carbon atoms. Its exceptional properties, such as a large theoretical specific surface area, high electrical conductivity, and remarkable thermal and chemical stability, make it a promising electrode material for enhancing supercapacitor performance.[220–224] Zhang et al.[220] successfully prepared a nanocomposite of reduced graphene oxide/lanthanum oxide (rGO/La_2O_3) through a straightforward reflux process. This rGO/La_2O_3 composite exhibited an impressive specific capacitance of 156.25 F/g at a current density of 0.1 A/g, as well as an outstanding cycling stability. Even after 500 charge-discharge cycles, the electrode material retained 78% of its initial charge-discharge efficiency. Zhao et al.[221] reported the preparation of nanostructured composites consisting of reduced graphene oxide wrapped around porous disk-like $Co_xNi_{1-x}(OH)_2$ structures using hydrazine hydrate reduction. The porous structure significantly enhances the permeation of electrolyte and cycling life for 20,000 cycles at 20 A g^{-1}. Li et al.[223] successfully prepared the layered $NiCo_2O_4$/RGO nanocomposite (Figure 8.19a), which showed an ultra-high specific capacity of 1,388 F g^{-1} at 0.5 A g^{-1} and an ultra-long cycle life with 90.2% capacity retention after 20,000 cycles at 5 A g^{-1}. They assembled an asymmetric supercapacitor with the $NiCo_2O_4$/RGO composites as the positive electrode and activated carbon as the negative electrode (Figure 8.19b), which achieved a high-power density of 375 W kg^{-1} and a high-energy density of 57 Wh kg^{-1} (Figure 8.19c–f). Additionally, the $NiCo_2O_4$/RGO/AC devices exhibited good stability (Figure 8.19g) and were able to power LEDs, suggesting potential applications in energy storage. Wang et al.[224] introduced a scalable synthesis method for producing mixed-valence manganese oxide nanoparticles anchored to reduced graphene oxide (rGO/MnO_x), which served as a high-performance electrode material for supercapacitors. The rGO/MnO_x displayed a high capacitive value of 202 F g^{-1} and an impressive areal-specific capacitance of 2.5 F cm^{-2} with exceptional long-term cycling life.

4 THE APPLICATION OF SUPERCAPACITORS

Supercapacitor possesses high-power density and long cyclic stability but low energy density compared to batteries. Therefore, applications that require a higher burst of energy in a shorter time and continuous cyclic operations are suited to supercapacitors. Currently, supercapacitors have been used in the fields of transportation, industry, communication, medical devices, and life.

4.1 Transportation

In electric vehicles, supercapacitors play a crucial role in starting, acceleration, and braking. When the vehicle starts, the extra energy stored in the supercapacitor can be discharged to ensure the acceleration performance of the electric vehicle. The supercapacitor also can receive more energy during braking for the next start and others. It can increase the cyclic life of the battery and improve battery performance. The use of ultracapacitors in railways and buses is another important application in the transport sector. The frequent braking of traditional fuel vehicles is bound to cause energy

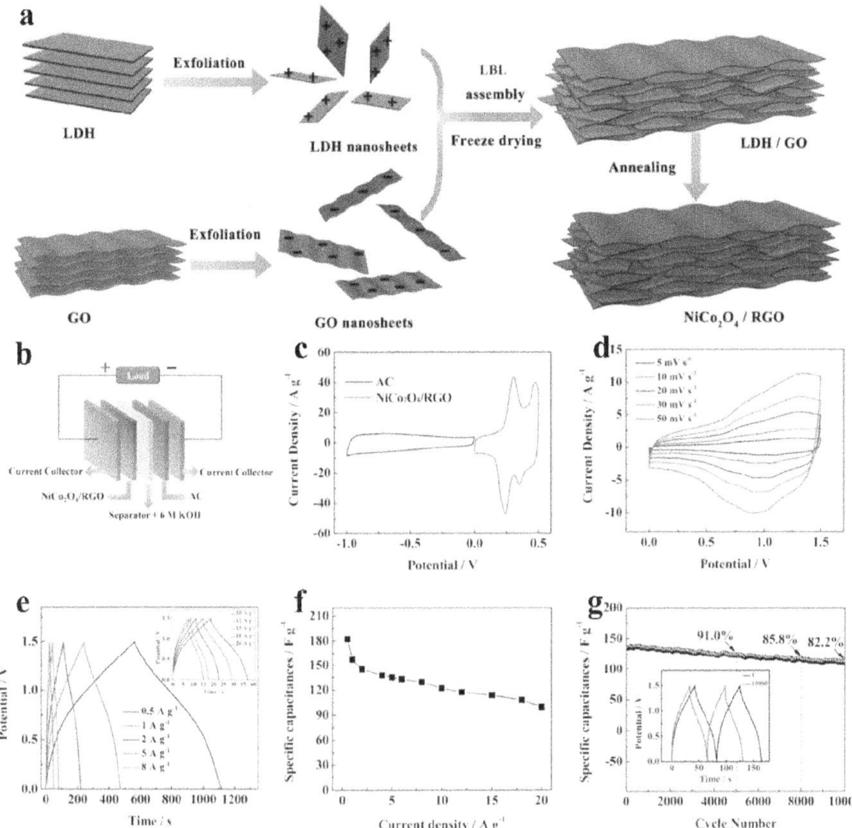

FIGURE 8.19 (a) Schematic of the fabrication of layered $NiCo_2O_4$/RGO composites. (b) Schematic illustration of the asymmetric supercapacitor. (c) CV curves of the $NiCo_2O_4$/RGO and AC electrodes at 20 mV s^{-1}. (d) CV curves at different scan rates from 5 to 50 mV s^{-1}. (e) GCD curves at different current densities. (f) Corresponding specific capacitance as a function of current density. (g) Cycling performance at a current density of 5 A g^{-1} for 10,000 cycles.[223]

wastage, which not only reduces the economy of the vehicle but also generates more harmful gases. These problems can be solved by using supercapacitor modules for the energy storage and power systems of buses. Meanwhile, ultracapacitors are also used in forklifts and excavators.

4.2 Industrial and Military/Aerospace Fields

As the rapid development of industrial modernization, the consumption of energy continuously increases. The usage of supercapacitors can efficiently save energy and enhance economic efficiency. In lifting cranes, a large amount of gravitational potential energy is released at the lowering of heavy objects. During the lowering of heavy things, supercapacitors are ideal as energy recovery devices where the recovered energy can be used in the braking period of lifting cranes, resulting in fuel energy-saving up to 40%.

Supercapacitors can be used in combination with other energy storage devices. For example, wind/photovoltaic power generation has intermittent and unstable features. In combination with supercapacitors and wind/photovoltaic power generation, supercapacitors can be used as a backup power source to provide transient power compensation for improving the stability and reliability of power generation. In addition, stable power output is essential for the safe and economic operation of the power grid, as well as reducing energy consumption, etc.

Aeronautical equipment requires safe and reliable energy storage units even under harsh working conditions such as wide operating temperatures, longer cyclic life, and higher power density. Supercapacitors can meet the needs of aerospace equipment to a certain extent and can be used as power supply devices and emergency power for aerospace. In wireless systems, the safety of radio frequencies can be improved by the application of supercapacitors. Supercapacitors with fast charge-discharge characteristics can provide instantaneous high-power density for portable electronic devices. The new high-powered military equipment must combine supercapacitors with conventional high-energy batteries to meet the practical operational requirements. In addition, military trucks, armored vehicles, and electric vehicles are required to use a combination of battery and supercapacitor.

4.3 Electronic Communication

In communications applications, supercapacitors are feasible as energy storage devices for high-power pulse applications such as portable devices for global systems for mobile communications (GSM) and general packet radio services (GPRS).

4.4 Life

Now, the supercapacitor has become an essential component in daily life and has been applied in light/storage lamps, portable electronic devices, warning, and real-time counters. In many electric toys, the supercapacitor has the advantages of lightweight, small size, high-power density, and rapid start-up and has attracted more attention. In addition, supercapacitors are also often used in traffic lights, solar watches, traffic signals, cameras, etc. Flexible supercapacitors are also gradually being used in new electronic products such as foldable mobile phones and smart watches. In particular, printable flexible solid-state supercapacitors are expected to have great potential for application in the field of wearable electronic devices.

5 CONCLUSIONS AND PROSPECTS

In conclusion, supercapacitors are essential in energy storage systems for a variety of applications. Recent research advances of supercapacitors in the fundamentals, electrode materials, and applications have been reviewed in this context. The electrode materials played a significant role in mechanisms, properties, and cyclic stability. In general, the energy storage of carbon materials is based on the principle of EDLCs, where ions in the electrolyte can be adsorbed onto the surface of the carbon electrode materials, forming a layer of charge on the electrode surface. This process is reversible and occurs rapidly, accordingly exhibiting the ultra-high cyclic stability. Carbon materials have many unique properties, including high surface area, outstanding conductivity, and structural tunability, which allows for rapid charge transfer between the electrode and the electrolyte. In addition, to further enhance the performance, a great deal of effort has been put into the design of various strategies including increased specific surface area, doping/surface modification with functional groups, the construction of porous structure, composites, and the introduction of different electrolytes, etc. Due to the attractive mechanical features, the carbon-based materials also allow for the design of flexible and lightweight supercapacitors.

The energy storage of the transition compounds and conducting polymer is based on the mechanism of pseudocapacitance, where the redox reactions take place at the electrode/electrolyte interface, resulting in the transfer of charge between the electrode and the electrolyte. This process provides a much larger capacitance for enhancing the energy storage capability than the carbon materials. However, some transition metal compounds have a narrow voltage range which can limit their use in certain applications. Significantly, the Faradic redox reactions of the transition compounds and conducting polymer usually give rise to low cyclic stability compared with the carbon-based materials. The unique morphology, doping, carbon-encapsulated, and complex effects can efficiently promote the cyclic stability of the transition compounds and conducting polymer.

Based on the above discussion, impressive progress has been reached in the past decades. Many aspects still need to be further explored to achieve higher electrochemical performance. Here, some perspectives and opportunities for future research directions have been summarized as follows:

1. The charge storage mechanisms with advanced in situ techniques and theoretical computational simulation: The fundamental charge storage mechanism is favorable for understanding the reaction disciplines at the electrolyte/electrode interfaces and enhancing the capacitive performance. Now, in situ characterization techniques and theoretical computational simulation have been developed rapidly and applied to various fields in recent years. Especially, significant achievements in the mechanism research of catalysis have been reached. Thus, more attention should be paid to the combination among experimental results, in situ characterization techniques and theoretical calculations to achieve new breakthroughs in energy storage mechanisms in the future.
2. The development of new materials for higher energy density: compared to batteries, supercapacitors have an excellent power density but low energy density. The development of novel materials with improved energy density, power density, and stability is crucial for advancing supercapacitors. Currently, in addition to carbon materials and transition metal compounds, new materials such as MXenes, metal-organic frameworks, and covalent organic frameworks have been found and show potential in supercapacitors. However, the relatively low conductivity in metal-organic frameworks and

covalent organic frameworks or the low capacitance in MXenes limits their further applications. Thus, the design and development of new materials is crucial for advanced supercapacitors with higher energy density for practical application.

3. Electrolyte optimization: Electrolytes play a crucial role in the performance of supercapacitors including ion-conductivity, capacitance, cyclic life, operated voltage, and product cost. The choice of electrolyte is one of the key parameters in the efficiency, durability, and cost of high-performance supercapacitors. So far, aqueous and organic electrolytes are two fundamental types of electrolytes for supercapacitors. Aqueous electrolytes are sulfuric acid solutions or potassium hydroxide solutions, which are more corrosive and the working voltage of the single supercapacitors is only about 1.5 V. Organic electrolytes are organic solutions of electrolytes such as tetraethylammonium tetrafluoroborate, and the working voltage of single supercapacitors is above 2.5 V. However, there exists the easy volatilization of organic solvents, potential safety hazards, and environmental pollution, limiting their application in high-voltage and high-power supercapacitors. Recently, solid electrolytes are considered as a promising alternative to conventional liquid electrolytes for supercapacitors. They offer several advantages including improved safety, wide operating temperature and voltage ranges, and long cycle life. Current research on solid electrolytes for supercapacitors focuses on polymer electrolytes, ionic liquid polymer electrolytes, and inorganic solid electrolytes. However, many challenges remain including low ionic conductivity, complicated synthesis, and poor interface contact with electrodes. Solid electrolytes are an important direction for supercapacitor research and development.

ACKNOWLEDGMENTS

This work was supported by the National Natural Science Foundation of China (22202042), and the Guangdong Basic and Applied Basic Research Foundation (2021A1515110058 and 2022A1515140012).

REFERENCES

1. D.P. Chatterjee and A.K. Nandi, A review on the recent advances in hybrid supercapacitors, *J. Mater. Chem. A*, 2021, **9**, 15880–15918.
2. Y. Wang, Y. Song and Y. Xia, Electrochemical capacitors: mechanism, materials, systems, characterization and applications, *Chem. Soc. Rev.*, 2016, **45**, 5925–5950.
3. D. Zhou, P. Cheng, J. Luo, W. Xu, J. Li and D. Yuan, Facile synthesis of graphene@ $NiMoO_4$ nanosheet arrays on Ni foam for a high-performance asymmetric supercapacitor, *J. Mater. Sci.*, 2017, **52**, 13909–13919.
4. T. Bao, J. Wangand and C. Liu, Recent advances in epitaxial heterostructures for electrochemical applications, *Nanosc. Adv.*, 2023, **5**, 313–322.
5. M. Klimpel, M.V. Kovalenko and K.V. Kravchyk, Advances and challenges of aluminum-sulfur batteries, *Commun. Chem.*, 2022, **5**, 77.

6. L. Wang, Y. Zhang and P.G. Bruce, Batteries for wearables, *Natl Sci. Rev.*, 2023, **10**, nwac062.
7. Z. Hu, S. Li, P. Cheng, W. Yu, R. Li, X. Shao, W. Lin and D. Yuan, N,P-co-doped carbon nanowires prepared from bacterial cellulose for supercapacitor, *J. Mater. Sci.*, 2015, **51**, 2627–2633.
8. R. Li, Z. Hu, X. Shao, P. Cheng, S. Li, W. Yu, W. Lin and D. Yuan, Large Scale Synthesis of NiCo Layered Double Hydroxides for Superior Asymmetric Electrochemical Capacitor, *Sci. Rep.*, 2016, **6**, 18737.
9. A.G. Olabi, Q. Abbas, A. Al Makky and M.A. Abdelkareem, Supercapacitors as next generation energy storage devices: Properties and applications, *Energy*, 2022, **248**, 123617.
10. D.-G. Wang, Z. Liang, S. Gao, C. Qu and R. Zou, Metal-organic framework-based materials for hybrid supercapacitor application, *Coord. Chem. Rev.*, 2020, **404**, 213093.
11. W. Yu, W. Lin, X. Shao, Z. Hu, R. Li and D. Yuan, High performance supercapacitor based on Ni_3S_2/carbon nanofibers and carbon nanofibers electrodes derived from bacterial cellulose, *J. Power Sources*, 2014, **272**, 137–143.
12. P. Simon and Y. Gogotsi, Materials for electrochemical capacitors, *Nat. Mater.*, 2008, **7**, 845–854.
13. Y. Shao, M.F. El-Kady, J. Sun, Y. Li, Q. Zhang, M. Zhu, H. Wang, B. Dunn and R.B. Kaner, Design and Mechanisms of Asymmetric Supercapacitors, *Chem. Rev.*, 2018, **118**, 9233–9280.
14. Y. Gogotsi and R.M. Penner, Energy Storage in Nanomaterials-Capacitive, Pseudocapacitive, or Battery-like?, *ACS Nano*, 2018, **12**, 2081–2083.
15. V. Augustyn, P. Simon and B. Dunn, Pseudocapacitive oxide materials for high-rate electrochemical energy storage, *Energy Environ. Sci.*, 2014, **7**, 1597.
16. D. Galizzioli, F. Tantardini and S. Trasatti, Ruthenium dioxide: a new electrode material. I. Behaviour in acid solutions of inert electrolytes, *J. Appl. Electrochem.*, 1974, **4**, 57–67.
17. K.V. Kravchyk and M.V. Kovalenko, Perspective on design and technical challenges of Li-garnet solid-state batteries, *Sci. Technol. Adv. Mater.*, 2022, **23**, 2018919.
18. L. Li, Q. Zhang, B. He, R. Pan, Z. Wang, M. Chen, Z. Wang, K. Yin, Y. Yao, L. Wei and L. Sun, Advanced Multifunctional Aqueous Rechargeable Batteries Design: From Materials and Devices to Systems, *Adv. Mater.*, *Adv. Mater.*, 2022, **34**, 2104327.
19. S.D. Zhang, M.Y. Qi, S.J. Guo, Y.G. Sun, X.X. Tan, P.Z. Ma, J.Y. Li, R.Z. Yuan, A.M. Cao and L.J. Wan, Advancing to 4.6 V Review and Prospect in Developing High-Energy-Density $LiCoO_2$ Cathode for Lithium-Ion Batteries, *Small Methods*, 2022, **6**, 2200148.
20. D. Chao and H.J. Fan, Intercalation Pseudocapacitive Behavior Powers Aqueous Batteries, *Chem*, 2019, **5**, 1359–1361.
21. H.J. Kim, J.S. Yeon, H.R. Park, S.J. Lee, W.I. Kim, G. Jang and H.S. Park, Intercalation Pseudocapacitance of Cation-Exchanged Molybdenum-Based Polyoxometalate for the Fast and Stable Zinc-Ion Storage, *ACS Appl. Mater. Interfaces*, 2023, **15**, 9350–9361.
22. Y. Liu, S.P. Jiang and Z. Shao, Intercalation pseudocapacitance in electrochemical energy storage: recent advances in fundamental understanding and materials development, *Mater. Today Adv.*, 2020, **7**, 100072.
23. Z. Wang, R. Li, C. Su and K.P. Loh, Intercalated phases of transition metal dichalcogenides, *SmartMat*, 2020, **1**, e1013.
24. M. Yang and Z. Zhou, Recent Breakthroughs in Supercapacitors Boosted by Nitrogen-Rich Porous Carbon Materials, *Adv. Sci.*, 2017, **4**, 1600408.
25. Z. Zhai, L. Zhang, T. Du, B. Ren, Y. Xu, S. Wang, J. Miao and Z. Liu, A review of carbon materials for supercapacitors, *Mater. Design*, 2022, **221**, 111017.

26. L.L. Zhang and X.S. Zhao, Carbon-based materials as supercapacitor electrodes, *Chem. Soc. Rev.*, 2009, **38**, 2520–2531.
27. R.Z. Ma, J. Liang, B.Q. Wei, B. Zhang, C.L. Xu and D.H. Wu, Study of electrochemical capacitors utilizing carbon nanotube electrodes, *J. Power Sources*, 1999, **84**, 126–129.
28. C. Niu, E.K. Sichel, R. Hoch, D. Moy and H. Tennent, High power electrochemical capacitors based on carbon nanotube electrodes, *Appl. Phys. Lett.*, 1997, **70**, 1480–1482.
29. M.H. Ervin, B.S. Miller, B. Hanrahan, B. Mailly and T. Palacios, A comparison of single-wall carbon nanotube electrochemical capacitor electrode fabrication methods, *Electrochim. Acta*, 2012, **65**, 37–43.
30. K.H. An, W.S. Kim, Y.S. Park, J.M. Moon, D.J. Bae, S.C. Lim, Y.S. Lee and Y.H. Lee, Electrochemical Properties of High-Power Supercapacitors Using Single-Walled Carbon Nanotube Electrodes, *Adv. Funct. Mater.*, 2001, **11**, 387–392.
31. E. Frackowiak, K. Metenier, V. Bertagna and F. Beguin, Supercapacitor electrodes from multiwalled carbon nanotubes, *Appl. Phys. Lett.*, 2000, **77**, 2421–2423.
32. C.G. Liu, M. Liu, F. Li and H.M. Cheng, Frequency response characteristic of single-walled carbon nanotubes as supercapacitor electrode material, *Appl. Phys. Lett.*, 2008, **92**, 143108.
33. Q. Zhang, C. Cai, J. Qin and B. Wei, Tunable self-discharge process of carbon nanotube based supercapacitors, *Nano Energy*, 2014, **4**, 14–22.
34. J.H. Kim, Y.-i. Ko, Y.A. Kim, K.S. Kim and C.-M. Yang, Sulfur-doped carbon nanotubes as a conducting agent in supercapacitor electrodes, *J. Alloys Comp.*, 2021, **855**, 157282.
35. W. Lu, L. Qu, K. Henry and L. Dai, High performance electrochemical capacitors from aligned carbon nanotube electrodes and ionic liquid electrolytes, *J. Power Sources*, 2009, **189**, 1270–1277.
36. B. Xu, F. Wu, F. Wang, S. Chen, G.-P. Cao and Y.-S. Yang, Single-walled Carbon Nanotubes as Electrode Materials for Supercapacitors, *Chinese J. Chem.*, 2006, **24**, 1505–1508.
37. G. Wang, R. Liang, L. Liu and B. Zhong, Improving the specific capacitance of carbon nanotubes-based supercapacitors by combining introducing functional groups on carbon nanotubes with using redox-active electrolyte, *Electrochim. Acta*, 2014, **115**, 183–188.
38. M. Sevilla, L. Yu, L. Zhao, C.O. Ania and M.-M. Titiricic, Surface Modification of CNTs with N-Doped Carbon: An Effective Way of Enhancing Their Performance in Supercapacitors, *ACS Sustain. Chem. Eng.*, 2014, **2**, 1049–1055.
39. T. Zhu, J. Zhou, Z. Li, S. Li, W. Si and S. Zhuo, Hierarchical porous and N-doped carbon nanotubes derived from polyaniline for electrode materials in supercapacitors, *J. Mater. Chem. A*, 2014, **2**, 12545.
40. A.R. John and P. Arumugam, Open ended nitrogen-doped carbon nanotubes for the electrochemical storage of energy in a supercapacitor electrode, *J. Power Sources*, 2015, **277**, 387–392.
41. S. Zong, J. Du, A. Chen, X. Gao, K.O. Otun, X. Liu and L.L. Jewell, N-doped crumpled carbon nanotubes as advanced electrode material for supercapacitor, *J. Alloys Comp.*, 2022, **928**, 167222.
42. Z. Liu, S. Zhang, L. Wang, T. Wei, Z. Qiu and Z. Fan, High-efficiency utilization of carbon materials for supercapacitors, *Nano Select.*, 2020, **1**, 244–262.
43. Y. Zheng, K. Chen, K. Jiang, F. Zhang, G. Zhu and H. Xu, Progress of synthetic strategies and properties of heteroatoms-doped (N, P, S, O) carbon materials for supercapacitors, *J. Energy Stor.*, 2022, **56**, 105995.
44. J. Patiño, N. López-Salas, M.C. Gutiérrez, D. Carriazo, M.L. Ferrer and F.D. Monte, Phosphorus-doped carbon–carbon nanotube hierarchical monoliths as true three-dimensional electrodes in supercapacitor cells, *J. Mater. Chem. A*, 2016, **4**, 1251–1263.

45. W. Chen, Z. Zhao and X. Yu, Phosphorus-modulated controllably oxidized carbon nanotube architectures for the ultrahigh energy density of pseudocapacitive capacitors, *Electrochim. Acta*, 2020, **341**, 136044.
46. G. Ma, G. Ning and Q. Wei, S-doped carbon materials: Synthesis, properties and applications, *Carbon*, 2022, **195**, 328–340.
47. W. Yang, M. Ni, X. Ren, Y. Tian, N. Li, Y. Su and X. Zhang, Graphene in Supercapacitor Applications, *Curr. Opin. Coll. Interface Sci.*, 2015, **20**, 416–428.
48. M. Horn, B. Gupta, J. MacLeod, J. Liu and N. Motta, Graphene-based supercapacitor electrodes: Addressing challenges in mechanisms and materials, *Curr. Opin. Green Sustain. Chem.*, 2019, **17**, 42–48.
49. X. Shi, S. Zheng, Z.-S. Wu and X. Bao, Recent advances of graphene-based materials for high-performance and new-concept supercapacitors, *J. Energy Chem.*, 2018, **27**, 25–42.
50. S.R.C. Vivekchand, C.S. Rout, K.S. Subrahmanyam, A. Govindaraj and C.N.R. Rao, Graphene-based electrochemical supercapacitors, *J. Chem. Sci.*, 2008, **120**, 9–13.
51. M.D. Stoller, S. Park, Y. Zhu, J. An and R.S. Ruoff, Graphene-Based Ultracapacitors, *Nano Lett.*, 2008, **8**, 3498–3502.
52. Y. Zhu, S. Murali, M.D. Stoller, K.J. Ganesh, W. Cai, P.J. Ferreira, A. Pirkle, R.M. Wallace, K.A. Cychosz, M. Thommes, D. Su, E.A. Stach and R.S. Ruoff, Carbon-Based Supercapacitors Produced by Activation of Graphene, *Science*, 2011, **332**, 1537–1541.
53. C. Li, X. Zhang, K. Wang, X. Sun, G. Liu, J. Li, H. Tian, J. Li and Y. Ma, Scalable Self-Propagating High-Temperature Synthesis of Graphene for Supercapacitors with Superior Power Density and Cyclic Stability, *Adv. Mater.*, 2017, **29**,1604690.
54. S.W. Bokhari, A.H. Siddique, P.C. Sherrell, X. Yue, K.M. Karumbaiah, S. Wei, A.V. Ellis and W. Gao, Advances in graphene-based supercapacitor electrodes, *Energy Reports*, 2020, **6**, 2768–2784.
55. H. Cao, X. Zhou, Z. Qin and Z. Liu, Low-temperature preparation of nitrogen-doped graphene for supercapacitors, *Carbon*, 2013, **56**, 218–223.
56. A. Śliwak, B. Grzyb, N. Díez and G. Gryglewicz, Nitrogen-doped reduced graphene oxide as electrode material for high rate supercapacitors, *Appl. Surf. Sci.*, 2017, **399**, 265–271.
57. Y.-Z. Liu, Y.-F. Li, F.-Y. Su, L.-J. Xie, Q.-Q. Kong, X.-M. Li, J.-G. Gao and C.-M. Chen, Easy one-step synthesis of N-doped graphene for supercapacitors, *Energy Stor. Mater.*, 2016, **2**, 69–75.
58. Z. Zuo, Z. Jiang and A. Manthiram, Porous B-doped graphene inspired by Fried-Ice for supercapacitors and metal-free catalysts, *J. Mater. Chem. A*, 2013, **1**, 13476.
59. D.Y. Yeom, W. Jeon, N.D. Tu, S.Y. Yeo, S.S. Lee, B.J. Sung, H. Chang, J.A. Lim and H. Kim, High-concentration boron doping of graphene nanoplatelets by simple thermal annealing and their supercapacitive properties, *Sci. Rep.*, 2015, **5**, 9817.
60. S.S. Balaji, M. Karnan, J. Kamarsamam and M. Sathish, Synthesis of Boron-Doped Graphene by Supercritical Fluid Processing and its Application in Symmetric Supercapacitors using Various Electrolytes, *ChemElectroChem*, 2019, **6**, 1492–1499.
61. D.J. Tarimo, K.O. Oyedotun, A.A. Mirghni and N. Manyala, Sulphur-reduced graphene oxide composite with improved electrochemical performance for supercapacitor applications, *Inter. J. Hydrog. Energy*, 2020, **45**, 13189–13201.
62. Y. Kan, G. Ning and X. Ma, Sulfur-decorated nanomesh graphene for high-performance supercapacitors, *Chin. Chem. Lett.*, 2017, **28**, 2277–2280.
63. M. Seredych, K. Singh and T.J. Bandosz, Insight into the Capacitive Performance of Sulfur-Doped Nanoporous Carbons Modified by Addition of Graphene Phase, *Electroanalysis*, 2014, **26**, 109–120.
64. W.-S. Jeon, C.H. Kim, J.-H. Wee, J.H. Kim, Y.A. Kim and C.-M. Yang, Sulfur-doping effects on the supercapacitive behavior of porous spherical graphene electrode derived from layered double hydroxide template, *Appl. Surf. Sci.*, 2021, **558**, 149867.

65. M.B. Arvas, H. Gürsu, M. Gencten and Y. Sahin, Supercapacitor applications of novel phosphorus doped graphene-based electrodes, *J. Energy Storage*, 2022, **55**, 105766.
66. Y. Wen, B. Wang, C. Huang, L. Wang and D. Hulicova-Jurcakova, Synthesis of phosphorus-doped graphene and its wide potential window in aqueous supercapacitors, *Chemistry*, 2015, **21**, 80–85.
67. Z.S. Wu, K. Parvez, A. Winter, H. Vieker, X. Liu, S. Han, A. Turchanin, X. Feng and K. Mullen, Layer-by-layer assembled heteroatom-doped graphene films with ultrahigh volumetric capacitance and rate capability for micro-supercapacitors, *Adv. Mater.*, 2014, **26**, 4552–4558.
68. C. Wang, Y. Zhou, L. Sun, Q. Zhao, X. Zhang, P. Wan and J. Qiu, N/P-Codoped Thermally Reduced Graphene for High-Performance Supercapacitor Applications, *J. Phys. Chem. C*, 2013, **117**, 14912–14919.
69. T. Wang, L.X. Wang, D.L. Wu, W. Xia and D.Z. Jia, Interaction between nitrogen and sulfur in co-doped graphene and synergetic effect in supercapacitor, *Sci. Rep.*, 2015, **5**, 9591.
70. J. Bedia, M. Peñas-Garzón, A. Gómez-Avilés, J.J. Rodriguez and C. Belver, Review on Activated Carbons by Chemical Activation with $FeCl_3$, *C-J. Carbon Res.*, 2020, **6,** 21.
71. Y. Gao, Y.S. Zhou, M. Qian, X.N. He, J. Redepenning, P. Goodman, H.M. Li, L. Jiang and Y.F. Lu, Chemical activation of carbon nano-onions for high-rate supercapacitor electrodes, *Carbon*, 2013, **51**, 52–58.
72. E.Y.L. Teo, L. Muniandy, E.-P. Ng, F. Adam, A.R. Mohamed, R. Jose and K.F. Chong, High surface area activated carbon from rice husk as a high performance supercapacitor electrode, *Electrochim. Acta*, 2016, **192**, 110–119.
73. B. Li, F. Dai, Q. Xiao, L. Yang, J. Shen, C. Zhang and M. Cai, Nitrogen-doped activated carbon for a high energy hybrid supercapacitor, *Energy Environ. Sci.*, 2016, **9**, 102–106.
74. D.R. Lobato-Peralta, R. Amaro, D.M. Arias, A.K. Cuentas-Gallegos, O.A. Jaramillo-Quintero, P.J. Sebastian and P.U. Okoye, Activated carbon from wasp hive for aqueous electrolyte supercapacitor application, *J. Electroanal. Chem.*, 2021, **901**, 115777.
75. Jain, C. Xu, S. Jayaraman, R. Balasubramanian, J.Y. Lee and M.P. Srinivasan, Mesoporous activated carbons with enhanced porosity by optimal hydrothermal pre-treatment of biomass for supercapacitor applications, *Micropor. Mesopor. Mater.*, 2015, **218**, 55–61.
76. J. Cheng, S.-C. Hu, G.-T. Sun, K. Kang, M.-Q. Zhu and Z.-C. Geng, Comparison of activated carbons prepared by one-step and two-step chemical activation process based on cotton stalk for supercapacitors application, *Energy*, 2021, **215**, 119144.
77. A. Elmouwahidi, E. Bailón-García, A.F. Pérez-Cadenas, F.J. Maldonado-Hódar and F. Carrasco-Marín, Activated carbons from KOH and H_3PO_4-activation of olive residues and its application as supercapacitor electrodes, *Electrochim. Acta*, 2017, **229**, 219–228.
78. C. Jiang, G.A. Yakaboylu, T. Yumak, J.W. Zondlo, E.M. Sabolsky and J. Wang, Activated carbons prepared by indirect and direct CO_2 activation of lignocellulosic biomass for supercapacitor electrodes, *Renew. Energy*, 2020, **155**, 38–52.
79. E.G. Calvo, F. Lufrano, P. Staiti, A. Brigandì, A. Arenillas and J.A. Menéndez, Optimizing the electrochemical performance of aqueous symmetric supercapacitors based on an activated carbon xerogel, *J. Power Sources*, 2013, **241**, 776–782.
80. Y. Ding, J. Qi, R. Hou, B. Liu, S. Yao, J. Lang, J. Chen and B. Yang, Preparation of High-Performance Hierarchical Porous Activated Carbon via a Multistep Physical Activation Method for Supercapacitors, *Energy Fuels*, 2022, **36**, 5456–5464.
81. A. Śliwak, N. Díez, E. Miniach and G. Gryglewicz, Nitrogen-containing chitosan-based carbon as an electrode material for high-performance supercapacitors, *J. Appl. Electrochem.*, 2016, **46**, 667–677.
82. R. Farma, A. Putri, E. Taer, A. Awitdrus and A. Apriwandi, Synthesis of highly porous activated carbon nanofibers derived from bamboo waste materials for application in supercapacitor, *J. Mater. Sci. Mater. Electron.*, 2021, **32**, 7681–7691.

83. L. Yang, T. Huang, X. Jiang and W. Jiang, Effect of steam and CO_2 activation on characteristics and desulfurization performance of pyrolusite modified activated carbon, *Adsorption*, 2016, **22**, 1099–1107.
84. A. Alazmi, Synergistic effect of hydrothermal and physical activation approaches to fabricate activated carbon for energy storage applications, *Ceram. Inter.*, 2022, **48**, 22131–22140.
85. X. Yu, J. Lu, C. Zhan, R. Lv, Q. Liang, Z.-H. Huang, W. Shen and F. Kang, Synthesis of activated carbon nanospheres with hierarchical porous structure for high volumetric performance supercapacitors, *Electrochim. Acta*, 2015, **182**, 908–916.
86. Z. Xu, J. Chen, X. Zhang, Q. Song, J. Wu, L. Ding, C. Zhang, H. Zhu and H. Cui, Template-free preparation of nitrogen-doped activated carbon with porous architecture for high-performance supercapacitors, *Micropor. Mesopor. Mater.*, 2019, **276**, 280–291.
87. S. Zhu, P.-L. Taberna, N. Zhao and P. Simon, Salt-template synthesis of mesoporous carbon monolith for ionogel-based supercapacitors, *Electrochem. Commun.*, 2018, **96**, 6–10.
88. B. Xue, L. Jin, Z. Chen, Y. Zhu, Z. Wang, X. Liu and X. Wang, The template effect of silica in rice husk for efficient synthesis of the activated carbon based electrode material, *J. Alloys Compd*, 2019, **789**, 777–784.
89. J. Yin, W. Zhang, N.A. Alhebshi, N. Salah and H.N. Alshareef, Synthesis Strategies of Porous Carbon for Supercapacitor Applications, *Small Methods*, 2020, **4**, 1900853.
90. S. Yoon, S.M. Oh and C. Lee, Direct template synthesis of mesoporous carbon and its application to supercapacitor electrodes, *Mater. Res. Bull.*, 2009, **44**, 1663–1669.
91. Z. Fan, Y. Liu, J. Yan, G. Ning, Q. Wang, T. Wei, L. Zhi and F. Wei, Template-Directed Synthesis of Pillared-Porous Carbon Nanosheet Architectures: High-Performance Electrode Materials for Supercapacitors, *Adv. Energy Mater.*, 2012, **2**, 419–424.
92. S. Rawat, R.K. Mishra and T. Bhaskar, Biomass derived functional carbon materials for supercapacitor applications, *Chemosphere*, 2022, **286**, 131961.
93. Z. Bi, Q. Kong, Y. Cao, G. Sun, F. Su, X. Wei, X. Li, A. Ahmad, L. Xie and C.-M. Chen, Biomass-derived porous carbon materials with different dimensions for supercapacitor electrodes: a review, *J. Mater. Chem. A*, 2019, **7**, 16028–16045.
94. J. Niu, R. Shao, M. Liu, Y. Zan, M. Dou, J. Liu, Z. Zhang, Y. Huang and F. Wang, Porous Carbons Derived from Collagen-Enriched Biomass: Tailored Design, Synthesis, and Application in Electrochemical Energy Storage and Conversion, *Adv. Funct. Mater.*, 2019, **29**, 1905095.
95. Z. Wang, A.T. Smith, W. Wang and L. Sun, Versatile Nanostructures from Rice Husk Biomass for Energy Applications, *Angew. Chem. Int. Ed.*, 2018, **57**, 13722–13734.
96. D. Yan, C. Yu, X. Zhang, W. Qin, T. Lu, B. Hu, H. Li and L. Pan, Nitrogen-doped carbon microspheres derived from oatmeal as high capacity and superior long life anode material for sodium ion battery, *Electrochim. Acta*, 2016, **191**, 385–391.
97. Z. Li, Q. Liu, L. Sun, N. Li, X. Wang, Q. Wang, D. Zhang and B. Wang, Hydrothermal synthesis of 3D hierarchical ordered porous carbon from yam biowastes for enhanced supercapacitor performance, *Chem. Eng. Sci.*, 2022, **252**, 117514.
98. H. Ma, C. Li, M. Zhang, J.-D. Hong and G. Shi, Graphene oxide induced hydrothermal carbonization of egg proteins for high-performance supercapacitors, *J. Mater. Chem. A*, 2017, **5**, 17040–17047.
99. C. Long, X. Chen, L. Jiang, L. Zhi and Z. Fan, Porous layer-stacking carbon derived from in-built template in biomass for high volumetric performance supercapacitors, *Nano Energy*, 2015, **12**, 141–151.
100. J. Deng, M. Li and Y. Wang, Biomass-derived carbon: synthesis and applications in energy storage and conversion, *Green Chem.*, 2016, **18**, 4824–4854.
101. F. Kurosaki, H. Koyanaka, M. Tsujimoto and Y. Imamura, Shape-controlled multi-porous carbon with hierarchical micro–meso-macro pores synthesized by flash heating of wood biomass, *Carbon*, 2008, **46**, 850–857.

102. Z. Zhang, J. He, X. Tang, Y. Wang, B. Yang, K. Wang and D. Zhang, Supercapacitors based on a nitrogen doped hierarchical porous carbon fabricated by self-activation of biomass: excellent rate capability and cycle stability, *Carbon Lett.*, 2019,**29**, 585–594.
103. Z. Li, W. Lv, C. Zhang, B. Li, F. Kang and Q.-H. Yang, A sheet-like porous carbon for high-rate supercapacitors produced by the carbonization of an eggplant, *Carbon*, 2015, **92**, 11–14.
104. M. Biswal, A. Banerjee, M. Deo and S. Ogale, From dead leaves to high energy density supercapacitors, *Energy Environ. Sci.*, 2013, **6,** 1249.
105. Y. Wang, X. Lin, T. Liu, H. Chen, S. Chen, Z. Jiang, J. Liu, J. Huang and M. Liu, Wood-Derived Hierarchically Porous Electrodes for High-Performance All-Solid-State Supercapacitors, *Adv. Funct. Mater.*, 2018, **28**, 1806207.
106. K. Zou, Y. Deng, J. Chen, Y. Qian, Y. Yang, Y. Li and G. Chen, Hierarchically porous nitrogen-doped carbon derived from the activation of agriculture waste by potassium hydroxide and urea for high-performance supercapacitors, *J. Power Sources*, 2018, **378**, 579–588.
107. X. Liu, C. Ma, J. Li, B. Zielinska, R.J. Kalenczuk, X. Chen, P.K. Chu, T. Tang and E. Mijowska, Biomass-derived robust three-dimensional porous carbon for high volumetric performance supercapacitors, *J. Power Sources*, 2019, **412**, 1–9.
108. X. Yang, C. Li and Y. Chen, Hierarchical porous carbon with ultrahigh surface area from corn leaf for high-performance supercapacitors application, *J. Phys. D*, 2017, **50**, 055501.
109. K. Yang, J. Peng, H. Xia, L. Zhang, C. Srinivasakannan and S. Guo, Textural characteristics of activated carbon by single step CO_2 activation from coconut shells, *J. Taiwan Inst. Chem. Eng.*, 2010, **41**, 367–372.
110. M. Yu, Y. Han, J. Li and L. Wang, CO_2-activated porous carbon derived from cattail biomass for removal of malachite green dye and application as supercapacitors, *Chem. Eng. J.*, 2017, **317**, 493–502.
111. Y. Jin, K. Tian, L. Wei, X. Zhang and X. Guo, Hierarchical porous microspheres of activated carbon with a high surface area from spores for electrochemical double-layer capacitors, *J. Mater. Chem. A*, 2016, **4**, 15968–15979.
112. W. Li, B. Li, M. Shen, Q. Gao and J. Hou, Use of Gemini surfactant as emulsion interface microreactor for the synthesis of nitrogen-doped hollow carbon spheres for high-performance supercapacitors, *Chem. Eng. J.*, 2020, **384**, 123309.
113. D. Saha, Y. Li, Z. Bi, J. Chen, J.K. Keum, D.K. Hensley, H.A. Grappe, H.M. Meyer, 3rd, S. Dai, M.P. Paranthaman and A.K. Naskar, Studies on supercapacitor electrode material from activated lignin-derived mesoporous carbon, *Langmuir*, 2014, **30**, 900–910.
114. Y.S. Hu, P. Adelhelm, B.M. Smarsly, S. Hore, M. Antonietti and J. Maier, Synthesis of Hierarchically Porous Carbon Monoliths with Highly Ordered Microstructure and Their Application in Rechargeable Lithium Batteries with High-Rate Capability, *Adv. Funct. Mater.*, 2007, **17**, 1873–1878.
115. S. Zhang, K. Tian, B.-H. Cheng and H. Jiang, Preparation of N-Doped Supercapacitor Materials by Integrated Salt Templating and Silicon Hard Templating by Pyrolysis of Biomass Wastes, *ACS Sustain. Chem. Eng.*, 2017, **5**, 6682–6691.
116. C. Zhu and T. Akiyama, Cotton derived porous carbon via an MgO template method for high performance lithium ion battery anodes, *Green Chem.*, 2016, **18**, 2106–2114.
117. S. Yang, S. Wang, X. Liu and L. Li, Biomass derived interconnected hierarchical micro-meso-macro- porous carbon with ultrahigh capacitance for supercapacitors, *Carbon*, 2019, **147**, 540–549.
118. X. Chen, J. Zhang, B. Zhang, S. Dong, X. Guo, X. Mu and B. Fei, A novel hierarchical porous nitrogen-doped carbon derived from bamboo shoot for high performance supercapacitor, *Sci. Rep.*, 2017, **7**, 7362.

119. T. Cai, W. Xing, Z. Liu, J. Zeng, Q. Xue, S. Qiao and Z. Yan, Superhigh-rate capacitive performance of heteroatoms-doped double shell hollow carbon spheres, *Carbon*, 2015, **86**, 235–244.
120. X. Huang, S. Kim, M.S. Heo, J.E. Kim, H. Suh and I. Kim, Easy synthesis of hierarchical carbon spheres with superior capacitive performance in supercapacitors, *Langmuir*, 2013, **29**, 12266–12274.
121. Q. Li, R. Jiang, Y. Dou, Z. Wu, T. Huang, D. Feng, J. Yang, A. Yu and D. Zhao, Synthesis of mesoporous carbon spheres with a hierarchical pore structure for the electrochemical double-layer capacitor, *Carbon*, 2011, **49**, 1248–1257.
122. Y. Liu, L. Pan, T. Chen, X. Xu, T. Lu, Z. Sun and D.H.C. Chua, Porous carbon spheres via microwave-assisted synthesis for capacitive deionization, *Electrochim. Acta*, 2015, **151**, 489–496.
123. K.L. Van Aken, C.R. Pérez, Y. Oh, M. Beidaghi, Y. Joo Jeong, M.F. Islam and Y. Gogotsi, High rate capacitive performance of single-walled carbon nanotube aerogels, *Nano Energy*, 2015, **15**, 662–669.
124. R. W. Pekala, Organic aerogels from the polycondensation of resorcinol with formaldehyde, *J. Mater. Sci.* 1989, **24**, 3221–3227.
125. X. Wu, J. Zhou, W. Xing, G. Wang, H. Cui, S. Zhuo, Q. Xue, Z. Yan and S.Z. Qiao, High-rate capacitive performance of graphene aerogel with a superhigh C/O molar ratio, *J. Mater. Chem.*, 2012, **22**, 23186.
126. H. Probstle, M. Wiener and J. Fricke, Carbon Aerogels for Electrochemical Double Layer Capacitors, *J. Porous Mater.*, 2003, **10**, 213–222.
127. J. Lang, X. Yan, W. Liu, R. Wang and Q. Xue, Influence of nitric acid modification of ordered mesoporous carbon materials on their capacitive performances in different aqueous electrolytes, *J. Power Sources*, 2012, **204**, 220–229.
128. D. Li, L. Zhao, X. Cao, Z. Xiao, H. Nan and H. Qiu, Nickel-catalyzed formation of mesoporous carbon structure promoted capacitive performance of exhausted biochar, *Chem. Eng. J.*, 2021, **406**, 126856.
129. Q. Liang, L. Ye, Q. Xu, Z.-H. Huang, F. Kang and Q.-H. Yang, Graphitic carbon nitride nanosheet-assisted preparation of N-enriched mesoporous carbon nanofibers with improved capacitive performance, *Carbon*, 2015, **94**, 342–348.
130. T. Le, Y. Yang, Z. Huang and F. Kang, Preparation of microporous carbon nanofibers from polyimide by using polyvinyl pyrrolidone as template and their capacitive performance, *J. Power Sources*, 2015, **278**, 683–692.
131. F. Miao, C. Shao, X. Li, N. Lu, K. Wang, X. Zhang and Y. Liu, Polyaniline-coated electrospun carbon nanofibers with high mass loading and enhanced capacitive performance as freestanding electrodes for flexible solid-state supercapacitors, *Energy*, 2016, **95**, 233–241.
132. X. Gong, W. Luo, N. Guo, S. Zhang, L. Wang, D. Jia, L. Ai and S. Feng, Carbon nanofiber@ZIF-8 derived carbon nanosheet composites with a core–shell structure boosting capacitive deionization performance, *J. Mater. Chem. A*, 2021, **9**, 18604–18613.
133. Q. Yu, J. Lv, Z. Liu, M. Xu, W. Yang, K.A. Owusu, L. Mai, D. Zhao and L. Zhou, Macroscopic synthesis of ultrafine N-doped carbon nanofibers for superior capacitive energy storage, *Sci. Bull.*, 2019, **64**, 1617–1624.
134. H. Zhu, C. Wang, C. Li, L. Guan, H. Pan, M. Yan and Y. Jiang, Engineering capacitive contribution in nitrogen-doped carbon nanofiber films enabling high performance sodium storage, *Carbon*, 2018, **130**, 145–152.
135. X. Li, D. Du, Y. Zhang, W. Xing, Q. Xue and Z. Yan, Layered double hydroxides toward high-performance supercapacitors, *J. Mater. Chem. A*, 2017, **5**, 15460–15485.
136. S. Gao, Y. Sun, F. Lei, L. Liang, J. Liu, W. Bi, B. Pan and Y. Xie, Ultrahigh energy density realized by a single-layer beta-$Co(OH)_2$ all-solid-state asymmetric supercapacitor, *Angew. Chem. Int. Ed.*, 2014, **53**, 12789–12793.

137. M. Xie, S. Duan, Y. Shen, K. Fang, Y. Wang, M. Lin and X. Guo, In-Situ-Grown $Mg(OH)_2$-Derived Hybrid α-$Ni(OH)_2$ for Highly Stable Supercapacitor, *ACS Energy Lett.*, 2016, **1**, 814–819.
138. Y. Zhu, C. Huang, C. Li, M. Fan, K. Shu and H.C. Chen, Strong synergetic electrochemistry between transition metals of α phase Ni–Co–Mn hydroxide contributed superior performance for hybrid supercapacitors, *J. Power Sources*, 2019, **412**, 559–567.
139. H. Li, Y. Gao, C. Wang and G. Yang, A Simple Electrochemical Route to Access Amorphous Mixed-Metal Hydroxides for Supercapacitor Electrode Materials, *Adv. Energy Mater.*, 2015, **5**, 1401767.
140. H. Chen, L. Hu, M. Chen, Y. Yan and L. Wu, Nickel–Cobalt Layered Double Hydroxide Nanosheets for High-performance Supercapacitor Electrode Materials, *Adv. Funct. Mater.*, 2014, **24**, 934–942.
141. J. Kumar, R.R. Neiber, Z. Abbas, R.A. Soomro, A. BaQais, M.A. Amin and Z.M. El-Bahy, Hierarchical NiMn-LDH Hollow Spheres as a Promising Pseudocapacitive Electrode for Supercapacitor Application, *Micromachines*, 2023, **14**, 487.
142. X. Zhu, X. Li, H. Tao and M. Li, Preparation of Co_2Al layered double hydroxide nanosheet/Co_2Mn bimetallic hydroxide nanoneedle nanocomposites on nickel foam for supercapacitors, *J. Alloys Compd*, 2021, **851**, 156868.
143. H. Xiong, L. Liu, L. Fang, F. Wu, S. Zhang, H. Luo, C. Tong, B. Hu and M. Zhou, 3D self-supporting heterostructure NiCo-LDH/ZnO/CC electrode for flexible high-performance supercapacitor, *J. Alloys Compd*, 2021, **857**, 158275.
144. L. Zhu, W. Wu, Y. Zhu, W. Tang and Y. Wu, Composite of CoOOH Nanoplates with Multiwalled Carbon Nanotubes as Superior Cathode Material for Supercapacitors, *J. Phys. Chem. C*, 2015, **119**, 7069–7075.
145. D. Zhang, X. Kong, Y. Zhao, M. Jiang and X. Lei, CoOOH ultrathin nanoflake arrays aligned on nickel foam: fabrication and use in high-performance supercapacitor devices, *J. Mater. Chem. A*, 2016, **4**, 12833–12840.
146. K. Ding, X. Zhang, P. Yang and X. Cheng, A precursor-derived morphology-controlled synthesis method for mesoporous Co_3O_4 nanostructures towards supercapacitor application, *CrystEngComm*, 2016, **18**, 8253–8261.
147. C. Guan, W. Zhao, Y. Hu, Z. Lai, X. Li, S. Sun, H. Zhang, A.K. Cheetham and J. Wang, Cobalt oxide and N-doped carbon nanosheets derived from a single two-dimensional metal-organic framework precursor and their application in flexible asymmetric supercapacitors, *Nanosc. Horiz.*, 2017, **2**, 99–105.
148. P. Wu, S. Cheng, M. Yao, L. Yang, Y. Zhu, P. Liu, O. Xing, J. Zhou, M. Wang, H. Luo and M. Liu, A Low-Cost, Self-Standing $NiCo_2O_4$@CNT/CNT Multilayer Electrode for Flexible Asymmetric Solid-State Supercapacitors, *Adv. Funct. Mater.*, 2017, **27**, 1702160.
149. J.S. Wei, H. Ding, P. Zhang, Y.F. Song, J. Chen, Y.G. Wang and H.M. Xiong, Carbon Dots/$NiCo_2O_4$ Nanocomposites with Various Morphologies for High Performance Supercapacitors, *Small*, 2016, **12**, 5927–5934.
150. T. Larbi, L. Ben said, A. Ben daly, B. ouni, A. Labidi and M. Amlouk, Ethanol sensing properties and photocatalytic degradation of methylene blue by Mn_3O_4 , $NiMn_2O_4$ and alloys of Ni-manganates thin films, *J. Alloys Compd*, 2016, **686**, 168–175.
151. S. Maiti, A. Pramanik, T. Dhawa, M. Sreemany and S. Mahanty, Bi-metal organic framework derived nickel manganese oxide spinel for lithium-ion battery anode, *Mater. Sci. Eng. B*, 2018, **229**, 27–36.
152. M. Puratchimani, V. Venkatachalam, M. Rajapriya and K. Thamizharasan, High Performance of $NiMn_2O_4$ Nanostructured Electrode for Supercapacitor Applications, *Analyt. Bioanalyt. Electrochem.*, 2022, **14**, 968–979.
153. S.K. Muzakir, A.S. Samsudin and B. Sahraoui, Transition Metal Dichalcogenide for High-Performance Electrode of Supercapacitor, *Makara J. Technol.*, 2019, **22**, 123.

154. J. Theerthagiri, R.A. Senthil, P. Nithyadharseni, S.J. Lee, G. Durai, P. Kuppusami, J. Madhavan and M.Y. Choi, Recent progress and emerging challenges of transition metal sulfides based composite electrodes for electrochemical supercapacitive energy storage, *Ceram. Int.*, 2020, **46**, 14317–14345.
155. Z. Meng, L.I.U. Xiujun and Z. Qingyin, Research progress on the electrode material of transition metal sulfide in the supercapacitor, *J. Funct. Mater.*, 2021, **52**, 1086–1090.
156. H. Huo, Y. Zhao and C. Xu, 3D Ni_3S_2 nanosheet arrays supported on Ni foam for high-performance supercapacitor and non-enzymatic glucose detection, *J. Mater. Chem. A*, 2014, **2**, 15111.
157. S. Peng, L. Li, H. Tan, R. Cai, W. Shi, C. Li, S.G. Mhaisalkar, M. Srinivasan, S. Ramakrishna and Q. Yan, MS_2(M = Co and Ni) Hollow Spheres with Tunable Interiors for High-Performance Supercapacitors and Photovoltaics, *Adv. Funct. Mater.*, 2014, 24, 2155–2162.
158. A. Mohammadi Zardkhoshoui, S.S. Hosseiny Davarani and A.A. Asgharinezhad, Designing graphene-wrapped $NiCo_2Se_4$ microspheres with petal-like FeS_2 toward flexible asymmetric all-solid-state supercapacitors, *Dalton Trans.*, 2019, 48, 4274–4282.
159. Y. Liu, Y. Wen, Y. Zhang, X. Wu, H. Li, H. Chen, J. Huang, G. Liu and S. Peng, Reduced $CoNi_2S_4$ nanosheets decorated by sulfur vacancies with enhanced electrochemical performance for asymmetric supercapacitors, *Sci. China Mater.*, 2020, **63**, 1216–1226.
160. K. Zhou, W. Zhou, L. Yang, J. Lu, S. Cheng, W. Mai, Z. Tang, L. Li and S. Chen, Ultrahigh-Performance Pseudocapacitor Electrodes Based on Transition Metal Phosphide Nanosheets Array via Phosphorization: A General and Effective Approach, *Adv. Funct. Mater.*, 2015, **25**, 7530–7538.
161. J. Gu, L. Sun, Y. Zhang, Q. Zhang, X. Li, H. Si, Y. Shi, C. Sun, Y. Gong and Y. Zhang, MOF-derived Ni-doped CoP@C grown on CNTs for high-performance supercapacitors, *Chem. Eng. J.*, 2020, **385**, 123454.
162. Y. Lan, H. Zhao, Y. Zong, X. Li, Y. Sun, J. Feng, Y. Wang, X. Zheng and Y. Du, Phosphorization boosts the capacitance of mixed metal nanosheet arrays for high performance supercapacitor electrodes, *Nanoscale*, 2018, **10**, 11775–11781.
163. N.S. Arul and J.I. Han, Facile hydrothermal synthesis of hexapod-like two dimensional dichalcogenide $NiSe_2$ for supercapacitor, *Mater. Lett.*, 2016, **181**, 345–349.
164. H. Peng, G. Ma, K. Sun, Z. Zhang, J. Li, X. Zhou and Z. Lei, A novel aqueous asymmetric supercapacitor based on petal-like cobalt selenide nanosheets and nitrogen-doped porous carbon networks electrodes, *J. Power Sources*, 2015, **297**, 351–358.
165. C. Gong, M. Huang, J. Zhang, M. Lai, L. Fan, J. Lin and J. Wu, Facile synthesis of Ni0.85Se on Ni foam for high-performance asymmetric capacitors, *RSC Adv.*, 2015, **5**, 81474–81481.
166. Banerjee, S. Bhatnagar, K.K. Upadhyay, P. Yadav and S. Ogale, Hollow $Co_{0.85}Se$ nanowire array on carbon fiber paper for high rate pseudocapacitor, *ACS Appl. Mater. Interfaces*, 2014, 6, 18844–18852.
167. K. Guo, S. Cui, H. Hou, W. Chen and L. Mi, Hierarchical ternary Ni-Co-Se nanowires for high-performance supercapacitor device design, *Dalton Trans.*, 2016, **45**, 19458–19465.
168. M. Jaymand, Recent progress in chemical modification of polyaniline, *Prog. Polym. Sci.*, 2013, **38**, 1287–1306.
169. R. Jain, N. Jadon and A. Pawaiya, Polypyrrole based next generation electrochemical sensors and biosensors: A review, *Trends Analyt. Chem.*, 2017, **97**, 363–373.
170. M. Jaymand, M. Hatamzadeh and Y. Omidi, Modification of polythiophene by the incorporation of processable polymeric chains: Recent progress in synthesis and applications, *Prog. Polym. Sci.*, 2015, **47**, 26–69.

171. I.E. Jacobs, Y. Lin, Y. Huang, X. Ren, D. Simatos, C. Chen, D. Tjhe, M. Statz, L. Lai, P.A. Finn, W.G. Neal, G. D'Avino, V. Lemaur, S. Fratini, D. Beljonne, J. Strzalka, C.B. Nielsen, S. Barlow, S.R. Marder, I. McCulloch and H. Sirringhaus, High-Efficiency Ion-Exchange Doping of Conducting Polymers, *Adv. Mater.*, 2022, **34**, 2102988.
172. H. Lu, X. Li and Q. Lei, Conjugated Conductive Polymer Materials and its Applications: A Mini-Review, *Front. Chem.*, 2021, **9**, 732132.
173. H. Li, J. Wang, Q. Chu, Z. Wang, F. Zhang and S. Wang, Theoretical and experimental specific capacitance of polyaniline in sulfuric acid, *J. Power Sources*, 2009, **190**, 578–586.
174. Y. Wang, X. Yang, L. Qiu and D. Li, Revisiting the capacitance of polyaniline by using graphene hydrogel films as a substrate: the importance of nano-architecturing, *Energy Environ. Sci.*, 2013, **6**, 477–481.
175. B.K. Kuila, B. Nandan, M. Bohme, A. Janke and M. Stamm, Vertically oriented arrays of polyaniline nanorods and their super electrochemical properties, *Chem. Commun.*, 2009, **38**, 5749–5751.
176. H. Wang, Q. Hao, X. Yang, L. Lu and X. Wang, Effect of graphene oxide on the properties of its composite with polyaniline, *ACS Appl. Mater. Interfaces*, 2010, **2**, 821–828.
177. J. Xu, K. Wang, S.Z. Zu, B.H. Han and Z. Wei, Hierarchical nanocomposites of polyaniline nanowire arrays on graphene oxide sheets with synergistic effect for energy storage, *ACS Nano*, 2010, **4**, 5019–5026.
178. Z. Niu, L. Liu, L. Zhang, Q. Shao, W. Zhou, X. Chen and S. Xie, A universal strategy to prepare functional porous graphene hybrid architectures, *Adv. Mater.*, 2014, **26**, 3681–3687.
179. H. Guo, W. He, Y. Lu and X. Zhang, Self-crosslinked polyaniline hydrogel electrodes for electrochemical energy storage, *Carbon*, 2015, **92**, 133–141.
180. Y. Liu, Y. Ma, S. Guang, F. Ke and H. Xu, Polyaniline-graphene composites with a three-dimensional array-based nanostructure for high-performance supercapacitors, *Carbon*, 2015, **83**, 79–89.
181. K. Zhang, L.L. Zhang, X.S. Zhao and J. Wu, Graphene/Polyaniline Nanofiber Composites as Supercapacitor Electrodes, *Chem. Mater.*, 2010, **22**, 1392–1401.
182. L. Pan, G. Yu, D. Zhai, H.R. Lee, W. Zhao, N. Liu, H. Wang, B.C. Tee, Y. Shi, Y. Cui and Z. Bao, Hierarchical nanostructured conducting polymer hydrogel with high electrochemical activity, Proc. Natl. Acad. *Proc. Natl Acad. Sci. U. S. A.*, 2012, **109**, 9287–9292.
183. K. Wang, X. Zhang, C. Li, X. Sun, Q. Meng, Y. Ma and Z. Wei, Chemically Crosslinked Hydrogel Film Leads to Integrated Flexible Supercapacitors with Superior Performance, *Adv. Mater.*, 2015, **27**, 7451–7457.
184. Q. Wen, K. He, C. Wang, B. Wang, S. Yu, C. Hao and K. Chen, Clip-like polyaniline nanofibers synthesized by an insitu chemical oxidative polymerization and its strong electrorheological behavior, *Synth. Met.*, 2018, **239**, 1–12.
185. Y. Zhao, H. Wei, M. Arowo, X. Yan, W. Wu, J. Chen, Y. Wang and Z. Guo, Electrochemical energy storage by polyaniline nanofibers: high gravity assisted oxidative polymerization vs. rapid mixing chemical oxidative polymerization, *Phys. Chem. Chem. Phys.*, 2015, **17**, 1498–1502.
186. Z. Zakaria, N.F.A. Halim, M.H.V. Schleusingen, A.K.M.S. Islam, U. Hashim and M.N. Ahmad, Effect of Hydrochloric Acid Concentration on Morphology of Polyaniline Nanofibers Synthesized by Rapid Mixing Polymerization, *J. Nanomat.*, 2015, **2015**, 1–6.
187. S. Bhandari and D. Khastgir, Corrosion-free electrochemical synthesis of polyaniline using Cu counter electrode in acidic medium, *Int. J. Polym. Mater. Polym. Biomater.*, 2016, **65**, 543–549.

188. C. Kaneda, Y. Sueyasu, E. Tanaka and M. Atobe, Electrochemical synthesis of microporous polyaniline films using foam templates prepared by ultrasonication, *Ultrason. Sonochem.*, 2020, **64**, 104991.
189. Y.H. Navale, S.T. Navale, M.A. Chougule, S.M. Ingole, F.J. Stadler, R.S. Mane, M. Naushad and V.B. Patil, Electrochemical synthesis and potential electrochemical energy storage performance of nodule-type polyaniline, *J. Colloid. Interface Sci.*, 2017, **487**, 458–464.
190. F. de Salas, I. Pardo, H.J. Salavagione, P. Aza, E. Amougi, J. Vind, A.T. Martinez and S. Camarero, Advanced Synthesis of Conductive Polyaniline Using Laccase as Biocatalyst, *PLoS One*, 2016, **11**, 0164958.
191. A.L. Cholli, M. Thiyagarajan, J. Kumar and V.S. Parmar, Biocatalytic approaches for synthesis of conducting polyaniline nanoparticles, *Pure Appl. Chem.*, 2005, **77**, 339–344.
192. J. Chen, L. Bai, M. Yang, H. Guo and Y. Xu, Biocatalyzed synthesis of conducting polyaniline in reverse microemulsions, *Synth. Met.*, 2014, **187**, 108–112.
193. L. Yuan, C. Wan, X. Ye and F. Wu, Facial Synthesis of Silver-incorporated Conductive Polypyrrole Submicron Spheres for Supercapacitors, *Electrochim. Acta*, 2016, **213**, 115–123.
194. G.A. Grijalva-Bustamante, R.V. Quevedo-Robles, T. del Castillo-Castro, M.M. Castillo-Ortega, J.C. Encinas, D.E. Rodríguez-Félix, T.E. Lara-Ceniceros, D. Fernández-Quiroz, J. Lizardi-Mendoza and L. Armenta-Villegas, A novel bile salt-assisted synthesis of colloidal polypyrrole nanoparticles, *Colloids Surf. A Physicochem. Eng. Aspects*, 2020, **600**, 124961.
195. H.J. Jang, B.J. Shin, E.Y. Jung, G.T. Bae, J.Y. Kim and H.-S. Tae, Polypyrrole film synthesis via solution plasma polymerization of liquid pyrrole, *Appl. Surf. Sci.*, 2023, **608**, 155129.
196. O.I. Istakova, D.V. Konev, A.T. Glazkov, T.O. Medvedeva, E.V. Zolotukhina and M.A. Vorotyntsev, Electrochemical synthesis of polypyrrole in powder form, *J. Solid State Electrochem.*, 2018, **23**, 251–258.
197. E. Asghari, H. Ashassi-Sorkhabi, G.-R. Charmi, S. Jabbari and B. Rezaei-Moghadam, A facile electrochemical strategy for synthesis of 3D nanodimensional polypyrrole structures using self-assembled layers of pyrrole monomers, *Prog. Org. Coat.*, 2016, **101**, 130–141.
198. B. Zhang, P. Zhou, Y. Xu, J. Lin, H. Li, Y. Bai, J. Zhu, S. Mao and J. Wang, Gravity-assisted synthesis of micro/nano-structured polypyrrole for supercapacitors, *Chem. Eng. J.*, 2017, **330**, 1060–1067.
199. Y. Bian, C. Yang, Q. Gu, X. Zhu, Y. Wang and X. Zhang, Controlled syntheses of polythiophene nanoparticles with plain and hollow nanostructures templated from unimolecular micelles, *J. Polym. Sci.*, 2019, **57**, 1550–1555.
200. T. Kitao, M.W.A. MacLean, B. Le Ouay, Y. Sasaki, M. Tsujimoto, S. Kitagawa and T. Uemura, Preparation of polythiophene microrods with ordered chain alignment using nanoporous coordination template, *Polym. Chem.*, 2017, **8**, 5077–5081.
201. K. Yoshino, S. Hayashi and R.-I. Sugimoto, Preparation and Properties of Conducting Heterocyclic Polymer Films by Chemical Method, Japan. *J. Appl. Phys.*, 1984, **23**, L899.
202. A. Gök, M. Omastová and A.G. Yavuz, Synthesis and characterization of polythiophenes prepared in the presence of surfactants, *Synth. Met.*, 2007, **157**, 23–29.
203. X.G. Li, J. Li and M.R. Huang, Facile optimal synthesis of inherently electroconductive polythiophene nanoparticles, *Chem.*, 2009, **15**, 6446–6455.
204. N. Ballav and M. Biswas, Preparation and evaluation of a nanocomposite of polythiophene with Al_2O_3, *Poly. Int.*, 2003, **52**, 179–184.

205. H. Vijeth, S.P. Ashokkumar, L. Yesappa, M. Niranjana, M. Vandana and H. Devendrappa, Camphor sulfonic acid assisted synthesis of polythiophene composite for high energy density all-solid-state symmetric supercapacitor, *J. Mater. Sci. Mater. Electron.*, 2019, **30**, 7471–7484.
206. Q. Wang, Y. Ma, X. Liang, D. Zhang and M. Miao, Flexible supercapacitors based on carbon nanotube-MnO_2 nanocomposite film electrode, *Chem. Eng. J.*, 2019, **371**, 145–153.
207. T. Lv, Y. Yao, N. Li and T. Chen, Highly Stretchable Supercapacitors Based on Aligned Carbon Nanotube/Molybdenum Disulfide Composites, *Angew. Chem. Int. Ed.*, 2016, **55**, 9191–9195.
208. Y.-F. Wang, S.-X. Zhao, L. Yu, X.-X. Zheng, Q.-L. Wu and G.-Z. Cao, Design of multiple electrode structures based on nano Ni_3S_2 and carbon nanotubes for high performance supercapacitors, *J. Mater. Chem. A*, 2019, **7**, 7406–7414.
209. Y.-H. Lai, S. Gupta, C.-H. Hsiao, C.-Y. Lee and N.-H. Tai, Multilayered nickel oxide/carbon nanotube composite paper electrodes for asymmetric supercapacitors, *Electrochim. Acta*, 2020, **354**, 136744.
210. H. Li, Z. Li, Z. Wu, M. Sun, S. Han, C. Cai, W. Shen and Y. Fu, Nanocomposites of Cobalt Sulfide Embedded Carbon Nanotubes with Enhanced Supercapacitor Performance, *J. Electrochem. Soc.*, 2019, **166**, A1031–A1037.
211. X. Hou, T. Peng, Q. Yu, R. Luo, X. Liu, Y. Zhang, Y. Wang, Y. Guo, J.-K. Kim and Y. Luo, Facile Synthesis of Holothurian-Like γ-MnS/Carbon Nanotube Nanocomposites for Flexible All-Solid-State Supercapacitors, *ChemNanoMat*, 2017, **3**, 551–559.
212. Q. Liu, J. Zhu, L. Zhang and Y. Qiu, Recent advances in energy materials by electrospinning, *Renew. Sust. Energ. Rev.*, 2018, **81**, 1825–1858.
213. M. Mirjalili and S. Zohoori, Review for application of electrospinning and electrospun nanofibers technology in textile industry, *J. Nanostruct. Chem.*, 2016, **6**, 207–213.
214. A.S. Levitt, M. Alhabeb, C.B. Hatter, A. Sarycheva, G. Dion and Y. Gogotsi, Electrospun MXene/carbon nanofibers as supercapacitor electrodes, *J. Mater. Chem. A*, 2019, **7**, 269–277.
215. E. Samuel, B. Joshi, H.S. Jo, Y.I. Kim, S. An, M.T. Swihart, J.M. Yun, K.H. Kim and S.S. Yoon, Carbon nanofibers decorated with FeO nanoparticles as a flexible electrode material for symmetric supercapacitors, *Chem. Eng. J.*, 2017, **328**, 776–784.
216. J. Tan, Y. Han, L. He, Y. Dong, X. Xu, D. Liu, H. Yan, Q. Yu, C. Huang and L. Mai, In situ nitrogen-doped mesoporous carbon nanofibers as flexible freestanding electrodes for high-performance supercapacitors, *J. Mater. Chem. A*, 2017, **5**, 23620–23627.
217. S. Liu, H. Du, K. Liu, M.-G. Ma, Y.-E. Kwon, C. Si, X.-X. Ji, S.-E. Choi and X. Zhang, Flexible and porous Co_3O_4-carbon nanofibers as binder-free electrodes for supercapacitors, *Adv. Compos. Hybrid Mater.*, 2021, **4**, 1367–1383.
218. W. Xu, L. Liu and W. Weng, High-performance supercapacitor based on MnO/carbon nanofiber composite in extended potential windows, *Electrochim. Acta*, 2021, **370**, 137713.
219. B. Üstün, H. Aydın, S.N. Koç and Ü. Kurtan, Amorphous ZnO@S-doped carbon composite nanofiber for use in asymmetric supercapacitors, *Diam. Relat. Mater.*, 2023, **136**, 110048.
220. J. Zhang, Z. Zhang, Y. Jiao, H. Yang, Y. Li, J. Zhang and P. Gao, The graphene/lanthanum oxide nanocomposites as electrode materials of supercapacitors, *J. Power Sources*, 2019, **419**, 99–105.
221. B. Zhao, D. Chen, X. Xiong, B. Song, R. Hu, Q. Zhang, B.H. Rainwater, G.H. Waller, D. Zhen, Y. Ding, Y. Chen, C. Qu, D. Dang, C.-P. Wong and M. Liu, A high-energy, long cycle-life hybrid supercapacitor based on graphene composite electrodes, *Energy Storage Mater.*, 2017, **7**, 32–39.

222. F. Bahmani, S.H. Kazemi, Y. Wu, L. Liu, Y. Xu and Y. Lei, $CuMnO_2$-reduced graphene oxide nanocomposite as a free-standing electrode for high-performance supercapacitors, *Chem. Eng. J.*, 2019, **375**, 121966.
223. Q. Li, C. Lu, C. Chen, L. Xie, Y. Liu, Y. Li, Q. Kong and H. Wang, Layered $NiCo_2O_4$/reduced graphene oxide composite as an advanced electrode for supercapacitor, *Energy Storage Mater.*, 2017, **8**, 59–67.
224. Y. Wang, W. Lai, N. Wang, Z. Jiang, X. Wang, P. Zou, Z. Lin, H.J. Fan, F. Kang, C.-P. Wong and C. Yang, A reduced graphene oxide/mixed-valence manganese oxide composite electrode for tailorable and surface mountable supercapacitors with high capacitance and super-long life, *Energy Environ. Sci.*, 2017, **10**, 941–949.

9 Preparation, Structure Control, and Electrochemical Performance of Electrospun Carbon Nanofiber (CNF)

Yan Song, Xiaodong Tian and Tao Yang

1 INTRODUCTION

The globalization and industrial revolution have remarkably accelerated the consumption speed of fossil fuels and aggravated environmental pollution as well as health hazards, in order to build a more sustainable and green world, scientists and policy makers have focused on developing efficacious, clean, and renewable energy sources, such as solar and wind power to fulfill energy demands. However, given the seasonal characteristics, regionalism, and discontinuity, it is of vital importance for searching reliable green energy storage systems to make them favorable to use directly in the industry. Thereinto, supercapacitors (SCs) and secondary chargeable batteries have become the fast-growing technologies due to their efficient energy storage capability for power and energy supply.[1, 2] Taking the increasing demand for higher energy and power density along with longer lifespan into consideration, tremendous efforts have been devoted to explore next-generation energy storage devices. As we all know, the performance of these energy storage systems mainly depends on the characteristics of the electrode materials.

Carbon materials with various microtextures are widely investigated as electrode materials in terms of high electrical conductivity, tunable pore structure, controlled microcrystalline structure, easy availability, nontoxic, stability, and environmental friendliness.[3] However, most of the state-of-the-art carbon materials, especially powders and particles, suffer from limited ion-accessible areas, which restrict the storage capability under large loads and worsen power output.[4] Therefore, it is crucial to prepare carbon materials with high ion-accessible areas and charge diffusion pathways to achieve high capacity at a large charging rate.

DOI: 10.1201/9781003387831-9

Owing to the large high surface-to-volume ratio, short ion/electron transport kinetics, and facile stress relaxation processes, one-dimensional (1D) nanomaterials, such as nanowires, nanorods, nanofibers, and nanotubes, have attracted much interest. Among them, carbon nanofibers (CNFs) present a higher aspect ratio, superior access to electrolytes, better conductivity along the longitudinal direction, and compatibility with other electrode materials. In this line, various strategies, such as chemical vapor deposition (CVD),[5] hydrothermal/solvothermal,[6] electrospinning,[7] self-assembly,[8] molten-salt method,[9] etc., have been adopted to prepare CNFs. Thereinto, electrospinning is regarded as the most facile and controllable method to prepare CNFs due to the following advantages: (1) tuneable morphology and functional property; (2) adjustable specific surface area and pore interconnectivity; (3) unidirectional electron flow and free-standing without the involvement of conductive agents and insulated binders when used as electrodes for energy storage devices; (4) easy tuning of properties through post-electrospinning treatments (hydrothermal/solvothermal method, CVD, electrodeposition, etc.)[10–13] In this chapter, we will introduce how the electrospinning parameters affect the structure of the as-prepared CNFs and their applications in typical electrochemical energy storage systems, such as SCs, lithium-ion batteries (LIBs), sodium-ion batteries (SIBs), potassium ion batteries (PIBs) and other advanced energy systems.

2 THE FACTORS AFFECTING THE PREPARATION OF ELECTROSPUN CARBON NANOFIBERS

There are many factors that affect the morphology and components of electrospun fibers. Both the electrospun process and post-processing have effects on the structure and morphology of electrospun CNFs. The solution parameters, electrospun operating parameters, and environmental parameters during electrospinning play an important role in regulating the structure and property of pristine nanofibers, while post-treatment processes make sense to accommodate the properties of final CNFs (Figure 9.1). In this section, the impacts of parameters in electrospinning and post-treatment on morphology and microstructure were fully summarized.

2.1 ELECTROSPINNING PARAMETERS

Given the diverse combinations of spinning precursors and receiving modes during the electrospinning process, the morphology and composition of the electrospun fibers are highly adjustable. Three basic aspects (precursor solution, operating parameter, and setup geometries) dominate the nanofiber regulation. For example, various continuous nanofibers with different surface states and components can be achieved by blending different components with a certain ratio in electrospinning solutions, during which polymer type, conductivity, viscosity, and solvent influence the quality of the as-prepared nanofibers. Most of the soluble polymers including natural biopolymers and synthetic polymers can be used for electrospinning if the molecular weight is high enough. These polymer solutions usually have non-Newtonian flow behavior. Natural biopolymers and synthetic polymers with proper molecular weight,

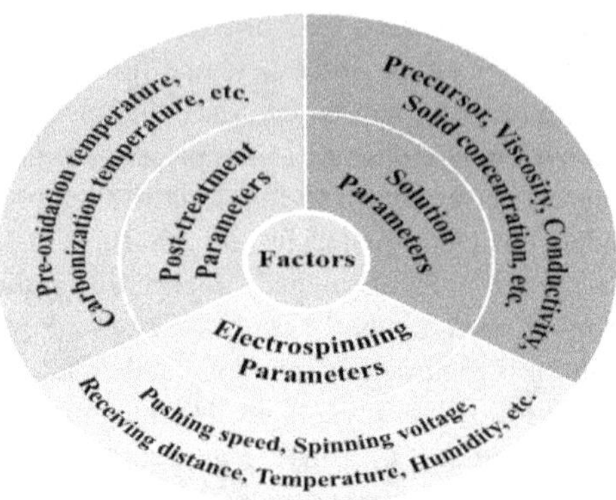

FIGURE 9.1 The factors influencing the fiber formation during electrospinning process.

such as silk fibroin,[14] chitosan,[15] polyacrylonitrile (PAN),[16] poly(vinylidene fluoride) (PVDF),[17] etc., have also been electrospun into nanofibers, leading to the diversity of the structure and properties of the electrospun fibers. However, conductive polymers with relatively low molecular weights and rigid backbones usually cannot be fibers due to poor solubility and inadequate chain entanglement in the solutions. Additionally, high solution conductivity makes it difficult to form a stable jet. Thus, only a limited number of conductive polymers have been electrospun into fibers by adjusting the parameters.[18–20] The viscosity of the precursor solution, which is related to the concentration and molecular weight of the polymer, determines the length and continuity of as-electrospun fibers.[21] The continuous fiber cannot be formed if the viscosity is really low due to the low polymer chain entanglements, while an unstable flow rate occurs when a very high viscosity of the solution is adopted.[5] McKee and his co-workers studied the dependence of the fiber diameter on the zero shear rate viscosity (η_0) and normalized concentration. They found that an increase in η_0 indicated a larger number of entanglement couplings, thereby generating larger electrospun fibers.[22] Yun et al. also demonstrated that the diameter of fibers decreases as the polymer concentration decreases.[23]

As for the electrospinning process, the applied voltage, feeding rate, temperature, humidity, collecting condition, etc., indirectly affect the morphology of these electrospun nanofibers because of the fact that these parameters usually have an appropriate range. To be specific, both high and low voltage can lead to more defects like beads and inhomogeneity of the as-spun nanofibers due to the mismatch of the interaction strength between jet and electric field, which could greatly reduce the properties of nanofibers. Generally, beads will occur in fibers when low voltage is used. Fine fibers can be achieved when the voltage increases in a certain range. Feeding rate is an important factor that affects the geometry and diameter of the as-electrospun fiber. Fiber diameters will be enlarged when the feeding rate increases. However, when the feeding rate is too high to over a threshold, too much solution is ejected without enough stretch, and hence nonuniform fiber or beads form. Whereas, when the feed rate is too low, needle-blocking issues will arise because of the evaporation of the solvent. Apart from the feeding rate, the distance between the spinneret and the collector is another factor which determines the unstable time of the jet before being deposited on the collector. To ensure full extension and sodiation of the jet, a long distance is required, resulting in thin fibers. The humidity and temperature during electrospinning also influence the morphology of the fibers by affecting the evaporation rate of the solvent, which in turn affects the solidification rate of the jet. Therefore, they should be controlled in a proper range in order to prepare high-quality nanofibers. For instance, a high environmental humidity will lead to water condensation on the fiber surface and generate more pores, which is favorable for ion adsorption when used as electrodes for energy storage devices.[24–26] Thin fibers with rapid ion efficiency and short ion/electron transport pathways can be fabricated when cutting down the humidity. However, the extension of the jet will be hindered if the humidity is too low due to the quick evaporation of the solvent solution. Similarly, the temperature should also be in a reasonable range. Although high temperature is beneficial to reduce the viscosity of the spinning precursor solution, it accelerates solvent evaporation and limits jet extension.

Except for the factors mentioned above, the needle and shape of the collector can also tailor the morphology and arranged state of nanofibers. Hollow and multichannel nanofibers instead of solid fibers can be fabricated by altering the needle tip design, and the mechanism of channel formation will be discussed in the following part (Figure 9.2a,b).[27, 28] By changing the collectors, nanofibers with different orientations and configurations can be obtained. Several spinneret collectors including metal plates, rotating drums, wheel-like bobbins, etc. are used to collect the fiber mats (Figure 9.2c).[29] Low-cost metal foil collector is mostly used to get the disordered mats.[32] Electrospun fibers could be aligned parallel to each other with high density when a drum rotating at a high speed or a rotating wheel-like bobbin is used as the collector.[33] Liquid solution can also be used as a collector to collect the as-electrospun nanofibers. As shown in Figure 9.2d, liquid nitrogen was used as a collector to gather the nanofibers. By regulating the post-treatment after electrospinning, fibers with different surfaces could be obained.[30] In addition to the use of a rotating drum and wheel-like bobbin, the combined use of conductors and insulators as collectors can also achieve the adjustment of the fiber arrangement. As illustrated in Figure 9.2e, two parallel gold electrodes are separated by an insulating gap of variable width.[31] It can be found that electrospun nanofibers are lined up across the gap. The uniaxially aligned array can also be achieved by changing the substrate with different bulk resistivity.

2.2 Post-treatment parameters

To further improve the properties of electrospun fibers in some features, the as-obtained electrospun fibers usually go through some thermal modifications. Various atmospheres such as air,[34] Ar,[32] N_2,[35] C_2H_2, H_2,[36] or the mixture of them[37] have been applied to prepare CNFs with different structures and components during which the decomposition of polymer and metallic precursor conversion has occurred. At the same time, the crystallization, surface functionalization, and structural development could be adjusted, resulting in various electrochemical performances. For example, the blending solution of polyureasilazane (PSN)/polyvinylpyrrolidone

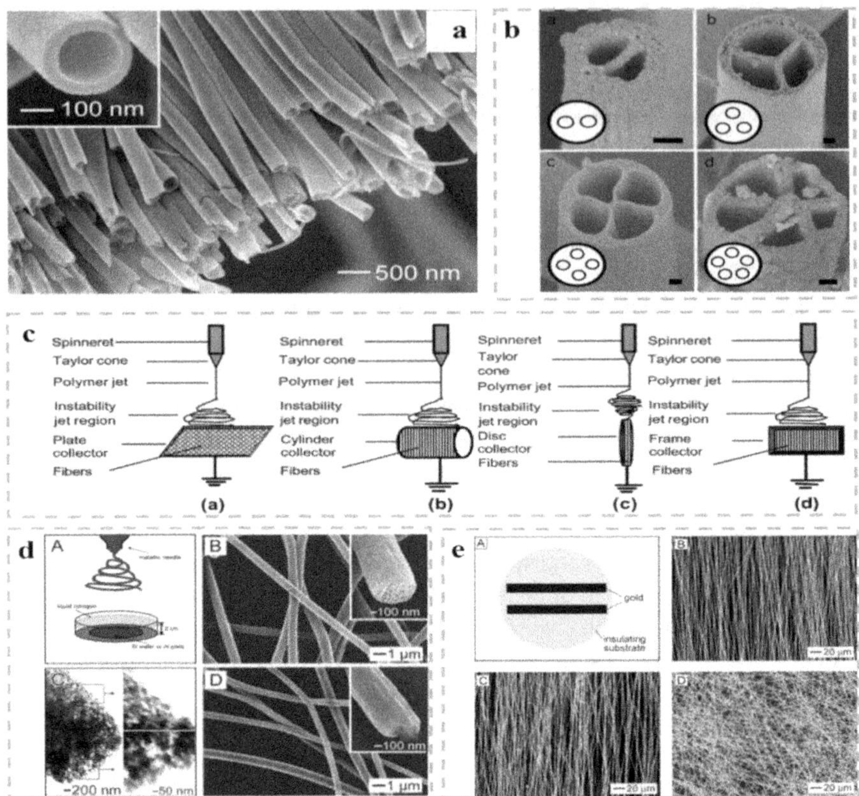

FIGURE 9.2 (a) SEM image of a uniaxially aligned array of anatase hollow fibers,[27] Reproduced with permission from ref.[27]. Copyright 2004, American Chemical Society; (b) SEM images of multichannel tubes with variable diameter and channel number,[28] reproduced with permission from ref.[28]. Copyright 2007, American Chemical Society; (c) types of nanofiber-collecting devices,[29] reproduced with permission from ref.[29]. Copyright 2007, Mary Ann Liebert, Inc.; (d) the electrospinning setup with the liquid nitrogen bath collector and the corresponding nanofibers,[30] reproduced with permission from ref.[30]. Copyright 2006, American Chemical Society; (e) schematic illustration of a test pattern that comprised two gold electrodes and nanofibers deposited on different insulating solid substrates of quartz wafer, polystyrene sheet and glass slide,[31] reproduced with permission from ref.[31]. Copyright 2004, John Wiley and Sons.

(PVP) was electrospun into nanofibers first. Afterwards, the as-electrospun fibers were cured in air at 200°C for 2 h followed by pyrolyzed under N_2 at 800°C to obtain porous hollow CNFs. In this work, PVP was used as a sacrificial template and porator. The gas release during the decomposition of PSN also contributes to the pore generation.[38] Chen and co-workers fabricated N-doped hollow carbon nanotube-carbon nanofiber (CNT-CNF) hybrid material by using poly(methyl methacrylate) (PMMA) solution as the inner fluid and a mixture of nickel acetate ($Ni(Ac)_2 \cdot 4H_2O$ and PAN as the outer fluid via co-electrospinning technique. A mixture gas of Ar/H_2 was used to form CNTs on the surface of CNFs, during which Ni acted as the catalyst for the growth of CNTs.[39] Similar to this work, Yu et al. fabricated root-whisker structured CNTs-CNFs network by altering PMMA as PVP.[37]

Apart from CVD, other technologies such as the hydrothermal method, sol-gel, and microwave were also combined with electrospinning to tailor the structural and compositional feature of the as-obtained materials. For example, hydrothermal-assisted electrospinning strategy was used to prepare worm-like $NiMoO_4$ decorated CNFs for SC and LIB appilication.[40] Porous TiO_2 nanofibers with good photocatalytic activities could be fabricated by collaborating sol-gel with electrospinning method. Microwave-assisted oxidation was introduced to regulate the surface state of electrospun CNFs. By changing the microwave treatment parameters, the structure, electron, and electrochemical properties of as-obtained CNFs could be adjusted.[41]

Clearly, as a basic technology, electrospinning can be easily integrated with other techniques to achieve competitive hybrid composites with unique structure and composition for energy storage applications. Notably, the electrospun CNFs can not only be directly used as electrodes but also can be used as a substrate for other materials' decoration, leading to composites with adjustable electrical, chemical, photonic, magnetic properties. In other words, there will be thousands of methods and means that can be used to regulate the performance of electrospun CNF-based materials. In the following section, the methods of electrospun CNF's structural regulation will be summarized.

3 THE METHODS FOR ELECTROSPUN CARBON NANOFIBERS' STRUCTURAL REGULATION

Briefly, novel strategies and design rules have been explored to boost CNFs' electrochemical performance via pore regulation, microcrystalline modulation, and surface chemistry adjustment (Figure 9.3). Pore size and distribution influence the electrochemical performance most via changing effective ion-accessible areas for energy storage. Generally, the materials with similar pore size to that of solvated ion of the used electrolyte and a hierarchically porous structure can maximize the energy storage performance.[42] As for microcrystalline, it mainly affects the electrical conductivity, which is an important factor for high power output. The ion-accessible area can also be influenced by wettability. Proper surface chemical composition is favorable for the enhancement of wettability as well as extra capacity contribution. In this section, we refine the structural regulation into three parts: pore structure modulation, microcrystalline structure modulation, and surface chemical structure adjustment.

3.1 Porous Structure

Ion diffusion, adsorption, and insertion are really important factors during energy storage. Proper porous structure, especially multi-scale pore structure, is highly desired due to the potential of the enhancement of effective ion-accessible area. In this part, we will examine some representative approaches, including the etching approach, template method, and compounding method, to create different types of pores and their incorporation into electrospun CNFs for enhanced charge storage performance will be examined.

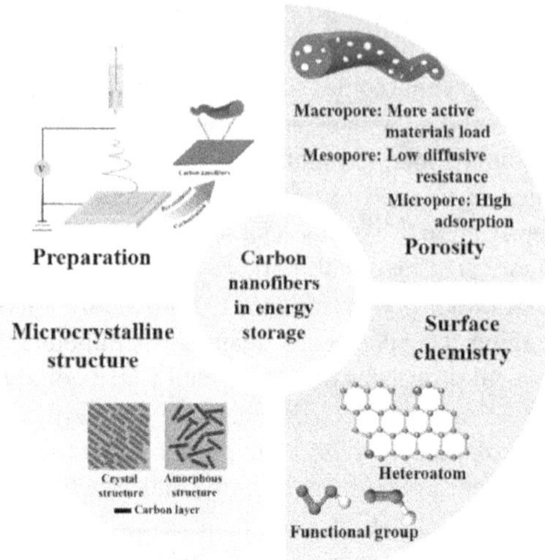

FIGURE 9.3 The factors influencing the electrochemical performance of the as-obtained CNFs.

3.1.1 Etching approach

In most cases, the surface area generated directly by the dehydration and dehydrogenation of organic compounds during the pyrolysis process is too low for efficient energy storage. To increase the specific surface area, an etching approach is usually adopted. Various carbon corrosion reagents, such as oxidizing gases (air, CO_2, and H_2O steam), acids, and alkalis, are frequently implemented to augment the ion-accessible surface area and capacity. It is known to all, a well-controlled etching treatment can give rise to a hierarchical porous structure. Herein, typically used etching approaches were summarized.

The etching reaction of electrospun CNFs using air, CO_2, and H_2O steam is usually regarded as a physical process during which carbon atoms are partially oxidized by these oxidants at high temperatures to release gases and generate pores. The activation effect of N_2, H_2O, and CO_2 on PAN-based CNFs has been deeply studied by Kim. The specific surface area and pore volume along with average diameter of the activated CNFs (ACNFs) are heavily dependent on the activation condition. The results show that the oxidation reactivity with the carbon of these activation agents follows in the order of $H_2O > CO_2 > N_2$. Compared with the inert N_2 and CO_2, the diameter activated by H_2O reduced the most. Pore size could be tailored from 0.64 to 0.81 nm and the specific surface area ranged from 404 to 1,624 $m^2 g^{-1}$.[43]

Electrospun-activated CNFs with hollow core/highly mesoporous shell structures were fabricated by combining the use of co-electrospinning and H_2O steam activation. By changing preparation parameters, a tri-modal pore distribution with three distinctive peaks at 4, 25, and 38 nm was achieved. The total specific surface, pore volume, average pore diameter, and mesopore average diameter are 1,191 $m^2 g^{-1}$, 0.99 $cm^3 g^{-1}$, 3.3 nm, and 17.9 nm.[44] Porous CNFs with a suitable pore size and porosity were also produced by electrospinning PAN/PVP blending followed by phase separation, carbonization, and CO_2 activation treatments. The specific surface area, total pore capacity, and meso/macropore capacity increased from $355.2\,m^2g^{-1}$, $0.134\,cm^3g^{-1}$, $0.112\,cm^3g^{-1}$ to $1,256.2\,m^2g^{-1}$, $0.232\,cm^3g^{-1}$, and $0.197\,cm^3g^{-1}$, respectively, indicating the positive role of CO_2 activation in enlarging porosity.[45]

As another important etching approach, chemical activation is often performed by immersing the as-spun polymer fibers or CNFs into an aqueous solution of activating agents, followed by carbonization and washing treatment.[38] In order to achieve similar porosity with low-dosage activating agents, a one-step carbonization/activation strategy was also explored. In this means, a small dose of activating agent was introduced into electrospinning solution, and activated CNFs could be obtained through pyrolysis during which activation takes place synchronously.[32] KOH is extensively used among various chemical reagents, the activation mechanism by KOH is complex and dependent on both the reactivity of carbon sources and the experimental parameters.[46] A commonly acknowledged understanding is the expansion of the carbon lattice caused by the metallic K intercalation, which originates from the dehydration and reduction of KOH. Ma et al. investigated the influence of the amount and mode of KOH addition (i.e. adding KOH into electrospinning solution directly, immersing the as-spun polymer fibers into KOH solution) on the development of pore structure evolution of CNFs.[47, 48] They discovered that when the concentration of KOH solution increased from 0 to 20 wt%, the specific surface area and total pore volume improved greatly no matter which KOH addition mode was chosen. Compared with the method of adding KOH into precursor solution (ACNF), the immersion mode (ICNF) affect the porosity better. Specifically, the specific surface area and total pore volume of the activated CNFs enlarged from $597\,m^2g^{-1}$ and $0.27\,cm^3g^{-1}$ for ACNF to $1,317\,m^2g^{-1}$ and $0.69\,cm^3g^{-1}$ for ICNF, respectively. As a result, ICNF delivers superior overall electrochemical performance than ACNF when used as SC electrode. Similarly, the electrochemical property of the porous CNFs can also be optimized via activating by NaOH,[38] $NaHCO_3$,[49] $ZnCl_2$,[50] H_3PO_4,[51] etc.

As outlined above, physical activation can flourish the porosity effectively. However, this strategy is usually plagued by low carbon yield and high energy consumption, while chemical activation is associated with issues of special equipment with strong corrosion resistance and extra cost for the removal of aggressive chemical reagents, thus limiting the widespread applications. Besides, only using the etching method to regulate the pore structure, the prepared CNFs tend to be mostly microporous, and it is difficult to achieve the precise regulation of the pore structure. Therefore, it is necessary to develop new pore-forming approaches to realize the directional design and regulation of the pore structure.

3.1.2 Template approaches

In comparison with the activation approach, the template method is perceived as one of the most popular strategies to precisely control the pore structure of CNFs, through tailoring either the type or the dosage of templates. Typically, a template in powder form is mixed evenly into the polymer solution used for electrospinning and the porous CNFs can then be obtained via the carbonization treatment, followed by removing the templates. The commonly used templates can be divided into inorganic and organic species.

Silica (SiO_2) is often selected as the inorganic template for fabricating hierarchical porous carbon nanostructures, owing to its wide availability, low cost in large quantity, and facile regulation to various morphologies and sizes.[52, 53] Hierarchical porous PAN-based CNFs were prepared by using a SiO_2 template and KOH-activated agent. Mesoporous volume changes ranging from 0.10 to 0.93 $cm^3 g^{-1}$ and total volume increases from 0.84 to 1.53 $cm^3 g^{-1}$ when the mass ratio of SiO_2 and PAN increases from 0% to 50%. Nevertheless, the removal of SiO_2 templates usually needs HF or hot concentrated alkaline solutions with acute toxicity and strong corrosivity. In comparison, many inorganic salts can be dissolved in the polymer precursor solutions, they are more versatile than SiO_2 templates for the engineering of porous CNFs. Based on the reaction types during the pyrolysis process, inorganic salt templates can be divided into stable templates and decomposable templates. NaCl, KCl, and K_2S with high thermal stability are typical stable templates.[54–56] For the latter, $Mg(NO_3)_2$,[57] $Fe(NO_3)_3$,[58] $NaHCO_3$,[49] etc., have been used as decomposable metal salt templates prepare CNF with mesoporous structure. Taking $NaHCO_3$ as an example, it decomposes into Na_2CO_3, CO_2, and H_2O at low temperatures, Na_2O occurs along with the increasing temperature (800°C). The mesoporous CNFs were obtained by immersing the carbonized nanofibers into water to remove the soluble Na_2O.[49] A high specific surface area of 724 $m^2 g^{-1}$ can be achieved. N-doped CNFs were fabricated by using an $Mg(OH)_2$/PAN precursor solution. During carbonization, $Mg(OH)_2$ was transferred into MgO, and the as-formed MgO template can be easily removed by HCl. The pore size of the resultant CNFs is mainly located at 2–4 nm, and the specific surface area can reach up to 926.4 $m^2 g^{-1}$, which was 4.5 times higher than that of the pristine CNFs without the addition of $Mg(OH)_2$.[59] g-C_3N_4 can also be used as a sacrificial template to prepare CNFs with mesoporous structure.[60]

Unlike inorganic hard templates, most of the organic templates not only play the performing role but also serve as the carbon resources. Till now, numerous polymer hybrids with different carbon yields, such as PAN/PMMA,[61] chitosan/PEO (polyethylene oxide),[15] PEA (poly(ether amide))/PVP,[62] PAN/pitch,[63] phenolic resin/PVA (polyvinyl alcohol), etc., have been used to prepare porous CNFs, during which the polymer with low carbon yield severed as sacrifice template. As a special polymer, PVP can not only be used as porogen but also as a carbon source. Since PVP can be dissolved in water, expect for pyrolyzation, hydrothermal, immersion in water can also be utilized to convert as-spun nanofibers into CNFs with controlled porous structures. Figure 9.4 illustrates the CNFs with different morphologies prepared by changing the emission way of PVP. As shown in Figure 9.4a, N,F co-doped mesoporous CNFs (NFMCNFs) were prepared by combining the use of electrospinning, hydrothermal, carbonization, and vacuum plasma treatment.[64] The relative content of PVP plays an important role in the pore structure evolution. The basic polymer frame of the PAN fibers swelled up during the PVP extraction and a lot of porous structures formed. As the content of PVP increases, the surface modification progresses more efficiently and dynamically. However, partial collapse occurred when the PVP ratio against PAN was increased to above 3. The PVP in as-spun PAN/PVP nanofibers was extracted by immersing into the deionized water with the assistant of ultrasonic-assisted vibration before carbonization. As depicted in Figure 9.4b, no intra-fiber pores can be detected from PAN and PAN/PVP electrospun nanofibers. However, after removing PVP, the PAN porous nanofibers with pore sizes ranging from several nm to hundred nm in diameter were obtained. The agglomeration of PVP in PAN/PVP blends should be the response to the enlarged pore size. Liu et al. prepared porous hollow CNFs by directly carbonizing the as-spun PSN/PVP nanofibers at 800°C (HCF800) followed by NaOH activation. The formation mechanism can be explained as follows: (i) A core-shell structure was formed in the PSN/PVP precursor nanofibers due to the difference of conductivity, in which PSN is mainly in the core, whereas the shell is PVP. (ii) Si in the PSN core can be removed by NaOH in the activation process. Besides, the NaOH activation can also introduce micro/mesopores into the fiber shell owing to its strong corrosion[38] (Figure 9.4c).

Nowadays, some fossil/biomass polymers, such as pitch,[65] lignin,[66] have been used to prepare CNFs, which can be regarded as an effective waste management approach. However, the spinnability and complicated pretreatments should be taken into consideration due to the low molecular weight. In this regard, some other organic compounds, such as terephthalic acid (TPA), have been explored as an additive for the maintaining of fibrous morphology. For example, cross-linked CNFs with good flexibility were fabricated by using TPA as a multi-functional additive. In this work, TPA can not only act as template to in situ create pores on/in the electrospun CNFs but also plays an important role in accelerating the polymerization of the pitch. When adopted as supercapacitor electrodes, the as-obtained porous CNFs exhibit high specific capacitance of 267 F g^{-1} at 0.2 A g^{-1} and outstanding rate capability (178 F g^{-1} at 100 A g^{-1}).[67]

Except for organic polymers, decomposable $Zn(Ac)_2$,[68] $Cu(Ac)_2$,[69] $Ni(Ac)_2$,[70] $Co(Ac)_2$,[71] $VO(acac)_2$,[72] etc., are frequently used as organic salt templates. Their pore-forming mechanism is quite similar to that of decomposable inorganic salts. Taking $Zn(Ac)_2$ as an example, the foldable CNFs could be produced by using $Zn(Ac)_2$–assisted electrospinning–preoxidation–carbonization strategy. ZnO was generated in situ during the decomposition of $Zn(Ac)_2$ in the carbonization process. The newly formed ZnO could oxidize the local carbon to form pores and defects, accompanied by the removal of ZnO itself though the formula of $ZnO + C \rightarrow Zn + CO_x$. Additionally, the utilization of $Zn(Ac)_2$ could enhance the mechanical flexibility of CNFs by relieving the stress concentration and creating porous structures.[68]

Metal-organic frameworks (MOFs) have been widely applied as organic templates for the construction of porous structures due to the controllable structure, and pore type since they were invented.[73, 74] Generally, MOFs can be used in two main synthesis approaches: (i) MOFs are deposited in situ or by polarity induction on the electrospun polymer nanofibers[75]; (ii) polymer/MOF solution is electrospun to form nanofibers followed by the carbonization and subsequent removal of templates.[76] For example, porous N-doped CNF assembled with interconnected carbon hollow nanoparticles was designed by carbonizing the electrospun zeolitic imidazolate framework (ZIF-8)/PAN composite precursor.[76] The majority of the pores fall in the range of 3–4 nm. Besides, the corresponding specific surface area can reach up to 417.9 m^2g^{-1}, which is larger than that of the N-doped carbon materials derived from ZIF-8 (223.1 m^2g^{-1}) and electrospun PAN nanofibers (8.7 m^2g^{-1}). More recently, a

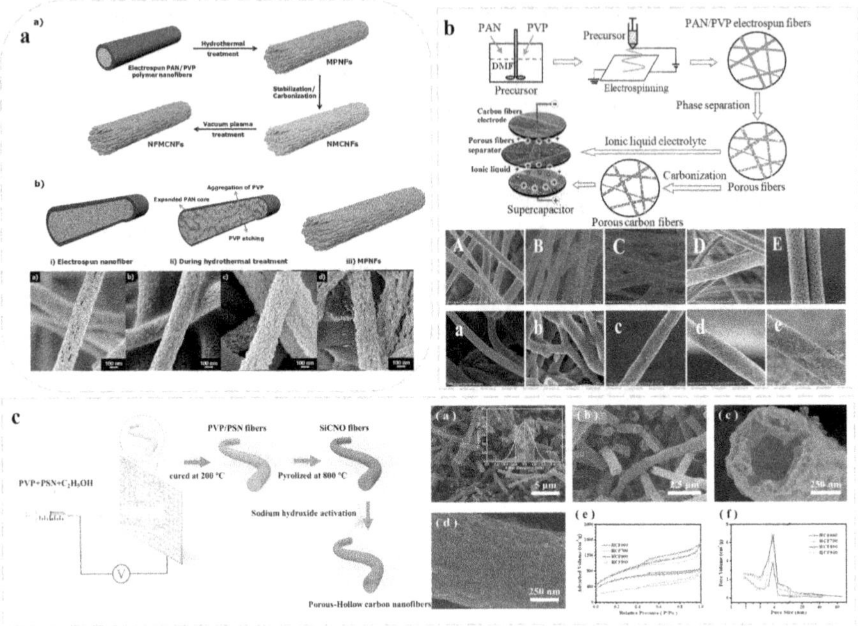

FIGURE 9.4 (a) Schematic illustration of the overall preparation process of NFMCNFs and SEM images of the MPNFs fabricated with various PAN/PVP mixed ratios in the sequence of 2:1, 1:1, 1:2 and 1:3, respectively,[64] reproduced with permission from ref.[64], Copyright 2017, Royal Society of Chemistry; (b) schematic diagrams of PAN-derived nanofiber supercapacitor and the corresponding SEM images of as-spun and carbonized nanofibers with different of PAN/PVP ratios (10:0, 10:1, 10:5 and 10:10),[45] reproduced with permission from ref.[45], Copyright 2018, Elsevier; (c) schematic illustration of the preparation of porous hollow carbon nanofibers and typical SEM images of the activated products of HCF800 under different magnifications and the associating textural structure,[38] reproduced with permission from ref.[38], Copyright 2020, American Chemical Society.

monolayer sheath of MOFs was decorated on the electrospun polymer nanofibers by Gong and co-workers.[75] Compared with PVA and polylactic acid (PLA), PAN nanofibers were more favorable for the formation of the monolayer sheath of MOF particles with excellent uniformity in size and morphology owing to the existence of a superior polar side group (–CN). This general approach could also be expanded to the growth of other types of MOFs on the electrospun PAN nanofibers, such as Cu-BTC (trimesic acid), Ni-2-MIM (2-methylimidazole), Ni-BTC, and Mn-BTC. Excellent electrochemical performance could be achieved after transforming these PAN@ZIF-8 nanofibers into N-doped porous CNFs with a conductive carbon core and a monolayer coating of hollow carbon frames.

3.1.3 Other special approaches

It has been demonstrated that designing the nanostructure of materials could greatly expand their application. As mentioned in Section 2.1, coaxial electrospinning is widely used to prepare hollow micro/nanofibers. In general, the hollow fibers can be obtained either by core extration[38] or by core decomposition.[77] The former approach uses special solvents to selectively dissolve only the core materials, just like He's work.[45] While the other method uses heat to remove the core materials. Based on a similar mechanism, multichanneled structure[28] could be achieved by altering the number of capillaries in the spinneret. For example, as shown in Figure 9.2b, a mixed solution of $Ti(OiPr)_4$ and PVP and commercially available paraffin oil served as an outer liquid and inner liquid, respectively. A spinneret with different numbers of capillaries was used to prepare fibers with various channels. By removing the organics of as-spun products through calcination, multichanneled TiO_2 was obtained.[28]

Except for the multichannel structures, more complex nanostructures, including fiber-in-tube,[78] tube-in-tube,[79] and even modified fiber-in-tube,[80] were also fabricated afterward. The fibers with fiber-in-tube structures were prepared by using tricoaxial electrospinning. In this work, the paraffin oil emulsion serves as the sacrificial phase that separates the core and shell fluids ($Ti(OBu)_4$ sol) during spinning and results in as-spun nanofibers with a sandwich-like structure. After calcinating, fiber-in-tube structure was fabricated. The flow rate of the fluid plays an important role in structure regulation. As illustrated in Figure 9.5a, the diameter of the fiber inside the tube was enlarged when the flow rate of the inner fluid increased from 0.5 to 1.0 mL h^{-1}. By altering the flow rate of the middle fluid, the diameter of the channel could be changed.[81] Interestingly, such channels can also be filled with other substances. For instance, Si nanoparticles were filled into the fiber-in-tube structure by using styrene-co-acrylonitrile as the sacrificial carrier of Si nanoparticles.[80] In this work, a distinctive tri-layered structure consisting of a carbon cylindrical core, Si nanoparticles, and a carbon tubular shell was achieved. The average C-shell thickness and C-core diameter were 127.0 nm and 315.8 nm, respectively (Figure 9.5b). For energy storage applications, ion-accessible area is quite an important factor that influences the electrochemical property, and the nanofibers with tube-in-tube have attracted strong interest due to their abundant inner surface, which is a benefit for extra contact between the electrolyte and electrode, thus providing higher capacity (Figure 9.5c).

Compared with the solid or hollow nanofibers, although nanofibers with multichannel structure, or any other complex nanostructures have considerable advantages, such as larger surface-to-volume, independent channels for ion transfer or for functionalization, superior mechanical stability, etc., such complex structures were hard to practical control, which in turn impedes their extensive application.

3.2 Surface Chemistry

Chemical doping of carbon skeletons with heteroatoms, including but not limited to nitrogen (N),[83] sulfur (S),[67] oxygen (O),[41] phosphorous (P),[84] boron (B),[85] fluorine (F),[64] etc., by which the electronic structure, porosity, interlayer spacing and surface chemistry of the host CNFs could be manipulated, is regarded as another powerful strategy to enhance their electrochemical activity.[3, 10, 86]

Basically, N/B-doping is commonly identified as substitutional doping by the way of replacing carbon atoms in the sp^2 lattice with newcome atoms.[87] To be specific, the three valence electrons of B could result in higher charge carrier concentration and density of states at the Fermi level, acting as an electron acceptor. Given the lower

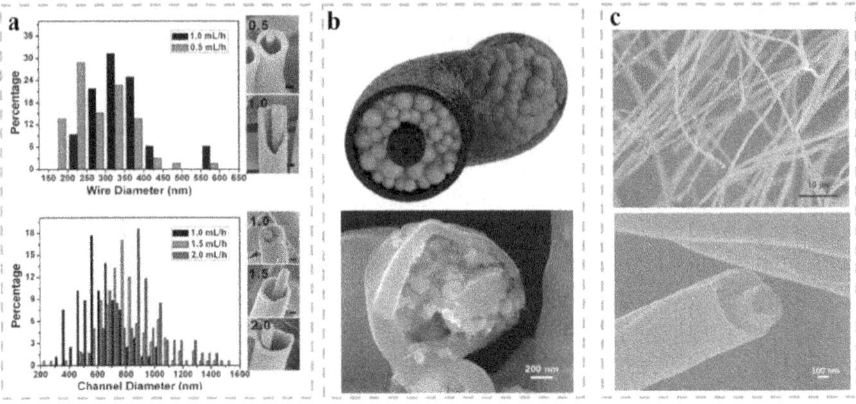

FIGURE 9.5 (a) The influence of the inner flow rate (up) and middle flow rate (down) on the size distribution and morphology regulation.[81] Reproduced with permission from ref.[81], Copyright 2010, American Chemical Society; (b) schematic diagram (up) and SEM image (down) of the well-prepared tri-layered C-core/Si-medium/C-shell nanofiber.[80] Reproduced with permission from ref.[80], Copyright 2014, Royal Society of Chemistry; (c) the nanofibers with tube-in-tube structure under different magnifications.[82] Reproduced with permission from ref.[82], Copyright 2014, IOP Publishing.

electronegativity of B (2.04) compared with C (2.55), B–C bond can be formed more easily than that of C–C bond. Besides, when B is doped into the carbon matrix, the π electrons will be redistributed. As a result, the C–C bonds will be softened and the C–O bond will be strengthened.[85] It is also demonstrated that the functional groups formed by N with C are more electrochemically stable than the C–O group and can serve as electron donors. The higher electronegativity of N (3.04) results in better electrolyte ion adsorption.[88, 89] F goes the similar way with N and B, and it can also lead to the charge redistribution among carbon atoms due to the electronegativity difference between them.[90] However, given the specificity of F sources and low doping level, there is an evident shortage of research on F-doped CNFs.[91] Compared with F_2 and HF, polyvinylidene fluoride (PVDF) is considered a stable and safe F-containing precursor. Gong et al. prepared N,F-co-doped microporous CNFs by electrospinning the mixture of PAN/PVP/PVDF (mass ratio = 1:1:1) followed by hydrothermal removal of PVP and carbonization. The maximum F content was only 2.81% by changing the PVDF dosage.[91] Therefore, more studies should be needed to regulate the F-doping concentration. As for P and S, there is no unified conclusion as to whether their doping is a substitution mechanism.[92, 93] However, there's no doubt that they play a vital role in adjusting the electrochemical properties of carbon materials.[94, 95] It has been reported that P-doping can enhance the structural/interfacial stability, facilitate the adsorption of alkali metal ions and improve the electrochemical reactivity of carbon.[96–98] Wu and his co-workers prepared P-doped hard carbon nanofibers by using PAN and H_3PO_4 blend as electrospinning solution. The homogenous distribution of P and the unique 1D nanofibers with macroporous morphology endow the as-obtained P-doped hard carbon nanofibers with excellent electrochemical performance when adopted as anode for sodium-ion storage.[84] Due to the large covalent radius of S, carbon skeleton formed cannot maintain the planar structure when the substitution of S occurs in the sp^2 carbon framework.[99] The –C–S–C– group and S=O group could affect the π electronic structure and provide extra pseudocapacitive performance. Additionally, S-doping can also enlarge the interlayer spacing, which is good for the accommodation of more electrolyte ions. And the reduced specific surface area is also essential for the optimization of the coulombic efficiency.[100] Thiourea,[101] $(NH_4)_2SO_4$,[67] KS_2,[54] sublimed sulfur[100] have been adopted as S precursor. Unfortunately, the doping of S is accompanied by a pungent odor, which limits the applied studies to some extent. Additionally, compared with single-heteroatom doping, multi-heteroatom doping in CNFs could generate a synergistic effect of different dopants, leading to an enhanced performance.

Generally, there are two methods for producing nonmetallic element-doped CNFs: in situ method[62] and ex-situ method (post-thermal treatment method).[83] The in situ modulation approach can also be regarded as a self-doping strategy, where the precursors containing both carbon and heteroatom elements, such as PAN,[59] polybenzimidazol (PBI),[102] are used in the synthesis process. Another situation, i.e., carbon and heteroatom precursors are subjected together to prepare heteroatom-doped carbon, should also be considered as in situ method. In this approach, the type and content of heteroatoms are more influenced by the precursor. Thus, to achieve a high doping level, some precursors with high heteroatom content like g-C_3N_4 ,[60] melamine,[103] red phosphorus (Red P),[104] etc., were incorporated into the electrospinning process. For example, the N content in g-C_3N_4 involved CNF (NMCNFs) is about 8.6 wt.%, which is much higher than that in pure PAN-based CNFs (1.0 wt.%). The results show that NMCNFs have much more pyridinic N and quaternary/graphitic N than pyrrolic N and pyridinic N oxide, which is beneficial for the improvement of the electrical conductivity of the CNFs.[60] Unlike in situ doping, ex-situ doping is usually used to introduce heteroatoms into the existing carbon skeleton by post-treatment of the already obtained carbon materials, including but not least, chemical etching,[83] microwave irradiation,[41] plasma treatment,[105] etc. Our group increases N content in as-obtained NCNFs by post NH_3 activation.[83] Pitifully, NH_3 gas with a pungent smell is chemically aggressive and poisonous, which would need extra control as far as safety is concerned. Vacuum plasma approach was adopted by using C_4F_8 as F source to fabricate F-doped CNFs (NFMCNFs). The F content in CNFs can achieve up to 10.9%, semi-ionic and covalent C–F bonds could be deconvoluted.[64] Recently, the hydrothermal method was also applied to make high F-doping levels possible. For example, NH_4F[106] and $NaBF_4$[107] were used as F sources to decorate graphene and carbon nanotube fiber bundles via hydrothermal strategy. High F content as well as good overall performance can be achieved. These works pave the way to explore unique F-doped CNFs with more safe, effective method and F sources.

3.3 Microcrystalline Structure

A high conductivity is another important factor for the application of CNFs in energy storage devices. Although conventional CNFs are conductive, they are not conductive enough for high-performance devices. It is well known that conductivity is closely related to the microcrystalline structure, in order words, the degree of graphitization, which is related to the ratio of G and D band intensity (I_G/I_D) calculated from Raman spectra. To enhance the conductivity of carbon materials, high-temperature carbonization and graphitization are usually needed, which is very energy-consuming. Furthermore, high-temperature treatment is not conducive to the maintenance of fiber film flexibility and high heteroatom doping levels. Therefore, it is of great importance to find a feasible way to improve the conductivity of CNFs under relatively low temperatures.

Alternatively, incorporating carbon materials or specific metals, nonmetals, and their compounds into electrospinning solution, constructing unique cross-linking structured CNFs have been acclaimed as effective methods to improve the conductivity of CNFs prepared under mild thermal conditions. In this part, various strategies that can increase ordered structure in CNFs will be introduced.

3.3.1 Integration with conductive materials

The introduction of CNTs,[108] graphene,[109] graphene oxide,[110] carbon black (CB), and other types of conductive materials[111] into the electrospun CNFs can not only promote the electrical conductivity but can also affect the overall architecture, hence boosting the electrochemical properties. They can be added directly into the electrospinning solution with the help of proper solvent or through post-modification of the CNFs. Typically, the PAN/CB blend complex was electrospun into CNFs by combining the use of air stabilization and carbonization process. The results demonstrated that the introduction of CB additives increased the electrical conductivity over 4.5 times.[112] CNTs were also introduced into the electrospinning solution with the help of sodium dodecyl sulfate. About two times increase in the electrical conductivity was achieved due to the induction effect of the transition of amorphous to a graphitic structure. The I_D/I_G ratio decreased from 1.39 to 0.97. Deng confirmed that multiwalled CNFs could help reduce the activation energy of the oxidative stabilization of cellulose nanofibers. They could also increase the crystallite site, structural order, and electrical conductivity of the activated CNFs. The positive role in the formation of highly ordered graphitic interphase around the embedded CNTs was also evidenced by Cai et al.[113] Recently, Zhao et al. reported a simple strategy to integrate a continuous conductive network and a well-developed porous structure into CNFs by adding graphene quantum dots (GQDs).[114] The uniformly embedded GQDs play a crucial bifunctional role in constructing an entire reinforcing phase and conductive network. Compared with the pristine CNFs, the GQD-reinforced activated CNF exhibits a significantly improved conductivity and strength of 5.5 and 2.5 times, respectively. Despite some gratifying progress that has been made during the past years, a higher content of additives usually leads to a poor electrospinnability of the precursor solution. Thus, proper post-chemical modification of electrospun CNFs is highly needed.

In this line, 3D-activated carbon nanowhisker-wrapped graphitized CNFs were fabricated by He et al. Thanks to the optimized 3D electrical conductivity network (5.4 S cm^{-1}), the SC based on the as-prepared CNFs displays excellent rate performance and power density.[115] Vertically aligned CNTs (VACNTs) were directly grown on the electrospun CNF by using $Ni(Ac)_2$ as a catalyst.[116] The structure and morphology of VACNTs/CNFs could be precisely tuned and controlled by adjusting the percentage of reactants. Besides, the desired VACNTs/CNFs could not only possess high electric conductivity for efficient charge transport but could also increase surface area for accessing more electrolyte ions.

3.3.2 Catalytic graphitization

Catalytic graphitization is another effective strategy for the preparation of carbon materials with a high degree of graphitization under mild thermal conditions. Till now, there are two major graphitization mechanisms, namely the dissolution–precipitation mechanism and the carbide conversion mechanism.[117–119] The former always relates to transition metals, such as Fe, Co, and Ni. During high-temperature treatment, they can dissolve amorphous carbon to generate solid solutions, and then the catalytic graphitization reactions tend to take place according to the dissolution–precipitation mechanism. The other one always takes place between high-valence metals (such as Ti, V, and Mo) and carbon based on the formation of carbides, which will then decompose into metal vapors and graphitic carbons at high temperatures.[87]

Hollow-tunneled graphitic CNFs through Ni-diffusion-induced graphitization were prepared by Chen and his co-workers.[70] In this work, $Ni(Ac)_2$ was used as Ni source. Ni formed by the decomposition of $Ni(Ac)_2$. The Ni-induced graphitization by diffusing Ni nanoparticles from graphitic carbon spheres into N-doped amorphous carbon nanofibers, thereby turning amorphous carbon into graphitic carbon and producing a hollow-tunnel structure in electrospun carbon/Ni nanofibers. After being treated with acid, a novel architecture of hollow-tunneled graphitic CNFs was created. The low interlayer spacing (0.34 nm) and high ratio of I_G/I_D (1.84) compared with controlling CNFs (0.43 nm and 0.97) indicate a great increase in the degree of graphitic crystalline structure owing to the use of Ni catalyst. $FeCl_3$[120] and iron (III) acetylacetonate[121] can also be added to the polymer solution. The catalytic effects of the Fe species accelerate the graphitization of CNFs at a relatively low temperature as evidenced by the larger I_G/I_D values. As a consequence, the CNFs with satisfactory electrical conductivity can be obtained.

In addition to the above metal-based species, some nonmetals have also been demonstrated to be graphitization catalysts for CNFs. For example, B and Si are the canonical candidates to improve the electrical conductivity via the typical carbide conversion mechanism.[85, 117] Given that a high temperature of over 2,000°C is needed when they are expected to improve the degree graphitization effectively, which in turn hinders their further research, especially in the field that flexibly is needed. However, B-doping under mild temperatures can also give rise to the strengthening in overall electrochemical performance by extra pseudocapacitance.

3.3.3 Construction of the cross-linked structure

As we all known, the electrical conductivity of carbon materials is related to the arrangement of carbon atoms in microcrystalline structure. Therefore, the strategies that can regulate microcrystalline structure could be used to adjust electrical conductivity. Cross-linked CNF (CLCF), in which no interface between two adjacent nanofibers could be found, was obtained by controlling the evaporation rate of the solvent. The CLCF electrode shows smaller internal resistance drops than that of conventional CNFs at the given current density, indicating a faster electronic response.[122] This work opens a new view to modulate the electrical conductivity of

CNFs by constructing cross-linked structures. Inspired by this study, more literatures had been made to obtain CNFs with proper cross-linked structures via altering pre-oxidation parameters of PAN, and some positive results were achieved.[123, 124] However, the state-of-the-art of the as-obtained cross-linked CNFs suffer from the low specific surface area, and thus, further improvement is encouraged.

Guo prepared CLCFs by taking advantage of the strong coordination of PVP with $Cu(NO_3)_2$, which acts as a cross-linking agent. The lower electrical resistivity of CLCFs (19 Ω cm) in comparison with traditional CNFs (52 Ω cm) indicates that the cross-linked structure is helpful to improve the conductivity of the film.[125] The researchers from Institute of Coal Chemistry (Chinese Academy of Sciences) use HNO_3 and TPA to intervene in the pre-oxidation process and the CNFs with well-balanced cross-linked structure and good flexibility could be achieved. When used as electrodes for energy storage, attractive rate performances and power output can be achieved.[65, 126, 127]

4 THE ELECTROCHEMICAL APPLICATION OF ELECTROSPUN CARBON NANOFIBERS

Energy issues and environmental pollution problems have become one of the most prominent global dilemmas, in contrast to upgrading the existing coal chemical system, pioneering impactful electrochemical energy storage systems hold great promise toward a sustainable energy world. As mentioned before, by changing the component of the precursor solution or structural design strategy, CNFs with various nanostructures and compounds could provide a more efficient ion-accessible area, short ion/electron transport length, and tunable surface state for energy storage. Owing to the versatility of electrospinning, this technique has been widely used to fabricate important components of SCs, LIBs, SIBs, etc., to realize improvements in electrochemical activity. In this section, we review the progress of the application of CNFs in some typical energy storage devices.

4.1 Supercapacitor Application

As one of the most promising energy storage devices, especially in fields that need prompt energy, SCs have gathered tremendous interest for their impressive power density, excellent rate capability, as well as long lifespan. More importantly, the adaptability of SCs to the wide temperature range has also further expanded their applications. The commonly used CNF-based materials include pure CNFs and CNF-based composites (such as CNF/metal compounds, CNF/conductive polymers), which follow the electrostatic absorbance mechanism and the Faradic electrochemical storage mechanism.

As typical carbonaceous materials based on the electrostatic absorbance mechanism, the ion-accessible area for ion adsorption is essential. Therefore, various activated CNFs with high specific surface area, good electronic channels had been fabricated by adjusting preparation strategy. Ma et al. studied the influence of the effective electrochemical surface area caused by KOH activation approach on the electrochemical performance. In their works, certain amount of KOH was introduced into the solution[47] or coated on the as-spun fiber's surface,[48] and then one-step carbonization/activation was conducted. Under the same dosage of KOH utilization, the specific surface areas and specific capacity of the activated CNFs were 597 vs. 1,317 $m^2 g^{-1}$ and 256 vs. 362 F g^{-1}, respectively. The improved ion-accessible area should be responsible for the enhancement of specific capacitance. Nevertheless, the capacity under high current load was quite unsatisfied. Thus, heteroatom doping, hierarchical porous construction, and conductivity regulation were carried out to optimize the overall performance.[128–130] And some remarkable results had been achieved. However, due to the energy storage mechanism of electrostatic adsorption, the energy density of the CNF electrode is difficult to continue to increase significantly, and for this reason, the researchers combined it with the pseudocapacitive material aiming to further improvement of energy density. For example, bi-metallic MOF-generated phosphides within and outside the hollow CNF electrode shows a high specific capacitance of 1,392 F g^{-1} at 1 A g^{-1} and good cycling stability, with capacitance retention of approximately 89% at 25 A g^{-1}. The enhanced electrochemical performance should be ascribed to the combination of Faradic reaction and electrostatic adsorption. Based on similar concept, metal oxide, sulfide, and conductive polymers were also cooperated with CNFs and the energy density of the as-assembled SCs improved signicantly.[11, 12, 40]

Inspired by the aforementioned works, one can conclude that in order to improve the electrochemical performance of the SCs, the electrospun CNFs should be designed according to the following fundamental requirements: (1) well-developed pore channel structures with abundant electrochemical active sites ensure the well-infiltration as well as diffusion and adsorption of electrolyte; (2) continuous conductive networks for fast electron transport; (3) chemical stable fiber skeleton with good mechanical strength and richful adsorption sites for pseudocapacitive materials loading and volume change.

4.2 Battery Application

Compared with SC, secondary batteries are widely studied in portable electronics and electric vehicle applications due to the merits of high energy density, no memory effect, and relatively slow self-discharge rate.[131] Due to the development process of the technology, various kinds of batteries, such as LIBs, SIBs, PIBs, etc., have been explored. The conventional micrometer-sized powder materials suffer from long ion diffusion distance, limited electronic transport efficiency and insufficient accommodation room for releasing strain, resulting in unsatisfied rate performance and cycling lifetime.[132] To address these issues, one of the promising strategy is optimizing the structure design and component via nano engineering. Electrospun CNFs can provide enough electrode-electrolyte interface, and short ion/electron transport length along the longitudinal direction. Moreover, by changing fabrication parameters, CNFs with various kinds of components and microstructures could be achieved. To be specific, the microstructure of CNFs could be easily regulated from amorphous to partially graphitic, which could provide more voids and adjustable interlayer spacing for capacitive metal ion storage and reversible plating/striping of metal clusters. Furthermore, heteroatom doping and compounding can also improve the overall performance. In the case of metal compounds or alloy-type materials, CNF is also a promising choice to solve many problems (huge volume change, unstable SEI, severe aggregation, etc.) facing them during cycling. In this section, CNF-based anode and cathode materials will be discussed, focusing on the effects of the structure design and component regulation on the electrochemical performance.

4.2.1 Cathode materials

Similar to EDLC application, CNF with high specific surface area is essential for battery cathode application. However, the utilization of CNF-based materials in cathode is afflicted by the unsatisfied capacity despite their good rate performance and long lifespan due to the adsorption mechanism. For example, activated CNF with abundant micropores displays no capacity fading even after 1,000 cycles at 2 A g^{-1}, pitifully, only a low discharge capacity of 80 mAh g^{-1} at 0.1 A g^{-1} was achieved in Li^+-storage.[133] Therefore, to further enlarge the application of CNF, a CNT/C/$LiFePO_4$ hybrid nanofibers were obtained. Owing to the conductive network and short Li^+ diffusion pathways provided by CNFs, the composite cathode delivers a high capacity of 169 mAh g^{-1} and good rate capability.[134] Similar results can also be found in other cathode materials (such as $LiCoO_2$, $LiMn_2O_4$, $LiNi_{0.5}Mn_{1.5}O_4$, V_2O_5, etc.).[135–137] For SIBs, the mostly studied cathode materials are phosphate-based materials.[138] Like the cathode materials of LIBs, the cathode materials used for SIBs also face the problems of low conductivity, structural instability, and slow ion dynamics. Liu et al. reported electrospun $Na_3V_2(PO_4)_3$/C nanofibers as stable cathode materials for SIBs. In this work, $Na_3V_2(PO_4)_3$ nanoparticles with 20–30 nm were uniformly encapsulated into CNFs. The unique 1D morphology connected with efficient electrochemical coupling and conductive network leads to superior rate capability. However, only 101 mAh g^{-1} could be achieved at 0.1 C in its initial cycle.[139] Reasonable structure design and microcrystalline regulation may be the breakthrough to improve the properties of $Na_3V_2(PO_4)_3$ cathode. In this line, budding willow branches shaped $Na_3V_2(PO_4)_3$/C nanofibers were synthesized with the help of PVP, in which $Na_3V_2(PO_4)_3$ was the preferential growth along the (113) plane. Such unique metrics played a vital role in improving the capacity, cycling stability, and rate performance due to improved active sites. The cathode exhibited an initial specific capacity of 106.8 mAh g^{-1} at 0.2 C, still stabling at 107.2 mAh g^{-1} after 125 cycles. Moreover, a high capacity retention of 95.7% was obtained when it cycled stepwise from 0.2 to 2 C.[140] The high theoretical specific capacity also make $Na_{6.24}Fe_{4.88}(P_2O_7)_4$ (117.4 mAh g^{-1})[141] and $NaVPO_4F$ (143 mAh g^{-1})[142] a competitive cathode candidate for SIBs. Except for the directly Li/Na-storage application, CNFs can also be used as scaffold or host to support S or modification layer or used as a scaffold for metal compounds deposition to further enhancing the energy density.[63, 143–146]

The larger ionic radius of K^+, compared with Li^+ and Na^+, causes significant challenges for PIB cathode development. The currently researched cathodes are layered transition metal oxides, Prussian blue analogs, polyanionics, and organic materials.[147–149] Poor interfacial stability is one of the most critical issues for the utilization of these cathodes and various modifications have been done to address this problem. Unfortunately, few attentions have been paid to the composites of CNF and these cathodes.

4.2.2 Anode materials

Compared with cathode, CNFs with tunable ordered structure have been widely used as anodes. For instance, by the introduction of air into the Ar flow during carbonization, flexible highly porous electrospun CNFs (HPCNFs) with a large number of micro/mesopores and higher disorder degrees were created. The as-obtained HPCNFs deliver high reversible capacity (1,780 mAh g^{-1} after 40 cycles at 50 mA g^{-1}), ultralong cycle life (1,550 mAh g^{-1} after 600 cycles at 0.5 A g^{-1}) and good rate performance (200 mAh g^{-1} at 25 A g^{-1}) when adopted as anode materials for LIBs. The outstanding overall electrochemical performance can be ascribed to the unique architecture, in which, pores provide good access to the electrolyte of the electrode surface and facilitate fast charge transfer across the electrode/electrolyte interface. 3D interconnected network decreases the connect resistance obviously.[150] Due to the larger ionic radius of Na^+, the electrode materials suffer from more serious volume fluctuation and sluggish kinetics as well as inferior cycling life in comparison with LIBs. Thus, the electrospun CNFs with large interlayer spacing are really needed. The flexible multichannel CNF film with a dilated interlayer spacing of 0.398 nm delivers a reversible and high specific capacity of 222 mAh g^{-1} at 0.1 A g^{-1} after 100 cycles. A high capacity of 139 mAh g^{-1} could be remained even after 700 cycles at a current density of 0.5 A g^{-1}.[151] To further improve the electrochemical performance of CNFs, a cross-linking strategy was proposed.[125] In this work, the strong coordination of PVP with $Cu(NO_3)_2$ should be responsible for the formation of the cross-linked structure. Compared with the CNF without cross-linked structure, the cross-linked CNF exhibited an excellent rate performance and long cycle stability when used as a binder-free anode for SIBs. The enhanced rate capability raised from cross-linked structure could also be found in LIB and PIB applications.[127]

It is well accepted that heteroatom doping is an effective way to regulate the electrical conductivity and increase lithium capacity via heteroatomic defects.[131] P-doped carbon nanosheets/nanofibers interwoven paper shows enhanced specific capacity, excellent rate capability, as well as superior cycling stability compared to the undoped sample.[152] An imidization strategy was introduced to prepare interconnected N-doped CNF. Thanks to the high N-doping level and the uniformly distributed microporous materials along with excellent structural stability, the as-prepared presented a stable capacity of 377 mAh g^{-1} at 0.1 A g^{-1} after 100 cycles and retained 154 mAh g^{-1} when the current load was up to 15 A g^{-1}. Additionally, a high capacity of 210 mAh g^{-1} could be maintained even after 7,000 cycles at 5 A g^{-1}. Taking advantage of the stability and good conductivity, CNFs were also incorporated with Si,[80] $CoSnO_3$,[153] Fe_3O_4,[120] titania[154] to solve the volume expansion and short lifetime issues facing in their application.

Recently, PIBs have also attracted great attention due to the following reasons: (1) The lowest reduction potential of K/K^+ redox couple in propylene carbonate (PC) compared to Li/Li^+ and Na/Na^+, endow PIB cell with a wider potential window, which is good for high energy density.[155] (2) The weaker Lewis acidity of K^+, resulting in smaller solvated ions, endowing solvated K^+ with higher conductivity and mobility.[156] CNFs, which are composed of short, randomly oriented and cross-linked turbostratic structure with an interlayer spacing of 0.375 nm, are prepared and applied as anode for PIBs. Due to the 3D conductive network, turbostratic structure as well as large interlayer spacing, the CNF electrode delivers an impressive rate and discernible cycling performance.[127] Adams et al. compared the effect of O-doping and N-doping on the K-storage performance. The results demonstrated that oxygen functionalization did not significantly increase the specific capacity as compared to N-doped CNF.[157] Liu and her co-workers confirmed that the extra capacity in N-doped CNFs was mainly generated in the whole voltage window range, which corresponded to both surface/interface K^+ adsorption due to the pyridinic N and pyrrolic N-induced atomic vacancies and layer-by-layer intercalation due to the effects of graphitic N.[158] Moreover, N-doping into CNF is favorable for the adsorption of P atoms into the CNF, which provide a promising way to further enhancement of K-storage by P-doping into N-doped CNF. And the results shows that unprecedented long cycle life with high reversible capacity (465 mAh g^{-1} at 2A g^{-1} after 800 cycles) as well as outstanding rate capability (342 mAh g^{-1} at 5 A g^{-1}) is realized.[159] Apart from heteroatom-doped CNF, the composites composed of CNF and other types of anode materials, such as Sb and its derivative,[160] transition metal compounds,[161–165] etc., were also explored and great progresses had been made due to the synergistic reaction between them.

The results mentioned above demonstrates that when CNFs are adopted as SC electrodes and battery cathodes, a well-balanced conductivity and defective structure with high specific surface area is essential for achieving high ion-accessible area, which is an important index for energy storage. When they are applied as the battery anode electrodes, the following mechanism are usually obeyed: (1) the rich defective sites improve the capacitive adsorption of metal ions, leading to improved capacity; (2) highly porous 1D morphology provides a short transport distance, benefiting to good rate performance; (3) the turbostratic and graphitic structure in CNFs can be used to store metal ions via intercalation, while microporous and voids are useful to capture metal clusters by reversible plating/striping; (4) the free-standing skeleton is favorable to further modification.

4.2.3 Interlayer materials

In addition to the conventional application in cathode and anode, the CNFs with high conductivity and surface area could be used as interlayer between electrodes and separator to immensely improve as-assembled battery performance by suppressing the transfer of polysulfides within the S cathode side and/or to inhibiting metal dendrites, which are of vital importance for the commercial production of metal-sulfur batteries and metal anode batteries. Given the fact that for the research and development of Na-S battery and K-S battery, Na/K metal anode systems are really limited, and in this section, Li-S and Li anode cells are introduced.

The activated CNF (ACNF)-filter-coated separator with tunable micropores have been developed and investigated as interlayer to inhibit the migrating polysulfides by taking advantage of the physical absorption and excellent ionic conductivity. Compared with conventional separator, the modified separator endows the Li-S cell with high electrochemical utilization with amazing polysulfide retention (capacity fade rate: 0.13% per cycle after 200 cycles at C/5 rate).[166] Inspired by this work, our group introduced oxygen-doped CNF nonwovens as interlayer to further accelerate the performance of the battery cell. Even at 2C, the cell achieved an ultralow capacity decay rate of 0.06%, demonstrating hydrophilic interlayers which are favorable for anchoring and accelerating reutilization of the trapped polysulfides.[63] Apart from pure CNF interlayer, CNFs with hybrid structure is also used to optimize the cycling stability of metal- sulfur batteries. Fe-embedded N-doped porous CNF (Fe-N-C) is obtained as interlayer between separator and S cathode by using $Fe(acac)_3/SiO_2/$GO/PAN as electrospinning precursor.[167] This unique layer has not only displayed a rapid Li-ion transport rate but is also regarded as current collector to reutilize captured S species due to the porous intercalated structure, excellent electrical conductivity and robust LiPS capture ability of the Fe-N-C. The Li-S battery displayed an excellent cycling stability of 0.053% capacity decay per cycle over 500 cycles. Even though ameliorated cycle stability could be achieved, the slow kinetics which might be caused by low porosity of interlayer leads to low reversible capacity at high current load. A bioinspired ZnO/PAN-based CNF interlayer achieved a high capacity of 776 mAh g^{-1} after 200 cycles at 1C, obtaining a 0.05% average capacity loss per cycle.[168] In this work, the well-organized ZnO nanowire assembled brush-like structure acted well in entrapping migrating polysulfides due to the robust chemical bonding. Ultrafine TiO_2 nanoparticles-decorated CNFs (CNF-T) provided abundant room to adjust the volume expansion of active substance as well as richful anchoring points for dissolved polysulfides, resulting in high initial reversible capacity of 935 mAh g^{-1} at 1C with a reverse capacity of 74.2% after 500 cycles.[169]

Owing to the superhigh energy density of Li anode, Li metal battery has aroused particular attention in recent years. However, the practical application of Li metal anode is hindered by the formation of Li dendrites, instable Li metal interface, and inevitable volume change.[170] Therefore, how to passivate lithium metal and realize controllable uniform lithium deposition is crucial. Various carbon-based materials have been used as host or protection layer to suppress Li dendrites.[171–174] Among them, free-standing CNF mats have captured particular attention due to the polymer-binder-free feature along with suitable internal and external radius ratios, which is pivotal to induce the uniform deposition of the lithium metal, leading to safe and long-life lithium metal batteries. Man et al. used PAN-based CNF as interlayer covering lithium metal anode to regulate the lithium plating/stripping behavior.[172] The results demonstrated that the CNF interlayer could be used as a reservoir with excellent electrolyte uptake ability to optimize the redistribution of Li^+ flux near Li metal anode, therefore prolonging dendrite-free Li plating/stripping process with low initial nucleation overpotential (30 mV) even at the current density of 1 mA cm^{-2}. Benefiting from these, the symmetrical cell of Li metal anode protected by CNF delivered long duration time (>1,200 h) at 1 mA cm^{-2}. Lithiophilic Co, N co-doped porous CNFs were achieved through pyrolysis of self-assembled Co/Zn MOF nanosheets on CNFs.[175] The synergistic effect between the simultaneously introduced Co and N defects not only boosts the Li affinity of pyridinic and pyrrolic N but also transfers the graphitic N into lithiophilic, providing crucial roles for uniform plating of Li. Meanwhile, the porous structure formed by Zn sublimation provides sufficient buffer space for internal stress caused by Li plating/stripping. As a result, the PCNF–Co/N@Li anode delivers a long lifespan over 1,400 h with a low voltage hysteresis. Ag nanoparticles,[176] MnS and MnF_2[177] were also anchored on CNFs to guide the uniform generation of smooth Li with low overpotential. Inspired by these works, Yang et al. used Cu nanoparticles-modified CNFs as interlayer to construct a stable Zn anode and promising results were achieved.[178] Therefore, electrospun CNF-based materials could also be regarded as a tempting option used as a host or interlayer in other battery systems with similar problems.

From the aforementioned studies, we can find that an effective CNF-based interlayer should have plentiful conductive pathways along with controlled porous architecture for efficient electron/ion transport and confinement of polysulfides or lithium dendrites. To be more exact, to minimize the interlayer thickness, the integration of the interlayer with the separator is urgently required.

5 DEVELOPMENT AND APPLICATION OF ELECTROSPUN CARBON NANOFIBERS IN FLEXIBLE ENERGY STORAGE DEVICES

Currently, the demands for energy storage devices have grown significantly with the thriving development of flexible and wearable electronics. One of the most important challenges lies in designing light weighted, thin, and flexible energy storage devices. During the past decade, many efforts have been made to realize mechanical flexibility of these energy storage devices by making each component more flexible. In other words, mechanical robustness and flexibility are the essential key parameters. That is, on the one hand, mechanical properties must meet the requirements for flexible device assembly and operation, increasing the interfacial stability; On the other hand, the mechanical properties of CNFs need to adapt to the dynamic bending and stretching behavior to certain strength, preventing the propagation of mechanical fractures and cracks during usage.

Similar to the commercial carbon fibers, the tensile modulus of electrospun CNFs is mainly related to the degree of graphitization, the size and configuration of graphitic domains, crystallite orientation, and is immediately or indirectly impacted by fiber diameter, orientation, and porosity.[87, 179] Therefore, to some extent, the regulation of pore structure, heteroatom doping and microcrystalline structure (conductivity) can be linked to the mechanical properties of the electrospun CNFs, either directly or indirectly. In comparison with the numerous endeavors that have been devoted to regulating energy storage performance, few attention was focused on the mechanical performances of the as-prepared CNFs.

In most cases, the as-spun mats produced by electrospinning process have good flexibility and bendability. After post-treatment, the organic fibers transform into CNFs, ceramic fibers, metal fibers, or the mix of them, which may decrease the mechanical robustness and flexibility to some extent. Such flexibility and bendability change are normally demonstrated in many works.[65, 67, 126] Therefore, the comprehensive regulation of the spinning parameters and the post-processing process are important for balancing the mechanical and electrochemical properties. It is well accepted that a good mechanical property could be achieved when the fiber displays a smaller diameter, good crystallinity, and less surface defect.[180] However, some others hold the view that the porous structure provides a room for stress release and makes it easy for CNFs to restore their original shape during the folding process.[181, 182] Thus, the effect of the porosity on the mechanical performance of CNFs needs further research. More recently, some reports pointed out that the introduction of

CNTs to induce the generation of highly ordered graphitic structure could enhance the mechanical stability.[108, 183] By adding 1 wt.% CNTs to CNFs, the tensile strength increased from 50 kPa to 100 kPa.[184] The construction of the cross-linked structure via adjusting pre-oxidation was also confirmed as a practical method to boost the mechanical performance of CNFs due to the successful regulation of the crystallinity.[67, 126] GQDs could also be used to create CNF with high mechanical performance based on the cross-linking reinforcing mechanism.[114, 185] Improving treated temperature is another way to augment the crystallinity. Zhou et al. found that the tensile strength and the Young's modulus of CNFs were improved by 67% and 45% when the temperature increased from 1,000 to 2,200°C.[186]

Although CNFs with appropriate designs have indeed shown a certain mechanical stability, there remains a long way to put them into practical application in large scales, more deeper insights of the regulation of CNFs' mechanical performance is still needed. Most of the state-of-the-art works present the mechanical performance by folding and bending, lacking stretching and elastic behavior. Moreover, quantitative characterizations of flexibility and bendability are still largely lacking. Tensile performances, including tensile strength, Young's modulus, and elongation, are really encouraged to be used widely.

6 SUMMARY AND PROSPECT

As a simple, versatile, efficient, and low-cost technique, electrospinning can produce various interesting continuous nanostructures to be applied as functional materials used in energy storage devices. By adjusting electrospinning parameters and proper post-treatment, composite nanofibers with unique structures, components, and robustness can be obtained. In this chapter, we have systematically summarized the recent innovative and significant strategies for morphology/microstructure/component controlled CNF-based materials for energy storage applications. What's more, the requirements of mechanical performance of these CNF-based materials for flexible and free-standing electrode are also highlighted. This review can enlighten readers about the methods and the corresponding mechanisms of the geometry and functionality of CNFs and promote the understanding of "structure-performance" relationship of them in energy storage application.

There's no doubt that great progresses have been made during the past years, and many newly emerged opportunities and challenges still remain to be overcome before commercialization of electrospun CNF-based materials:

1. The low production rate is the main block in view of industrial manufacture and commercial application. To date, multi-spinnerets and non-needle electrospinning systems have been realized to the large-scale production of electrospun nanofibers. However, these new machines are too expensive compared with conventional devices.
2. There is a long way to regulate the structure of the final CNF-based materials accurately and steadily for industrial production due to the multi-step control process. To achieve this goal, trade-off among the key performance parameters is really needed. From the perspective of the precise

pore controlling, heteroatom doping and microcrystalline regulation, more innovative strategies are necessary. Maybe integrating the electrospinning machine with artificial intelligence along with in situ and real-time observation and characterization will be a good choice. This way may allow the production process to focus on fine-tuning the synthesis and structural control with reduced carbon emission and tiny carbon footprints.
3. Another key issue lies in how to prepare the target CNF-based materials with mechanical robustness. As abovementioned, to obtain CNFs with good tensile, bending, shear properties, a trade-off between the physical properties of CNF and mechanical properties is necessary. Integrating with self-healing materials or conductive elastic polymers in the porous fiber skeletons might be effective for enhancing the mechanical performance.
4. The aim of scientific research is to be used in practice. In this regard, packing issue and actual working environments must be considered in studies, such as the simultaneous occurrence of bending, twisting, and stretching in real wearing states. Additionally, washability, breathability, waterproof, and safety of these products should also be fully considered. However, these issues do not gain much attention at present.

Despite these challenges, electrospinning still shows bright opportunities to produce nanofibers with designed structure and compositions to achieve high-performance advanced energy storage devices. Electrospinning is emerging as a powerful and versatile technique from an industrial perspective for the synthesis of nanofibers. We believe that with the rapid development of electrospinning, flexible energy storage devices with low-cost and long-life-span will be industrially used for large-scale applications in the foreseeable future.

REFERENCES

1. P. Simon, Y. Gogotsi and B. Dunn, Where do batteries end and supercapacitors begin, *Science*, 2014, **343**, 1210–1211.
2. A. Masias, J. Marcicki and W.A. Paxton, Opportunities and challenges of lithium ion batteries in automotive applications, *ACS Energy Lett.*, 2021, **6**, 621–630.
3. G. Wang, M. Yu and X. Feng, Carbon materials for ion-intercalation involved rechargeable battery technologies, *Chem. Soc. Rev.*, 2021, **50**, 2388–2443.
4. S. Dutta, A. Bhaumik and K.C.W. Wu, Hierarchically porous carbon derived from polymers and biomass: Effect of interconnected pores on energy applications, *Energy Environ. Sci.*, 2014, **7**, 3574–3592.
5. X. Li, Y. Chen, H. Huang, Y.-W. Mai and L. Zhou, Electrospun carbon-based nanostructured electrodes for advanced energy storage—A review, *Energy Storage Mater.*, 2016, **5**, 58–92.
6. L.-F. Chen, Z.-H. Huang, H.-W. Liang, W.-T. Yao, Z.-Y. Yu and S.-H. Yu, Flexible all-solid-state high-power supercapacitor fabricated with nitrogen-doped carbon nanofiber electrode material derived from bacterial cellulose, *Energy Environ. Sci.*, 2013, **6**, 3331–3338.
7. K. Dong, J. Liang, Y. Wang, Z. Xu, Q. Liu, Y. Luo, T. Li, L. Li, X. Shi, A.M. Asiri, Q. Li, D. Ma and X. Sun, Honeycomb carbon nanofibers: A superhydrophilic O_2 -entrapping electrocatalyst enables ultrahigh mass activity for the two-electron oxygen reduction reaction, *Angew. Chem. Int. Ed.*, 2021, **60**, 10583–10587.

8. H. Chen, T. Liu, J. Mou, W. Zhang, Z. Jiang, J. Liu, J. Huang and M. Liu, Free-standing N-self-doped carbon nanofiber aerogels for high-performance all-solid-state supercapacitors, *Nano Energy*, 2019, **63**, 103836.
9. Y. Zhong, T. Wang, M. Yan, X. Huang and X. Zhou, *Int. J. Biol. Macromol.*, 2022, **207**, 541–548.
10. X. Zhang, C. Jiang, J. Liang and W. Wu, Electrode materials and device architecture strategies for flexible supercapacitors in wearable energy storage, *J. Mater. Chem. A*, 2021, **9**, 8099–8128.
11. T. Lv, M. Liu, D. Zhu, L. Gan and T. Chen, Nanocarbon-based materials for flexible all-solid-state supercapacitors, *Adv. Mater.*, 2018, **30**, 1705489.
12. J. Liang, H. Zhao, L. Yue, G. Fan, T. Li, S. Lu, G. Chen, S. Gao, Abdullah M. Asiri and X. Sun, Recent advances in electrospun nanofibers for supercapacitors, *J. Mater. Chem. A*, 2020, **8**, 16747–16789.
13. X.W. Mao, T.A. Hatton and G.C. Rutledge, A review of electrospun carbon fibers as electrode materials for energy storage, *Curr. Org. Chem.*, 2013, **17**, 1390–1401.
14. H. He, Y. Zhang, W. Zhang, Y. Li, X. Zhu, P. Wang and D. Hu, Porous carbon nanofibers derived from silk fibroin through electrospinning as N-doped metal-free catalysts for hydrogen evolution reaction in acidic and alkaline solutions, *ACS Appl. Mater. Interfaces*, 2021, **14**, 834–849.
15. L. Szabó, X. Xu, K. Uto, J. Henzie, Y. Yamauchi, I. Ichinose and M. Ebara, Tailoring the structure of chitosan-based porous carbon nanofiber architectures toward efficient capacitive charge storage and capacitive deionization, *ACS Appl. Mater. Interfaces*, 2022, **14**, 4004–4021.
16. Z. Zhang, X. Li, C. Wang, S. Fu, Y. Liu and C. Shao, Polyacrylonitrile and carbon nanofibers with controllable nanoporous structures by electrospinning, *Marcomol. Mater. Eng.*, 2009, **294**, 673–678.
17. S.-S. Choia, Y.S. Leeb, C.W. Joob, S.G. Leeb, J.K. Parkc and K.-S. Hanc, Electrospun PVDF nanofiber web as polymer electrolyte or separator, *Electrochim. Acta*, 2004, **50**, 339–343.
18. L. Wang, G. Yang, S. Peng, J. Wang, W. Yan and S. Ramakrishna, One-dimensional nanomaterials toward electrochemical sodium-ion storage applications via electrospinning, *Energy Storage Mater.*, 2020, **25**, 443–476.
19. A. Laforgue, All-textile flexible supercapacitors using electrospun poly (3,4-ethylenedioxythiophene) nanofibers, *J. Power Sources*, 2011, **196**, 559–564.
20. S. Chaudhari, Y. Sharma, P.S. Archana, R. Jose, S. Ramakrishna, S. Mhaisalkar and M. Srinivasan, Electrospun polyaniline nanofibers web electrodes for supercapacitors, *J. Appl. Polym. Sci.*, 2013, **129**, 1660–1668.
21. N. Amiraliyan, M. Nouri and M.H. Kish, Effects of some electrospinning parameters on morphology of natural silk-based nanofibers, *J. Appl. Polym. Sci.*, 2009, **113**, 226–234.
22. M.G. McKee, G.L. Wilkes, R.H. Colby and T.E. Long, Correlations of solution rheology with electrospun fiber formation of linear and branched polyesters, *Macromolecules*, 2004, **37**, 1760–1767.
23. J.S. Yun, C.K. Park, J.H. Cho, J.-H. Paik, Y.H. Jeong, J.-H. Nam and K.-R. Hwang, The effect of PVP contents on the fiber morphology and piezoelectric characteristics of PZT nanofibers prepared by electrospinning, *Mater. Lett.*, 2014, **137**, 178–181.
24. L. Tian, D. Ji, S. Zhang, X. He, S. Ramakrishna and Q. Zhang, A humidity-induced nontemplating route toward hierarchical porous carbon fiber hybrid for efficient bifunctional oxygen catalysis, *Small*, 2020, **16**, 2001743.
25. M. Inagaki, Y. Yang and F. Kang, Carbon nanofibers prepared via electrospinning, *Adv. Mater.*, 2012, **24**, 2547–2566.
26. S. Cavaliere, S. Subianto, I. Savych, D.J. Jones and J. Rozière, Electrospinning: Designed architectures for energy conversion and storage devices, *Energy Environ. Sci.*, 2011, **4**, 4761–4785.

27. D. Li and Y. Xia, Direct fabrication of composite and ceramic hollow nanofibers by electrospinning, *Nano Lett.*, 2004, **4**, 933–938.
28. Y. Zhao, X. Cao and L. Jiang, Bio-mimic multichannel microtubes by a facile method, *J. Am. Chem. Soc.*, 2007, **129**, 764–765.
29. R. Murugan and S. Ramakrishna, Design strategies of tissue engineering scaffolds with controlled fiber orientation, *Tissue Eng.*, 2007, **13**, 1845–1866.
30. J.T. McCann, M. Marquez and Y. Xia, Highly porous fibers by electrospinning into a cryogenic liquid, *J. Am. Chem. Soc.*, 2006, **128**, 1436–1437.
31. D. Li, Y. Wang and Y. Xia, Electrospinning nanofibers as uniaxially aligned arrays and layer-by-layer stacked films, *Adv. Mater.*, 2004, **16**, 361–366.
32. X. Tian, N. Zhao, K. Wang, D. Xu, Y. Song, Q. Guo and L. Liu, Preparation and electrochemical characteristics of electrospun water-soluble resorcinol/phenol-formaldehyde resin-based carbon nanofibers, *RSC Adv.*, 2015, **5**, 40884–40891.
33. S.G. King, N.J. Terrill, A.J. Goodwin, R. Stevens, V. Stolojan and S.R.P. Silva, Probing of polymer to carbon nanotube surface interactions within highly aligned electrospun nanofibers for advanced composites, *Carbon*, 2018, **138**, 207–214.
34. Y. Han, J. Zou, Z. Li, W. Wang, Y. Jie, J. Ma, B. Tang, Q. Zhang, X. Cao, S. Xu and Z.L. Wang, Si@void@c nanofibers fabricated using a self-powered electrospinning system for lithium-ion batteries, *ACS Nano*, 2018, **12**, 4835–4843.
35. M.-X. Wang, Z.-H. Huang, F. Kang and K. Liang, Porous carbon nanofibers with narrow pore size distribution from electrospun phenolic resins, *Mater. Lett.*, 2011, **65**, 1875–1877.
36. G.-H. An and H.-J. Ahn, Activated porous carbon nanofibers using sn segregation for high-performance electrochemical capacitors, *Carbon*, 2013, **65**, 87–96.
37. H. Yu, L. Chen, W. Li, M. Dirican, Y. Liu and X. Zhang, Root-whisker structured 3D CNTs-CNFs network based on coaxial electrospinning: A free-standing anode in lithium-ion batteries, *J. Alloy Compd*, 2021, 158481.
38. Y. Liu, Q. Liu, L. Wang, X. Yang, W. Yang, J. Zheng and H. Hou, Advanced supercapacitors based on porous hollow carbon nanofiber electrodes with high specific capacitance and large energy density, *ACS Appl. Mater. Interfaces*, 2020, **12**, 4777–4786.
39. Y. Chen, X. Li, K. Park, J. Song, J. Hong, L. Zhou, Y.-W. Mai, H. Huang and J.B. Goodenough, Hollow carbon-nanotube/carbon-nanofiber hybrid anodes for Li-ion batteries, *J. Am. Chem. Soc.*, 2013, **135**, 16280–16283.
40. X. Tian, X. Li, T. Yang, K. Wang, H. Wang, Y. Song, Z. Liu and Q. Guo, Porous worm-like $NiMoO_4$ coaxially decorated electrospun carbon nanofiber as binder-free electrodes for high performance supercapacitors and lithium-ion batteries, *Appl. Surf. Sci.*, 2018, **434**, 49–56.
41. X. Mao, X. Yang, J. Wu, W. Tian, G.C. Rutledge and T.A. Hatton, Microwave-assisted oxidation of electrospun turbostratic carbon nanofibers for tailoring energy storage capabilities, *Chem. Mater.*, 2015, **27**, 4574–4585.
42. F. Zhang, Y. Si, J. Yu and B. Ding, Electrospun porous engineered nanofiber materials: A versatile medium for energy and environmental applications, *Chem. Eng. J.*, 2023, **456**, 140989.
43. C.H. Kim, C.-M. Yang, Y.A. Kim and K.S. Yang, Pore engineering of nanoporous carbon nanofibers toward enhanced supercapacitor performance, *Appl. Surf. Sci.*, 2019, **497**, 143693.
44. S.-H. Park, B.-K. Kim and W.-J. Lee, Electrospun activated carbon nanofibers with hollow core/highly mesoporous shell structure as counter electrodes for dye-sensitized solar cells, *J. Power Sources*, 2013, **239**, 122–127.
45. T. He, Y. Fu, X. Meng, X. Yu and X. Wang, A novel strategy for the high performance supercapacitor based on polyacrylonitrile-derived porous nanofibers as electrode and separator in ionic liquid electrolyte, *Electrochim. Acta*, 2018, **282**, 97–104.

46. X. Zhang, X. Tian, Y. Song, J. Wu, T. Yang and Z. Liu, High-performance activated carbon cathodes from green cokes for Zn-ion hybrid supercapacitors, *Fuel*, 2022, **310**, 122485.
47. C. Ma, Y. Song, J. Shi, D. Zhang, X. Zhai, M. Zhong, Q. Guo and L. Liu, Preparation and one-step activation of microporous carbon nanofibers for use as supercapacitor electrodes, *Carbon*, 2013, **51**, 290–300.
48. C. Ma, Y. Li, J. Shi, Y. Song and L. Liu, High-performance supercapacitor electrodes based on porous flexible carbon nanofiber paper treated by surface chemical etching, *Chem. Eng. J.*, 2014, **249**, 216–225.
49. K.K. Karthikeyan and P. Biji, A novel biphasic approach for direct fabrication of highly porous, flexible conducting carbon nanofiber mats from polyacrylonitrile (PAN)/ $NaHCO_3$ nanocomposite, *Micropor. Mesopor. Mater.*, 2016, **224**, 372–383.
50. C. Kim, B.T.N. Ngoc, K.S. Yang, M. Kojima, Y.A. Kim, Y.J. Kim, M. Endo and S.C. Yang, Self-sustained thin webs consisting of porous carbon nanofibers for supercapacitors via the electrospinning of polyacrylonitrile solutions containing zinc chloride, *Adv. Mater.*, 2007, **19**, 2341–2346.
51. J. Yan, K. Dong, Y. Zhang, X. Wang, A.A. Aboalhassan, J. Yu and B. Ding, Multifunctional flexible membranes from sponge-like porous carbon nanofibers with high conductivity, *Nat. Commun.*, 2019, **10**, 5584.
52. A.B. Fuertes, G. Lota, T.A. Centeno and E. Frackowiak, Templated mesoporous carbons for supercapacitor application, *Electrochim. Acta*, 2005, **50**, 2799–2805.
53. C. Ma, Y. Yu, Y.j. Li, J.l. Shi, Y. Song and L. Liu, Ion accumulation and diffusion behavior in micro-/meso-pores of carbon nanofibers, *J. Electrochem. Soc.*, 2014, **161**, A1330–A1337.
54. H. Liu, W. Song and A. Xing, In situ K_2S activated electrospun carbon nanofibers with hierarchical meso/microporous structures for supercapacitors, *RSC Adv.*, 2019, **9**, 33539–33548.
55. C. Zhan, Q. Xu, X. Yu, Q. Liang, Y. Bai, Z.-H. Huang and F. Kang, Nitrogen-rich hierarchical porous hollow carbon nanofibers for high-performance supercapacitor electrodes, *RSC Adv.*, 2016, **6**, 41473–41476.
56. R. Singhal and V. Kalra, Using common salt to impart pseudocapacitive functionalities to carbon nanofibers, *J. Mater. Chem. A*, 2015, **3**, 377–385.
57. C. Ma, J. Sheng, C. Ma, R. Wang, J. Liu, Z. Xie and J. Shi, High-performanced supercapacitor based mesoporous carbon nanofibers with oriented mesopores parallel to axial direction, *Chem. Eng. J.*, 2016, **304**, 587–593.
58. T. Wang, H. Wang, X. Chi, R. Li and J. Wang, Synthesis and microwave absorption properties of Fe-C nanofibers by electrospinning with disperse Fe nanoparticles parceled by carbon, *Carbon*, 2014, **74**, 312–318.
59. J. Tan, Y. Han, L. He, Y. Dong, X. Xu, D. Liu, H. Yan, Q. Yu, C. Huang and L. Mai, In situ nitrogen-doped mesoporous carbon nanofibers as flexible freestanding electrodes for high-performance supercapacitors, *J. Mater. Chem. A*, 2017, **5**, 23620–23627.
60. Q. Liang, L. Ye, Q. Xu, Z.-H. Huang, F. Kang and Q.-H. Yang, Graphitic carbon nitride nanosheet-assisted preparation of N-enriched mesoporous carbon nanofibers with improved capacitive performance, *Carbon*, 2015, **94**, 342–348.
61. T. Liu, J. Serrano, J. Elliott, X. Yang, W. Cathcart, Z. Wang, Z. He and G. Liu, Exceptional capacitive deionization rate and capacity by block copolymer-based porous carbon fibers, *Sci. Adv.*, 2020, **6**, eaaz0906.
62. Y.S. Kwon, G.-T. Park, J.-S. Lee, G.-H. Hwang and Y.G. Jeong, Poly(ether amide)-derived, nitrogen self-doped, and interfused carbon nanofibers as free-standing supercapacitor electrode materials, *ACS Appl. Energy Mater.*, 2021, **4**, 1517–1526.
63. T. Yang, X. Tian, Y. Song, S. Wu, J. Wu and Z. Liu, Oxygen-doped carbon nanofiber nonwovens as an effective interlayer towards accelerating electrochemical kinetics for lithium-sulfur battery, *Appl. Surf. Sci.*, 2023, **611**, 155690.

64. W. Na, J. Jun, J.W. Park, G. Lee and J. Jang, In situ nitrogen-doped mesoporous carbon nanofibers as flexible freestanding electrodes for high-performance supercapacitors, *J. Mater. Chem. A*, 2017, **5**, 17379–17387.
65. X. Li, X. Tian, T. Yang, Y. He, W. Liu, Y. Song and Z. Liu, Coal liquefaction residues based carbon nanofibers film prepared by electrospinning: An effective approach to coal waste management, *ACS Sustain. Chem. Eng.*, 2019, **7**, 5742–5750.
66. C. Ma, L. Wu, M. Dirican, H. Cheng, J. Li, Y. Song, J. Shi and X. Zhang, ZnO-assisted synthesis of lignin-based ultra-fine microporous carbon nanofibers for supercapacitors, *J. Colloid Interface Sci.*, 2021, **586**, 412–422.
67. X. Li, X. Du, Y. Li, X. Tian, H. Zheng and X. Li, Flexible and cross-linked N, S co-doped carbon nanofiber nonwovens derived from coal liquefaction residue for high performance supercapacitors, *J. Mater. Sci.*, 2022, **57**, 9357–9369.
68. R. Chen, Y. Hu, Z. Shen, P. Pan, X. He, K. Wu, X. Zhang and Z. Cheng, Facile fabrication of foldable electrospun polyacrylonitrile-based carbon nanofibers for flexible lithium-ion batteries, *J. Mater. Chem. A*, 2017, **5**, 12914–12921.
69. Y. Luan, G. Nie, X. Zhao, N. Qiao, X. Liu, H. Wang, X. Zhang, Y. Chen and Y.-Z. Long, The integration of SnO_2 dots and porous carbon nanofibers for flexible supercapacitors, *Electrochim. Acta*, 2019, **308**, 121–130.
70. Y. Chen, X. Li, X. Zhou, H. Yao, H. Huang, Y.-W. Mai and L. Zhou, Hollow-tunneled graphitic carbon nanofibers through Ni-diffusion-induced graphitization as high-performance anode materials, *Energy Environ. Sci.*, 2014, **7**, 2689–2696.
71. D. Tian, Y. Ao, W. Li, J. Xu and C. Wang, General fabrication of metal-organic frameworks on electrospun modified carbon nanofibers for high-performance asymmetric supercapacitors, *J. Colloid Interface Sci.*, 2021, **603**, 199–209.
72. L. Xu, P. Xiong, L. Zeng, Y. Fang, R. Liu, J. Liu, F. Luo, Q. Chen, M. Wei and Q. Qian, Electrospun $VSe_{1.5}$/CNF composite with excellent performance for alkali metal ion batteries, *Nanoscale*, 2019, **11**, 16308–16316.
73. E. Doustkhah, R. Hassandoost, A. Khataee, R. Luque and M.H.N. Assadi, Hard-templated metal–organic frameworks for advanced applications, *Chem. Soc. Rev.*, 2021, **50**, 2927–2953.
74. A. Indra, T. Song and U. Paik, Metal organic framework derived materials: Progress and prospects for the energy conversion and storage, *Adv. Mater.*, 2018, **30**, 1705146.
75. Y. Gong, R. Chen, H. Xu, C. Yu, X. Zhao, Y. Sun, Z. Hui, J. Zhou, J. An, Z. Du, G. Sun and W. Huang, Polarity-assisted formation of hollow-frame sheathed nitrogen-doped nanofibrous carbon for supercapacitors, *Nanoscale*, 2019, **11**, 2492–2500.
76. L.-F. Chen, Y. Lu, L. Yu and X.W. Lou, Designed formation of hollow particle-based nitrogen-doped carbon nanofibers for high-performance supercapacitors, *Energy Environ. Sci.*, 2017, **10**, 1777–1783.
77. Y. Chen, A. Amiri, J.G. Boyd and M. Naraghi, Promising trade-offs between energy storage and load bearing in carbon nanofibers as structural energy storage devices, *Adv. Funct. Mater.*, 2019, **29**, 1901425.
78. Y.J. Hong, J.-W. Yoon, J.-H. Lee and Y.C. Kang, A new concept for obtaining SnO_2 fiber-in-tube nanostructures with superior electrochemical properties, *Chem. Eur. J.*, 2015, **21**, 371–376.
79. D. Han and A.J. Steckl, Triaxial electrospun nanofiber membranes for controlled dual release of functional molecules, *ACS Appl. Mater. Interfaces*, 2013, **5**, 8241–8245.
80. B.-S. Lee, H.-S. Yang, H. Jung, S.-Y. Jeon, C. Jung, S.-W. Kim, J. Bae, C.-L. Choong, J. Im, U.I. Chung, J.-J. Park and W.-R. Yu, Novel multi-layered 1-D nanostructure exhibiting the theoretical capacity of silicon for a super-enhanced lithium-ion battery, *Nanoscale*, 2014, **6**, 5989–5998.
81. H.Y. Chen, N. Wang, J.C. Di, Y. Zhao, Y.L. Song and L. Jiang, Nanowire-in-microtube structured core/shell fibers via multifluidic coaxial electrospinning, *Langmuir*, 2010, **26**, 11291–11296.

82. B.-S. Lee, H.-S. Yang and W.-R. Yu, Fabrication of double-tubular carbon nanofibers using quadruple coaxial electrospinning, *Nanotechnology*, 2014, **25**, 465602.
83. X. Tian, N. Zhao, Y. Song, K. Wang, D. Xu, X. Li, Q. Guo and L. Liu, Synthesis of nitrogen-doped electrospun carbon nanofibers with superior performance as efficient supercapacitor electrodes in alkaline solution, *Electrochim. Acta*, 2015, **185**, 40–51.
84. F. Wu, R. Dong, Y. Bai, Y. Li, G. Chen, Z. Wang and C. Wu, Phosphorus-doped hard carbon nanofibers prepared by electrospinning as an anode in sodium-ion batteries, *ACS Appl. Mater. Interfaces*, 2018, **10**, 21335–21342.
85. S. Lee, S.Y. Cho, Y.S. Chung, Y.C. Choi and S. Lee, High electrical and thermal conductivities of a PAN-based carbon fiber via boron-assisted catalytic graphitization, *Carbon*, 2022, **199**, 70–79.
86. L. Yue, H. Zhao, Z. Wu, J. Liang, S. Lu, G. Chen, S. Gao, B. Zhong, X. Guo and X. Sun, Recent advances in electrospun one-dimensional carbon nanofiber structures/heterostructures as anode materials for sodium ion batteries, *J. Mater. Chem. A*, 2020, **8**, 11493–11510.
87. G. Nie, X. Zhao, Y. Luan, J. Jiang, Z. Kou and J. Wang, Key issues facing electrospun carbon nanofibers in energy applications: On-going approaches and challenges, *Nanoscale*, 2020, **12**, 13225–13248.
88. Z. Zhang, Z. Gao, Y. Zhang, Z. Yan, I. Kesse, W. Wei, X. Zhao and J. Xie, Hierarchical porous nitrogen-doped graphite from tissue paper as efficient electrode material for symmetric supercapacitor, *J. Power Sources*, 2021, **492**, 229670.
89. W. Fu, K. Zhang, M.-s. Chen, M. Zhang and Z. Shen, One-pot synthesis of N-doped hierarchical porous carbon for high-performance aqueous capacitors in a wide pH range, *J. Power Sources*, 2021, **491**, 229587.
90. X. Wang, Q. Zhao, B. Yang, Z. Li, Z. Bo, K.H. Lam, N.M. Adli, L. Lei, Z. Wen, G. Wu and Y. Hou, Emerging nanostructured carbon-based non-precious metal electrocatalysts for selective electrochemical CO_2 reduction to CO, *J. Mater. Chem. A*, 2019, **7**, 25191–25202.
91. T. Gong, R. Qi, X. Liu, H. Li and Y. Zhang, N, F-codoped microporous carbon nanofibers as efficient metal-free electrocatalysts for ORR, *Nano-Micro Lett.*, 2019, **11**, 9.
92. C. Hu and L. Dai, Doping of carbon materials for metal-free electrocatalysis, *Adv. Mater.*, 2019, **31**, 1804672.
93. H.D. Pham, K. Mahale, T.M.L. Hoang, S.G. Mundree, P. Gomez-Romero and D.P. Dubal, Dual carbon potassium-ion capacitors: Biomass-derived graphene-like carbon nanosheet cathodes, *ACS Appl. Mater. Interfaces*, 2020, **12**, 48518–48525.
94. N. Li, Q. Yang, Y. Wei, R. Rao, Y. Wang, M. Sha, X. Ma, L. Wang and Y. Qian, Phosphorus-doped hard carbon with controlled active groups and microstructure for high-performance sodium-ion batteries, *J. Mater. Chem. A*, 2020, **8**, 20486–20492.
95. J. Yang, X. Zhou, D. Wu, X. Zhao and Z. Zhou, S-doped N-rich carbon nanosheets with expanded interlayer distance as anode materials for sodium-ion batteries, *Adv. Mater.*, 2016, **29** (6), 1604108.
96. D. Hulicova-Jurcakova, A.M. Puziy, O.I. Poddubnaya, F. Suárez-García, J.M. Tascón and G.Q. Lu, Highly stable performance of supercapacitors from phosphorus-enriched carbons, *J. Am. Chem. Soc.*, 2009, **131**, 5026–5027.
97. Y. Qian, S. Jiang, Y. Li, Z. Yi, J. Zhou, T. Li, Y. Han, Y. Wang, J. Tian, N. Lin and Y. Qian, In situ revealing the electroactivity of P-O and P-C bonds in hard carbon for high-capacity and long-life Li/K-ion batteries, *Adv. Energy Mater.*, 2019, **9**, 1901676.
98. L. Quan, Y. Zhang, B.J. Crielaard, A. Dusad, S.M. Lele, C.J.F. Rijcken, J.M. Metselaar, H. Kostková, T. Etrych, K. Ulbrich, F. Kiessling, T.R. Mikuls, W.E. Hennink, G. Storm, T. Lammers and D. Wang, Nanomedicines for inflammatory arthritis: Head-to-head comparison of glucocorticoid-containing polymers, micelles, and liposomes, *ACS Nano*, 2014, **8**(1), 458–466.
99. W. Kiciński, M. Szala and M. Bystrzejewski, Sulfur-doped porous carbons: Synthesis and applications, *Carbon*, 2014, **68**, 1–32.

100. X. Sun, C. Wang, Y. Gong, L. Gu, Q. Chen and Y. Yu, A flexible sulfur-enriched nitrogen doped multichannel hollow carbon nanofibers film for high performance sodium storage, *Small*, 2018, **14**, 1802218.
101. Y. Li, G. Zhu, H. Huang, M. Xu, T. Lu and L. Pan, A N, S dual doping strategy via electrospinning to prepare hierarchically porous carbon polyhedra embedded carbon nanofibers for flexible supercapacitors, *J. Mater. Chem. A*, 2019, **7**, 9040–9050.
102. C. Kim, S.-H. Park, W.-J. Lee and K.-S. Yang, Characteristics of supercapaitor electrodes of PBI-based carbon nanofiber web prepared by electrospinning, *Electrochim. Acta*, 2004, **50**, 877–881.
103. C. Yang, M. Zhang, N. Kong, J. Lan, Y. Yu and X. Yang, Self-supported carbon nanofiber films with high-level nitrogen and phosphorus co-doping for advanced lithium-ion and sodium-ion capacitors, *ACS Sustain. Chem. Eng.*, 2019, **7**, 9291–9300.
104. Y. Liu, N. Zhang, X. Liu, C. Chen, L.-Z. Fan and L. Jiao, Red phosphorus nanoparticles embedded in porous N-doped carbon nanofibers as high-performance anode for sodium-ion batteries, *Energy Storage Mater.*, 2017, **9**, 170–178.
105. Z. Liu, Z. Zhao, Y. Wang, S. Dou, D. Yan, D. Liu, Z. Xia and S. Wang, In situ exfoliated, edge-rich, oxygen-functionalized graphene from carbon fibers for oxygen electrocatalysis, *Adv. Mater.*, 2017, **29**, 1606207.
106. D. Cui, Z. Zheng, X. Peng, T. Li, T. Sun and L. Yuan, Fluorine-doped SnO_2 nanoparticles anchored on reduced graphene oxide as a high-performance lithium ion battery anode, *J. Power Sources*, 2017, **362**, 20–26.
107. Q. Chen, L. Cao, H. Wang, Z. Zhou and P. Xiao, Oxygen/fluorine-functionalized flexible carbon electrodes for high-performance and anti-self-discharge Zn-ion hybrid capacitors, *J. Power Sources*, 2022, **538**, 231586.
108. Y. Liu, B. Luo, H. Liu, M. He, R. Wang, L. Wang, Z. Quan, J. Yu and X. Qin, 3D printed electrospun nanofiber-based pyramid-shaped solar vapor generator with hierarchical porous structure for efficient desalination, *Chem. Eng. J.*, 2023, **452**, 139402.
109. Z. Zhou and X.-F. Wu, Graphene-beaded carbon nanofibers for use in supercapacitor electrodes: Synthesis and electrochemical characterization, *J. Power Sources*, 2013, **222**, 410–416.
110. X. Tian, X. Li, T. Yang, K. Wang, H. Wang, Y. Song, Z. Liu, Q. Guo and C. Chen, Flexible carbon nanofiber mats with improved graphitic structure as scaffolds for efficient all-solid-state supercapacitor, *Electrochim. Acta*, 2017, **247**, 1060–1071.
111. F. Miao, C. Shao, X. Li, N. Lu, K. Wang, X. Zhang and Y. Liu, Flexible solid-state supercapacitors based on freestanding electrodes of electrospun polyacrylonitrile @ polyaniline core-shell nanofibers, *Electrochim. Acta*, 2015, **176**, 293–300.
112. J.S. Im, S.C. Kang, S.-H. Lee and Y.-S. Lee, Improved gas sensing of electrospun carbon fibers based on pore structure, conductivity and surface modification, *Carbon*, 2010, **48**, 2573–2581.
113. J. Cai and M. Naraghi, The formation of highly ordered graphitic interphase around embedded CNTs controls the mechanics of ultra-strong carbonized nanofibers, *Acta Mater.*, 2019, **162**, 46–54.
114. J. Zhao, J. Zhu, Y. Li, L. Wang, Y. Dong, Z. Jiang, C. Fan, Y. Cao, R. Sheng, A. Liu, S. Zhang, H. Song, D. Jia and Z. Fan, Graphene quantum dot reinforced electrospun carbon nanofiber fabrics with high surface area for ultrahigh rate supercapacitors, *ACS Appl. Mater. Interfaces*, 2020, **12**, 11669–11678.
115. S. He, L. Chen, C. Xie, H. Hu, S. Chen, M. Hanif and H. Hou, Supercapacitors based on 3d network of activated carbon nanowhiskers wrapped-on graphitized electrospun nanofibers, *J. Power Sources*, 2013, **243**, 880–886.
116. Y. Qiu, G. Li, Y. Hou, Z. Pan, H. Li, W. Li, M. Liu, F. Ye, X. Yang and Y. Zhang, Vertically aligned carbon nanotubes on carbon nanofibers: A hierarchical three-dimensional carbon nanostructure for high-energy flexible supercapacitors, *Chem. Mater.*, 2015, **27**, 1194–1200.

117. Y. Li, X. Tian, Y. Song, T. Yang, S. Wu and Z. Liu, Preparation of high-performance anthracite-based graphite anode materials and their lithium storage properties, *New Carbon Mater.*, 2022, **37**, 1163–1171.
118. T. Wang, Y. Wang, G. Cheng, C. Ma, X. Liu, J. Wang, W. Qiao and L. Ling, Catalytic graphitization of anthracite as an anode for lithium-ion batteries, *Energy Fuels*, 2020, **34**, 8911–8918.
119. J. Yan, M. Zhong, C. Yu, J. Zhang, M. Ma, L. Li, Q. Hao, F. Gao, Y. Tian, Y. Huang, W. Shen and S. Guo, Multilayer graphene sheets converted directly from anthracite in the presence of molten iron and their applications as anode for lithium ion batteries, *Synthet. Met.*, 2020, **263**, 116364.
120. Q. Wu, R. Zhao, X. Zhang, W. Li, R. Xu, G. Diao and M. Chen, Synthesis of flexible Fe_3O_4/C nanofibers with buffering volume expansion performance and their application in lithium-ion batteries, *J. Power Sources*, 2017, **359**, 7–16.
121. B. Zhang, Z.-L. Xu, Y.-B. He, S. Abouali, M. Akbari Garakani, E. Kamali Heidari, F. Kang and J.-K. Kim, Exceptional rate performance of functionalized carbon nanofiber anodes containing nanopores created by (Fe) sacrificial catalyst, *Nano Energy*, 2014, **4**, 88–96.
122. Y. Cheng, L. Huang, X. Xiao, B. Yao, L. Yuan, T. Li, Z. Hu, B. Wang, J. Wan and J. Zhou, Flexible and cross-linked N-doped carbon nanofiber network for high performance freestanding supercapacitor electrode, *Nano Energy*, 2015, **15**, 66–74.
123. Y. Cheng, L. Huang, X. Xiao, B. Yao, Z. Hu, T. Li, K. Liu and J. Zhou, Cross-linked carbon network with hierarchical porous structure for high performance solid-state electrochemical capacitor, *J. Power Sources*, 2016, **327**, 488–494.
124. G. Xue, J. Zhong, Y. Cheng and B. Wang, Facile fabrication of cross-linked carbon nanofiber via directly carbonizing electrospun polyacrylonitrile nanofiber as high performance scaffold for supercapacitors, *Electrochim. Acta*, 2016, **215**, 29–35.
125. X. Guo, X. Zhang, H. Song and J. Zhou, Electrospun cross-linked carbon nanofiber films as free-standing and binder-free anodes with superior rate performance and long-term cycling stability for sodium ion storage, *J. Mater. Chem. A*, 2017, **5**, 21343–21352.
126. X. Tian, Y. He, Y. Song, T. Yang, X. Li and Z. Liu, Flexible cross-linked electrospun carbon nanofiber mats derived from pitch as dual-functional materials for supercapacitors, *Energy Fuels*, 2020, **34**, 14975–14985.
127. X. Li, N. Sun, X. Tian, T. Yang, Y. Song, B. Xu and Z. Liu, Electrospun coal liquefaction residues/polyacrylonitrile composite carbon nanofiber nonwoven fabrics as high-performance electrodes for lithium/potassium batteries, *Energy Fuels*, 2020, **34**, 2445–2451.
128. C. Ma, Y. Song, J. Shi, D. Zhang, Q. Guo and L. Liu, Preparation and electrochemical performance of heteroatom-enriched electrospun carbon nanofibers from melamine formaldehyde resin, *J. Colloid Interface Sci.*, 2013, **395**, 217–23.
129. C. Ma, L. Wu, M. Dirican, H. Cheng, J. Li, Y. Song, J. Shi and X. Zhang, Carbon black-based porous sub-micron carbon fibers for flexible supercapacitors, *Appl. Surf. Sci.*, 2021, **537**, 147914.
130. C. Ma, Q. Fan, M. Dirican, N. Subjalearndee, H. Cheng, J. Li, Y. Song, J. Shi and X. Zhang, Rational design of meso-/micro-pores for enhancing ion transportation in highly-porous carbon nanofibers used as electrode for supercapacitors, *Appl. Surf. Sci.*, 2021, **545**, 148933.
131. L. Li, D. Zhang, J. Deng, Y. Gou, J. Fang, H. Cui, Y. Zhao and M. Cao, Carbon-based materials for fast charging lithium-ion batteries, *Carbon*, 2021, **183**, 721–734.
132. Y. He, B. Matthews, J. Wang, L. Song, X. Wang and G. Wu, Innovation and challenges in materials design for flexible rechargeable batteries: From 1D to 3D, *J. Mater. Chem. A*, 2018, **6**, 735–753.

133. R. Shi, C. Han, X. Xu, X. Qin, L. Xu, H. Li, J. Li, C.P. Wong and B. Li, Electrospun N-doped hierarchical porous carbon nanofiber with improved degree of graphitization for high-performance lithium ion capacitor, *Chem. Eur. J.*, 2018, **24**, 10460–10467.
134. O. Toprakci, H.A.K. Toprakci, L. Ji, G. Xu, Z. Lin and X. Zhang, Carbon nanotube-loaded electrospun $LiFePO_4$/carbon composite nanofibers as stable and binder-free cathodes for rechargeable lithium-ion batteries, *ACS Appl. Mater. Interfaces*, 2012, **4**, 1273–1280.
135. Q. Liu, J. Zhu, L. Zhang and Y. Qiu, Recent advances in energy materials by electrospinning, *Renew. Sust. Energ. Rev.*, 2018, **81**, 1825–1858.
136. J.-W. Jung, C.-L. Lee, S. Yu and I.-D. Kim, Electrospun nanofibers as a platform for advanced secondary batteries: A comprehensive review, *J. Mater. Chem. A*, 2016, **4**, 703–750.
137. W. Li, M. Li, K.R. Adair, X. Sun and Y. Yu, Carbon nanofiber-based nanostructures for lithium-ion and sodium-ion batteries, *J. Mater. Chem. A*, 2017, **5**, 13882–13906.
138. L. Deng, F.-D. Yu, Y. Xia, Y.-S. Jiang, X.-L. Sui, L. Zhao, X.-H. Meng, L.-F. Que and Z.-B. Wang, Stabilizing fluorine to achieve high-voltage and ultra-stable $Na_3V_2(PO_4)_3F_3$ cathode for sodium ion batteries, *Nano Energy*, 2021, **82**, 105659.
139. J. Liu, K. Tang, K. Song, P.A. van Aken, Y. Yu and J. Maier, Electrospun $Na_3V_2(PO_4)_3$/C nanofibers as stable cathode materials for sodium-ion batteries, *Nanoscale*, 2014, **6**, 5081–5086.
140. H. Li, Y. Bai, F. Wu, Y. Li and C. Wu, Budding willow branches shaped $Na_3V_2(PO_4)_3$/C nanofibers synthesized via an electrospinning technique and used as cathode material for sodium ion batteries, *J. Power Sources*, 2015, **273**, 784–792.
141. Y. Niu, M. Xu, C. Dai, B. Shen and C.M. Li, Electrospun graphene-wrapped Na6.24Fe4.88(P2O7)4 nanofibers as a high-performance cathode for sodium-ion batteries, *Phys. Chem. Chem. Phys.*, 2017, **19**, 17270–17277.
142. T. Jin, Y. Liu, Y. Li, K. Cao, X. Wang and L. Jiao, Electrospun NaVPO4F/C nanofibers as self-standing cathode material for ultralong cycle life Na-ion batteries, *Adv. Energy Mater.*, 2017, **7**, 1700087.
143. Y. Liu, X. Zhou, R. Liu, X. Li, Y. Bai, H. Xiao, Y. Wang and G. Yuan, Tailoring three-dimensional composite architecture for advanced zinc-ion batteries, *ACS Appl. Mater. Interfaces*, 2019, **11**, 19191–19199.
144. L. Wu, J. Lang, R. Wang, R. Guo and X. Yan, Electrospinning synthesis of mesoporous $MnCoNiO_x$@double-carbon nanofibers for sodium ion battery anode with pseudocapacitive behavior and long cycle life, *ACS Appl. Mater. Interfaces*, 2016, **8**, 34342–34352.
145. B. He, Z. Zhou, P. Man, Q. Zhang, C. Li, L. Xie, X. Wang, Q. Li and Y. Yao, V_2O_5 nanosheets supported on 3D N-doped carbon nanowall arrays as an advanced cathode for high energy and high power fiber-shaped zinc-ion batteries, *J. Mater. Chem. A*, 2019, **7**, 12979–12986.
146. S. Liu, B. Yang, J. Zhou and H. Song, Nitrogen-rich carbon-onion-constructed nanosheets: An ultrafast and ultrastable dual anode material for sodium and potassium storage, *J. Mater. Chem. A*, 2019, **7**, 18499–18509.
147. S. Dhir, S. Wheeler, I. Capone and M. Pasta, Outlook on K-ion batteries, *Chem*, 2020, **6**, 2442–2460.
148. Y. Luo, L. Wei, H. Geng, Y. Zhang, Y. Yang and C.C. Li, Amorphous Fe-V-O bimetallic oxides with tunable compositions towards rechargeable Zn-ion batteries with excellent low-temperature performance, *ACS Appl. Mater. Interfaces*, 2020, **12**, 11753–11760.
149. W. Ren, X. Chen and C. Zhao, Ultrafast aqueous potassium-ion batteries cathode for stable intermittent grid-scale energy storage, *Adv. Energy Mater.*, 2018, **8**, 1801413.
150. W. Li, M. Li, M. Wang, L. Zeng and Y. Yu, Electrospinning with partially carbonization in air: Highly porous carbon nanofibers optimized for high-performance flexible lithium-ion batteries, *Nano Energy*, 2015, **13**, 693–701.

151. B. Yuan, L. Zeng, X. Sun, Y. Yu and Q. Wang, Enhanced sodium storage performance in flexible free-standing multichannel carbon nanofibers with enlarged interlayer spacing, *Nano Res.*, 2018, **11**, 2256–2264.
152. D. Li, D. Wang, K. Rui, Z. Ma, L. Xie, J. Liu, Y. Zhang, R. Chen, Y. Yan, H. Lin, X. Xie, J. Zhu and W. Huang, Flexible phosphorus doped carbon nanosheets/nanofibers: Electrospun preparation and enhanced Li-storage properties as free-standing anodes for lithium ion batteries, *J. Power Sources*, 2018, **384**, 27–33.
153. B. Zhang, Y. Wang, H. Shen, J. Song, H. Gao, X. Yang, J. Yu, Z. Wu, W. Lei and Q. Hao, Hollow porous $CoSnO_x$ nanocubes encapsulated in one-dimensional N-doped carbon nanofibers as anode material for high-performance lithium storage, *ACS Appl. Mater. Interfaces*, 2021, **13**, 660–670.
154. P. Udomsanti, T. Vongsetskul, P. Limthongkul, P. Tangboriboonrat, K. Subannajui and P. Tammawat, Interpenetrating network of titania and carbon ultrafine fibers as hybrid anode materials for high performance sodium-ion batteries, *Electrochim. Acta*, 2017, **238**, 349–356.
155. R. Rajagopalan, Y. Tang, X. Ji, C. Jia and H. Wang, Advancements and challenges in potassium ion batteries: A comprehensive review, *Adv. Funct. Mater.*, 2020, **30**, 1909486.
156. S. Wu, Y. Song, C. Lu, T. Yang, S. Yuan, X. Tian and Z. Liu, An adsorption-insertion mechanism of potassium in soft carbon, *Small*, 2022, **18**, 2105275.
157. R.A. Adams, J.-M. Syu, Y. Zhao, C.-T. Lo, A. Varma and V.G. Pol, Binder-free N- and O-rich carbon nanofiber anodes for long cycle life K-ion batteries, *ACS Appl. Mater. Interfaces*, 2017, **9**, 17872–17881.
158. F. Liu, J. Meng, F. Xia, Z. Liu, H. Peng, C. Sun, L. Xu, G. Van Tendeloo, L. Mai and J. Wu, Origin of the extra capacity in nitrogen-doped porous carbon nanofibers for high-performance potassium ion batteries, *J. Mater. Chem. A*, 2020, **8**, 18079–18086.
159. Y. Wu, S. Hu, R. Xu, J. Wang, Z. Peng, Q. Zhang and Y. Yu, Boosting potassium-ion battery performance by encapsulating red phosphorus in free-standing nitrogen-doped porous hollow carbon nanofibers, *Nano Lett.*, 2019, 19(2), 1351–1358.
160. D. Liu, L. Yang, Z. Chen, G. Zou, H. Hou, J. Hu and X. Ji, Ultra-stable Sb confined into N-doped carbon fibers anodes for high-performance potassium-ion batteries, *Sci. Bull*, 2020, **65**, 1003–1012.
161. C. Atangana Etogo, H. Huang, H. Hong, G. Liu and L. Zhang, Metal-organic-frameworks-engaged formation of $Co_{0.85}Se$@C nanoboxes embedded in carbon nanofibers film for enhanced potassium-ion storage, *Energy Storage Mater.*, 2020, **24**, 167–176.
162. C. Li, Y. Zhang, J. Yuan, J. Hu, H. Dong, G. Li, D. Chen and Y. Li, Ultra-small few-layered $MoSe_2$ nanosheets encapsulated in nitrogen-doped porous carbon nanofibers to create large heterointerfaces for enhanced potassium-ion storage, *Appl. Surf. Sci.*, 2022, **601**, 154196.
163. J. Xu, C. Lai, L. Duan, Y. Zhang, Y. Xu, J. Bao and X. Zhou, Anchoring ultrafine CoP and CoSb nanoparticles into rich N-doped carbon nanofibers for efficient potassium storage, *Sci. Chin. Mater.*, 2022, **65**, 43–50.
164. Q. Luo, J. Wen, G. Liu, Z. Ye, Q. Wang, L. Liu and X. Yang, Sb_2Se_3/Sb embedded in carbon nanofibers as flexible and twistable anode for potassium-ion batteries, *J. Power Sources*, 2022, **545**, 231917.
165. J. Dong, J. Xiao, K. Cao, H. He, Y. Zhu, H. Liu and C. Chen, Encapsulation of $KTi_2(PO_4)_3$ nanoparticles in porous N-doped carbon nanofibers as a free-standing electrode for superior Na/K-storage performance, *J. Alloys Compd*, 2023, **937**, 168358.
166. S.-H. Chung, P. Han, R. Singhal, V. Kalra and A. Manthiram, Electrochemically stable rechargeable lithium–sulfur batteries with a microporous carbon nanofiber filter for polysulfide, *Adv. Energy Mater.*, 2015, **5**, 1500738.

167. X. Song, S. Wang, G. Chen, T. Gao, Y. Bao, L.-X. Ding and H. Wang, Fe-N-doped carbon nanofiber and graphene modified separator for lithium-sulfur batteries, *Chem. Eng. J.*, 2018, **333**, 564–571.
168. T. Zhao, Y. Ye, X. Peng, G. Divitini, H.-K. Kim, C.-Y. Lao, P.R. Coxon, K. Xi, Y. Liu, C. Ducati, R. Chen and R.V. Kumar, Advanced lithium-sulfur batteries enabled by a bio-inspired polysulfide adsorptive brush, *Adv. Funct. Mater.*, 2016, **26**, 8418–8426.
169. G. Liang, J. Wu, X. Qin, M. Liu, Q. Li, Y.-B. He, J.-K. Kim, B. Li and F. Kang, Ultrafine TiO_2 decorated carbon nanofibers as multifunctional interlayer for high-performance Lithium-Sulfur battery, *ACS Appl. Mater. Interfaces*, 2016, **8**, 23105–23113.
170. Y. Guo, H. Li and T. Zhai, Reviving lithium-metal anodes for next-generation high-energy batteries, *Adv. Mater.*, 2017, **29**, 1700007.
171. Z. Xu, L. Xu, Z. Xu, Z. Deng and X. Wang, N, O-codoped carbon nanosheet array enabling stable lithium metal anode, *Adv. Funct. Mater.*, 2021, **31**, 2102354.
172. J. Man, K. Liu, H. Zhang, Y. Du, J. Yin, X. Wang and J. Sun, Dendrite-free lithium metal anode enabled by ion/electron-conductive N-doped 3D carbon fiber interlayer, *J. Power Sources*, 2021, **489**, 229524.
173. T. Jiang, L. Sun, Y. Zhang, X. Zhang, H. Lin, K. Rui and J. Zhu, Lithiophilic interface dynamic engineering to inhibit Li dendrite growth for intrinsically safe Li-metal batteries, *Chem. Eng. J.*, 2023, **464**, 142555.
174. H. Zhao, H. Yin, Z. Fu, Z. Chi, L. Li, Q. Zhang, Z. Guo and L. Wang, Constructing bimetallic ZIF-derived Zn,Co-containing N-doped porous carbon nanocube as the Lithiophilic host to stabilize Li metal anodes in Li-O_2 batteries, *ChemSusChem*, 2022, **15**, e202200648.
175. C. Zhao, S. Xiong, H. Li, Z. Li, C. Qi, H. Yang, L. Wang, Y. Zhao andT. Liu, A dendrite-free composite Li metal anode enabled by lithiophilic Co, N codoped porous carbon nanofibers, *J. Power Sources*, 2021, **483**, 229188.
176. C. Yang, Y. Yao, S. He, H. Xie, E. Hitz and L. Hu, Ultrafine silver nanoparticles for seeded lithium deposition toward stable lithium metal anode, *Adv. Mater.*, 2017, **29**, 1702714.
177. Q. Qin, N. Deng, L. Wang, L. Zhang, Y. Jia, Z. Dai, Y. Liu, W. Kang and B. Cheng, Novel flexible Mn-based carbon nanofiber films as interlayers for stable lithium-metal battery, *Chem. Eng. J.*, 2019, **360**, 900–911.
178. S. Yang, Y. Li, H. Du, Y. Liu, Y. Xiang, L. Xiong, X. Wu and X. Wu, Copper nanoparticle-modified carbon nanofiber for seeded zinc deposition enables stable Zn metal anode, *ACS Sustainable Chem. Eng.*, 2022, **10**, 12630–12641.
179. X. Yang, Y. Chen, C. Zhang, G. Duan and S. Jiang, Electrospun carbon nanofibers and their reinforced composites: Preparation, modification, applications, and perspectives, *Compos. Part B: Eng.*, 2022, 110386.
180. N. Ning, M. Wang, G. Zhou, Y. Qiu and Y. Wei, Effect of polymer nanoparticle morphology on fracture toughness enhancement of carbon fiber reinforced epoxy composites, *Compos. Part B: Eng.*, 2022, **234**, 109749.
181. H. Liu, C.-Y. Cao, F.-F. Wei, P.-P. Huang, Y.-B. Sun, L. Jiang and W.-G. Song, Flexible macroporous carbon nanofiber film with high oil adsorption capacity, *J. Mater. Chem. A*, 2014, **2**, 3557–3562.
182. Y. Sun, R.B. Sills, X. Hu, Z.W. Seh, X. Xiao, H. Xu, W. Luo, H. Jin, Y. Xin, T. Li, Z. Zhang, J. Zhou, W. Cai, Y. Huang and Y. Cui, A bamboo-inspired nanostructure design for flexible, foldable, and twistable energy storage devices, *Nano Lett.*, 2015, **15**, 3899–3906.
183. G. Sui, S.S. Xue, H.T. Bi, Q. Yang and X.P. Yang, Desirable electrical and mechanical properties of continuous hybrid nano-scale carbon fibers containing highly aligned multi-walled carbon nanotubes, *Carbon*, 2013, **64**, 72–83.

184. Y.K. Kim, S.I. Cha, S.H. Hong and Y.J. Jeong, A new hybrid architecture consisting of highly mesoporous cnt/carbon nanofibers from starch, *J. Mater. Chem.*, 2012, **22**, 20554–20560.
185. J. Zhu, S. Zhang, L. Wang, D. Jia, M. Xu, Z. Zhao, J. Qiu and L. Jia, Engineering cross-linking by coal-based graphene quantum dots toward tough, flexible, and hydrophobic electrospun carbon nanofiber fabrics, *Carbon*, 2018, **129**, 54–62.
186. Z. Zhou, C. Lai, L. Zhang, Y. Qian, H. Hou, D.H. Reneker and H. Fong, Development of carbon nanofibers from aligned electrospun polyacrylonitrile nanofiber bundles and characterization of their microstructural, electrical, and mechanical properties, *Polymer*, 2009, **50**, 2999–3006.

10 Porous Carbon from Biomass for Supercapacitor

Jing Huang and Gang Wang

1 INTRODUCTION

The increasing world population, rapid urbanization, and industrialization have resulted in a significant rise in energy demand, which simultaneously could bring about several environmental issues including climate change and air pollution.[1] Securing environmentally friendly green energies to complement a portion of fossil fuels has attracted great interest all over the world in recent decades.[2] Green energy storage/conversion devices such as batteries, supercapacitors, solar cells, and fuel cells have been vigorously precipitated to prevail the challenges from environmental pollution and excessive consumption of fossil fuels.[3]

Supercapacitors (SCs) can store and deliver charges within a much shorter time and over longer cycles than batteries owing to different operating mechanisms and constituting materials' properties. However, SCs provide less energy than batteries, which restricts industrial applications. Therefore, it is necessary to resolve the performance and stability issues of existing SCs by improving the properties of the energy storage materials.[4]

The surge in consumption of fossil fuel stipulates research to switch from petroleum-derived fuels to renewable resources, such as biomass. Biomass has been exploited for various applications such as catalysis, gas separation, dye-sensitized solar cells, optical devices, medicine and bioscience, energy storage, and conversion.[2] Importantly, biomass is abundantly available currently, providing over 10% of the global energy supply, which has been reckoned the world's fourth energy source (about 10–14%) after conventional fossil oil, coal, and natural gas.[5] In addition, the International Energy Agency (IEA) evaluates that almost half of the energy demand by 2,050 should be produced from biomass.[6] Moreover, large-scale recycling of biomass can help achieve the goal of "carbon neutrality."[7]

2 CARBON MATERIALS DERIVED FROM BIOMASS

At present, most nanocarbon materials are mainly derived from conventional non-renewable resources under the conditions of high-pressure and harsh chemicals. Although the yield of activated carbon produced from fossil fuels is very

DOI: 10.1201/9781003387831-10

high, the methods associated with its production lead to polluting environment and expensiveness.[8]

Biomass resources as a carbon precursor could not only refrain from the release of harmful substances in the synthesis process but also effectively overcome environmental and energy problems. Moreover, the natural and interconnected structure presenting in biomass improves the porous properties of activated carbons, and further contributes to the dual benefits of economy and environmental protection and to the sustainable development of mankind.[9]

2.1 Classification of biomass precursors

Biomass for the preparation of porous carbons is usually classified as forest materials and residues, agricultural crops and residues, industrial biowastes, marine wastes, and domestic biowastes[10] (Figure 10.1). The following sections provide multiple biomasses that are produced in each classification and their general compositions.

2.1.1 Forest materials and residues

Biomass derived from trees (logs and timbers) and their residues (barks, saw-timber, and wood chips) contribute to the major sources of forestry biomass. Energy crops as another source of forest biomasses comprise forestry (Poplar, Willow), herbaceous (Switch grass, Sorghum, and Jatropha), and aquatic (Water hyacinth, Algae).[11] Forest resources have several applications including wood production, soil conservation, and water resources as well as social or recreational requirements.[12] For instance,

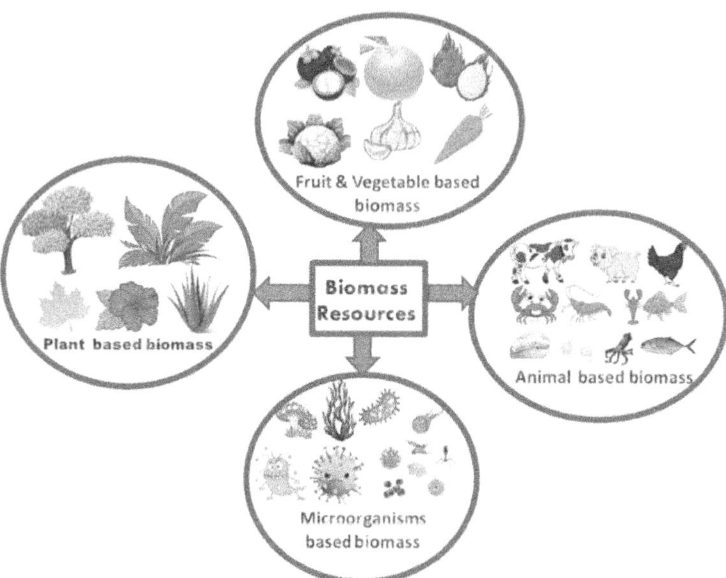

FIGURE 10.1 Schematic diagram of different types of biomass resources.[10]

Yu et al. has fabricated wood-derived porous carbon and employed as an electrode material for supercapacitors.[11]

2.1.2 Agricultural crops and residues

Energy crops as the plant-based agricultural biomass are specifically cultivated to provide feedstock for bioenergy and biomaterial production, including timothy grass, switch grass, elephant grass, and Miscanthus. Moreover, energy crops possess advantages such as superior energy density, fast growth, low maintenance cost, easily feeding, which is beneficial to biofuels and green chemicals production.[13] Additionally, agricultural residues have been classified as crops and processing residues. Crop residue refers to the materials left behind on cultivated land after harvesting the crops, such as wheat straw, soybean straw, corn straw, canola hull, canola straw, etc. And then, processing residues represent materials or by-products generated in the procedure of harvested crops to usable forms, including bran, husks, citrus peel, onion peel, garlic peel, shells, etc.[14] Several studies have reported the thermochemical and biological conversion of crop residues into biochart and activated carbon for industrial applications such as wastewater treatment, soil remediation and catalysis, energy, and energy carriers.[15] For instance, Biswal et al. have utilized dead neem leaves to fabricate carbon material with naturally exposed pores in the presence of minimal amount of calcium and magnesium as porogens.[16]

2.1.3 Industrial biowastes

Biowastes from the industries ground on the biosources, including food waste, wood processing waste, pulp industries etc. Peels, seeds, and sacks as biowaste produced in the process of juice or pulp processing industries, are naturally full of elements including calcium, magnesium, phosphorous, potassium and phenolic compounds. Additionally, a great deal of animal biomasses such as animal skin and bones, waste leather, and fish scales are rich source of collagen proteins, which could be acted as self-templates and source of nitrogen to prepare nitrogen-doped porous carbonaceous materials.[17] Wang et al. have utilized fish scale as precursor to fabricate thin layered carbon doped with multiple heteroatom (N, O, S), which indicates the porous nanosheet framework (thickness of 3–5 nm) with large micropore ratio, and further are exploited as the electrodes for supercapacitors.[18]

2.1.4 Microorganism-based biomass

The microorganisms are classified as bacteria, algae, and fungus, which are composed of proteins, lipids, carbohydrates, nucleic, and amino acids. As the major microorganism-based biomass, the fruiting part of the mushroom is more often utilized to fabricate active carbons owing to the presence of rich nitrogen content.[19] Ping Cheng et al. have reported a hierarchically porous carbon with natural shiitake mushroom as precursor, which demonstrates a specific surface area of 2,988 $m^2 g^{-1}$ and pore volume of 1.76 $cm^3 g^{-1}$.[20] Bacterial cellulose (BC) derived from fermentation has drawn much attention owing to its wide availability, cost-effective, renewability, and porous nature. Bingjun Zhu et al. have fabricated value-added carbons with algae biomass as precursor, which have been used as a high-performance "green" carbon-based supercapacitor electrode material.[21]

2.1.5 Municipal waste

Municipal waste refers to garbage, which are solid materials and classified as organic and inorganic waste on account of their chemical and biological properties. And then, municipal waste could be further subdivided into food residue, paper-based residue, textile, wood waste, glass, trash cans, metal scraps, etc. The organic solid waste as a resource could be employed to generate energy in view of waste-to-energy concept through several procedure such as incineration and carbonization.[22] Yu et al. have utilized human urine as precursor to prepare porous heteroatom-doped carbon for the supercapacitors.[23] Polyethylene Terephthalate (PET) as the plastic waste is non-biodegradable and rich in carbon content (62.5 wt%), which could be re-processed into carbon material for energy storage.[24]

2.2 Types of porous carbon materials from biomass

2.2.1 Activated porous carbon

Although the conventional carbon materials such as CNT and graphene as well as the AC derived from MOFs indicate comparable or superior electrochemical performance to activated carbon, the intricate preparation process and non-renewable precursors hinder their wide application. Researchers shift toward biomass-derived activated carbon owing to its abundance and eco-friendliness, cost-effectiveness (Table 10.1).[25]

Biomass-derived active carbon can indicate multiple dimensions for frameworks, such as spherical, tubular, honeycomb, or graphene-like carbon. Pore framework including porosity and size distribution play an important role on energy storage power of active carbon materials for supercapacitors. The multi-scale pores are desirable, owing to macropores (larger than 50 nm) for the ion-buffering reservoir, mesopores (2–50 nm) for ion transportation, and micropores (less than 2 nm) for the enhancement of charge storage. Recently, researches have paid attention to fabricate hierarchical porous carbon with different dimension via adjusting experimental conditions such as the type of active reagent, the temperature of activation.[26–28]

2.2.2 Carbon fibers

Carbon fibers/carbon nanofibers (CFs/CNFs) with small size (<100 nm) are categorized into platelet, fishbone, ribbon, stacked cup, and thickened CNFs in view of the structural arrangement of their graphitic planes[29] (Figure 10.2). CNFs have been intensely utilized as conductive support matrices for energy storage devices owing to high electrical conductivity, mechanical stability, light weight, and large-scale synthesis at low cost.[30] Generally, biomass-derived carbon nanofibers are fabricated through immersing in reagents and then mixing with polymer solution. Researchers have demonstrated that the blending biomass with polymer contributes to the synthesis of carbon nanofibers with cross-linked porous framework and further the enhancement of electrochemical performance. Tao et al. have utilized walnut shell as biomass precursor to fabricate carbon nanofiber with small diameter about 280 nm through electrospinning process.[31]

TABLE 10.1
Comparison of various biowaste derived activated carbon as an electrode material[25]

Source of biowaste derived AC	Activation method	Surface area of AC ($m^2 g^{-1}$)	Specific capacitance ($F g^{-1}$)	Electrolyte	Cycle stability
Coconut shell	Steam	1,532	228 at 5 mV s^{-1}	6M KOH	93% after 3,000 cycles
Peanut shell	$ZnCl_2$	1,549	340 at 1 A g^{-1}	1M H_2SO_4	95.3% after 10,000 cycles
Palm kernel shell	KOH	462.1	210 at 0.5 A g^{-1}	1M KOH	95% after 1,000 cycles
Watermelon rind	KOH	2,277	333.4 at 1 A g^{-1}	6M KOH	96.8% after 10,000 cycles
Tea leaves	KOH	2,841	330 at 1 A g^{-1}	2M KOH	92% after 2,000 cycles
Tea waste buds	KOH	1,610	332 at 1 A g^{-1}	6M KOH	97.8% after 10,0000 cycles
Sugarcane bagasse	KOH	1,939.9	298 at 1 A g^{-1}	1M H_2SO_4	94.5% after 5,000 cycles
Withered rose flower	KOH/KNO_3	1,980	350 at 1 A g^{-1}	6M KOH	96.5% after 15,000 cycles
Jujube fruit	NaOH	1,135	460 at 1 A g^{-1}	6M KOH	92.2% after 130,000 cycles
American Poplar fruit waste	KOH	942	423 at 1 A g^{-1}	6M KOH	97% after 200,000 cycles
Corncob	KOH	800	390 at 0.5 A g^{-1}	1M H_2SO_4	94% after 5,000 cycles
Plastic waste (polyethylene terephthalate)	KOH	2,326	169 at 0.2 A g^{-1}	6M KOH	90.6% after 5,000 cycles
Rice husk	KOH	3,145	367 at 5 mV s^{-1}	6M KOH	≈100% after 30,000
Soybean pods	NaOH	2,612	352.6 at 0.5 A g^{-1}	1M Na_2SO_4	94.2% after 50,000
Orange peel	KOH	2,160	460 at 1 A g^{-1}	1M H_2SO_4	98% after 10,000 cycles
Mangosteen peels	NaOH	2,623	357 at 1 A g^{-1}	6M KOH	94.5% after 130,000 cycles

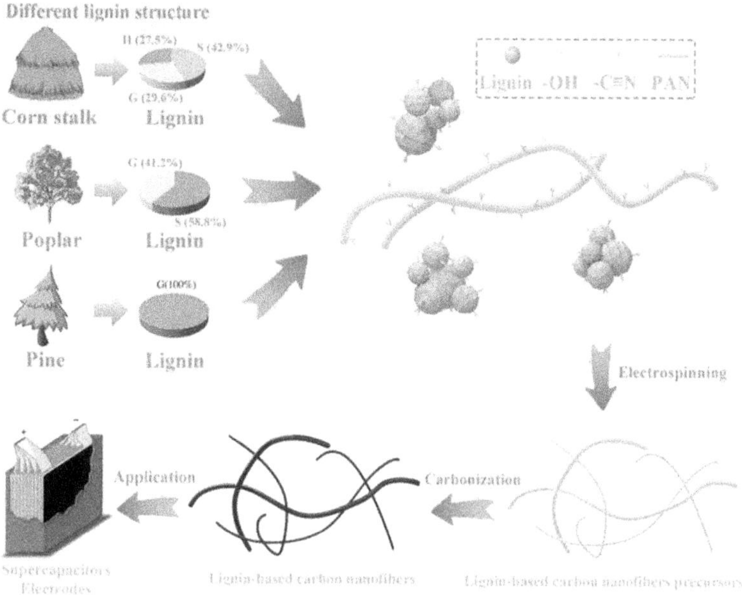

FIGURE 10.2 Schematic diagram of the synthesis of carbon nanofibers based on different lignin for supercapacitor.[29]

2.2.3 Carbon dots

The CDs (carbon-dots) are a class of carbon particles with nanoscale structure (<10 nm), which are composed of a nanocrystalline core and an amorphous shell with versatile surface functional groups. On account of the carbon cores, CDs could be classified as graphene quantum dots (GQDs), carbon nanodots (CNDs), and polymer dots (PDs)[32] (Figure 10.3). The diversity of CDs is mainly ascribed to the diversity of carbon precursors and synthetic strategies.[33] Based on the defects of conventional approaches such as low quantum yield and limited solubility, researchers have paid attention to developing carbon quantum dots derived from green carbon precursors. Biomass and their wastes have been applied to prepare the scalable, cost-effective carbon-based QDs (quantum dots) with superior optical feature, which have been applied in various optoelectronic fields, such as LEDs, solar cells, energy conversion, and photodetectors. For instance, Huang et al. have utilized sugarcane molasses as the precursor to fabricate biocompatible blue luminescent C-QDs (QY of 5.8%), which are employed as bioimaging agents and sensor for Sunset Yellow.[34]

2.2.4 Carbon-based gels

Carbon aerogels are a special type of low-density microcellular foams, which consist of interconnected carbon particles. Generally, carbon aerogels are provided with the advantages, such as high surface area assembly, good graphitized texture, controllable pore size, interconnected fibrous structure, rapid charge transport pathways, and excellent thermal conductivity, which are suitable for energy storage devices, electromagnetic shielding materials, and absorbers.[35] Li et al. have fabricated carbon

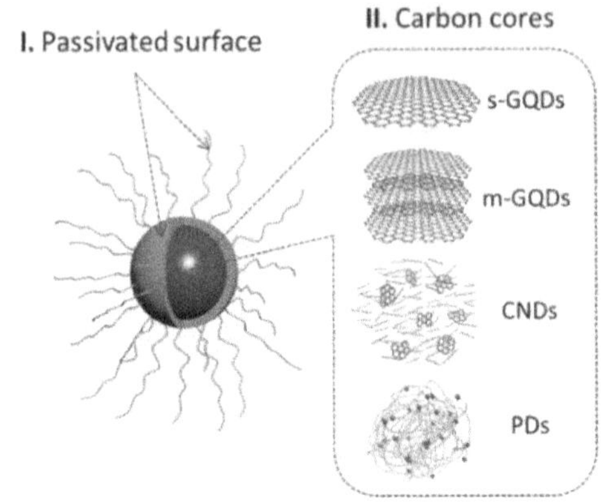

FIGURE 10.3 Schematic diagram of the structure of CDs.[33]

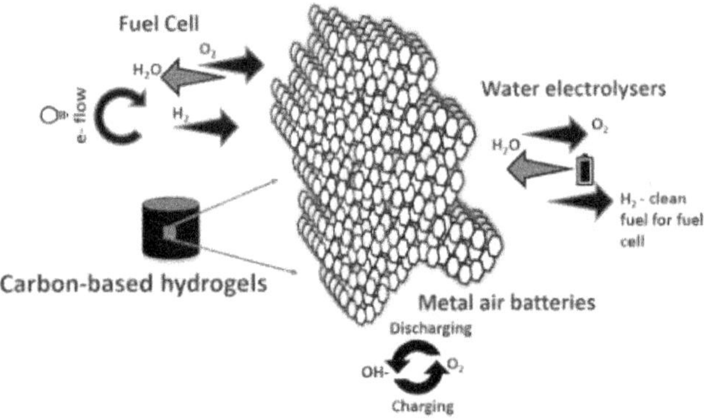

FIGURE 10.4 Schematic diagram showing carbon-based hydrogels.[37]

aerogels from various types of melon (winter melon, watermelon, and pumpkin) via a hydrothermal carbonization and post-pyrolysis, followed by introducing PW (paraffin wax) into the carbon aerogels.[36]

Hydrogels are a class of materials which are strongly hydrophilic, water-filled, three-dimensional, and cross-linked networks[37] (Figure 10.4). Carbon-based hydrogels have been broadly classified into three types: graphene-based, polymer-based, and biomass-derived hydrogels. The conductive framework contributes to creating low-resistance pathways for electron mobility and the porous structure facilitates ion diffusion freely.[38] Wu et al. have employed watermelon to prepare spongy hydrogels composed of carbon nanospheres and nanofibers via hydrothermal reaction.[39]

2.3 Preparation Methods

Porous carbons derived from non-renewable fossil feedstock sources are usually fabricated under harsh reaction conditions, which is environmental unfriendly and uneconomical. Therefore, it is a crucial topic of research to develop versatile functional biomass-derived carbons under cost-effective, sustainable conditions, such as carbonization, templating, activation, doping, and mechanochemically assisted syntheses.[40]

2.3.1 Carbonization

Carbonization refers to the treatment of carbonaceous precursor at high temperature in the presence of inert medium (such as N_2, argon) that gets rid of the volatile organics to fabricate pure carbon structure with small amount of heteroatoms and hydrogen. Pyrolysis and hydrothermal carbonization (HTC) are two predominant approaches to transform biomass into carbon.[41]

2.3.1.1 Pyrolysis

Pyrolysis refers to the conversion of biomass into carbonaceous products via thermal degradation of the chemical components in an inert medium at elevated temperatures. During the conversion of biomass into porous carbons, several reactions are involved such as dehydration, condensation, cross-linking depolymerization, fragmentation, and isomerization reactions. The ultimate products of the pyrolysis include gases, bio-oil, and biochar. Every product proportions will rely on the pyrolysis conditions, such as temperature, pressure, residence time, etc.[42]

Panmand et al. have reported the perforated grapheme (PG) with bougainvillea flower as raw material for supercapacitor via a simple carbonization step, which demonstrates high specific capacitance of 316.2 F g^{-1} at 2.28 A g^{-1}, and 93% of capacitance retention over 10,000 cycles.[43]

Liu et al. have utilized perilla frutescens (PF) leaves as precursor to fabricate O/N-co-doped porous carbon nanosheet through direct pyrolysis, which exhibits the surface area of 655 $m^2 g^{-1}$ and the pore volume of 0.44 $cm^3 g^{-1}$. In addition, the assembled symmetric supercapacitor indicates a high volumetric energy density of 14.8 Wh L^{-1} at 490 W L^{-1} and a high stability of 96.1% capacitance retention over 10,000 cycles.[44]

2.3.1.2 Hydrothermal carbonization

Hydrothermal carbonization (HTC) as a thermochemical process represents wet pyrolysis for the conversion of biomass precursors to a solid biochar or a structured carbon, which is usually carried out in a pressured liquid medium at a relatively mild temperature (100–240°C).[45] During the procedure of HTC, pretreated biomass dispersed in a liquid medium in a pressure-tight container (such as Teflon autoclave) would go through one of the following (or combined) stages: hydrolysis, dehydration, decarboxylation, polymerization, and aromatization.[46] Simultaneously, liquid medium usually serves as a solvent and a catalyst that contribute to the hydrolysis and cleavage of the chemical component of biomass.[47]

Currently, abundant porous carbon have been fabricated through HTC with versatile biomass as precursors, such as saccharides (glucose, sucrose, and starch), cellulose, algae, peanut shell, and camphor leaves.[48] However, HTC-derived porous carbon materials usually demonstrate a low specific surface and poor porosity as well as unordered structure in nature, which involve further activation or carbonization and graphitization to modify its chemical and physical properties.[49] Yong-Qing Zhao et al. have synthesized electrode porous carbon derived from tobacco rods via HTC and KOH activation, which demonstrates high capacitance of 286.6 F g^{-1} at 0.5 A g^{-1}, superior rate performance of 212.1 F g^{-1} at 30 A g^{-1}, and excellent cyclic stability of 96% capacitance retention over 10,000 cycles.[50]

2.3.2 Physical activation

Physical activation refers to carbonization at high temperature with gaseous activators such as CO_2, steam, air, O_2, water vapor, and the mixed atmosphere.[51] In the process of physical activation, three stages are involved: Firstly, the activated gas reacts with the biomass precursor to open the closed pores and expose crystallites' surface; Secondly, the original pores are enlarged; Thirdly, new active sites formed on the surface proceed to react with the activated gas to produce new pores.

The corresponding chemical reactions are demonstrated as follows (Eqns 10.1–10.4):

$$C+2H_2O \rightarrow 2H_2+CO_2 \tag{10.1}$$

$$C+H_2O \rightarrow H_2+CO \tag{10.2}$$

$$C+CO_2 \rightarrow CO \tag{10.3}$$

$$C+O_2 \rightarrow CO_2 \tag{10.4}$$

Although the effect of physical activation seems not very well such as the relatively low SSA according to chemical activation, physical activation is still more popular and effective for large-scale production owing to the developed porosity together with the environmental benign nature and cost-effective.[52]

Jiang et al. have reported the porous carbon with willow as precursor and CO_2 as activator, which indicates well-developed channels and SSA of 750.7 $m^2 g^{-1}$. Yu et al. have utilized CO_2 as the activator to prepare cattail-derived porous carbon, which demonstrates SSA of 441.12 $m^2 g^{-1}$ and pore volume of 0.33 $cm^3 g^{-1}$ as well as the specific capacitance of 126.5 F g^{-1} at 0.5 A g^{-1}.[53]

2.3.3 Chemical activation

Chemical activation is denoted as approach for fabricating porous carbon with chemical reagents as the activator, including $ZnCl_2$, KOH, NaOH, H_3PO_4, KCl, and HNO_3. Generally, chemical activation indicates many superiority factors, such as low-energy consumption, flexible control of the porosity, short activation time, high specific surface area, and high yield. However, the disadvantages of chemical activation are also obvious, including the special equipment for corrosion resistance, and the expensive activators and the treatment of wastewater and gas in the synthetic process.[54]

2.3.3.1 Alkaline activation

The activation of biomass with KOH is divided into two categories: the first is one-step carbonization activation and the second is step-by-step carbonization activation. The one-step carbonization activation refers to high temperature activation after directly mixing the biomass precursor and KOH. The step-by-step activation means pre-treating the biomass precursor before carbonization activation.

The pre-treatment of biomass precursor materials comprises hydrothermal carbonization, swelling, pre-carbonization, etc.[55] Currently, one commonly accepted assumption is that the expansion of the carbon lattice caused by the metallic K intercalation, which originates from the strong reaction ability of K and C. The mainly involving chemical reactions equations are below (Eqns 10.5–10.10)[56]:

$$2KOH = K_2O + H_2O \tag{10.5}$$
$$C + H_2O = CO + H_2 \tag{10.6}$$
$$CO + H_2O = CO_2 + H_2 \tag{10.7}$$
$$CO_2 + K_2O = K_2CO_3 \tag{10.8}$$
$$K_2O + H_2 = 2K + H_2O \tag{10.9}$$
$$K_2O + C = 2K + CO \tag{10.10}$$

Quan-Hong Yang et al. have reported a novel 3D honeycomb-like porous carbon (HLPC) with pomelo peels as the precursor and KOH as activation agent. The as-prepared HLPC demonstrates a specific surface area of 2,725 $m^2 g^{-1}$, and high specific capacitance of 342 F g^{-1} at 0.2 A g^{-1} as well as good stability with 98% retention over 1,000 cycles.[57]

Jing Huang et al. have utilized pomelo seeds as precursor to produce porous carbon through the activation of KOH, which exhibits a high specific capacitance of ~845 F g^{-1} at 1 A g^{-1} and an ultrahigh stability of ~93.8% even after 10,000 cycles.[58]

In general, the alkaline activation agents such as KOH and NaOH possess the advantages of high efficiency and simple experimental procedure. However, the usage of alkaline activation agents results in high corrosion of the equipment and the activation process are usually carried out at high temperature, which is not in consistent with the concept of green economy.[59]

2.3.3.2 Salt activation

Salt as an activation agent not only agrees well with the concept of green chemistry owing to low corrosion of equipment and environmental friendliness but also indicates comparable performance to alkaline activator. Presently, the salt activators that are used commonly with superior properties as follows: K_2CO_3, $KHCO_3$, $ZnCl_2$, Na_2CO_3, $FeCl_3$, etc.

K_2CO_3 with similar activation mechanism to KOH demonstrates a catalytic graphitization effect on the material. Xia et al. have utilized coconut shell as a precursor to prepare three-dimensional porous graphene sheets (3DPGLS) with K_2CO_3 as activation agent, which indicate the specific surface area of 1,506.19 $m^2 g^{-1}$ and high conductivity of 32.14 S cm^{-1}.[60]

Vediyappan Veeramani's group have fabricated grapheme sheet like porous carbon derived from Bougainvillea spectabilis with $ZnCl_2$ as activator, which indicates high surface area of 1, 197 $m^2 g^{-1}$and the specific capacitance of 233 F g^{-1} at 1.6 A g^{-1}. However, the application of $ZnCl_2$ activation is strictly restricted owing to a severe environment problem of waste water treatment.[61]

Zhang and his coworkers have reported highly porous graphene with K_2FeO_4 as activator and biomass-loofah as precursor, which demonstrates a high specific capacitance of 1308.2 F g^{-1} at 1 A g^{-1} and excellent cycling stability of 91.6% retention over 10,000 cycles.[62]

Wang et al. have utilized waste pine bark as precursor to fabricate ultrathin graphene-like PCNs with potassium formate as activation agent for supercapacitor electrode, which displays superior energy densities of 22.8 Wh kg^{-1} at an ultrahigh power density of 30.375 kW kg^{-1} and high rate performance with specific capacitance retention of 76.3% at 40 Ag^{-1}.[63]

2.3.3.3 Acid activation

The activation process of H_3PO_4 involves five categories as follows: hydrolysis, dehydration, aromatization, cross-linking with biopolymers, and pore formation. In the process of activation, H_3PO_4 cross-link with lignocellulosic structures in the form of phosphate ester to refrain from cell wall shrinking and further leave abundant voids. Generally, the biomass-derived porous carbons with H_3PO_4 as activation agent possess the specific surface area lower 1,500 m^2g^{-1}, and mesoporous texture as well as a large pore volume in the range of 1–1.5 cm^3g^{-1}, which are suitable for OEB, LIB, and EDLC.[64]

Chen and coworkers has reported the new activated carbon derived from cotton stalk via H_3PO_4 activation, which displays the specific surface area of 1,481 m^2g^{-1}. The assembled supercapacitor demonstrates the specific capacitance of 114 F g^{-1} at 0.5 A g^{-1} and cycling stability of 4.7% capacitance loss over 500 cycles in 1 M Et_4NBF_4 electrolyte.[65]

Jing Huang et al. have prepared porous carbon from pomelo valves by the hydrothermal activation with H_3PO_4 and simple carbonization, which exhibits a high specific capacitance of 966.4 F g^{-1} at 1 A g^{-1} and an ultrahigh stability of 95.6% even after 10,000 cycles.[66]

2.3.4 Template methods

The template method represents the fabrication of multifunctional nanomaterials with adjustable structure such as morphologies and pore size distribution in the presence of sacrificial templates with specific features. Template methods are generally classified as hard and soft templating strategies.[67]

2.3.4.1 Hard template method

Hard templates (HT) refer to a kind of rigid material with a defined nanoporous structure, such as silicas, zeolites, and metal oxides (e.g., MgO, Fe_2O_3, and ZnO) to be used for the synthesis of highly porous ordered or disordered carbons (Figure 10.5).[68] Typically, the method includes the following steps: synthesis of the template, combination of the carbon precursor with the template, pyrolysis of the composite, and finally removal of the template.[69]

Zhang et al. have produced a porous carbon foam with chitosan as the raw material and silica as the template, which demonstrates the specific surface area of 2,905 m^2g^{-1} and excellent gravimetric capacitance of 374.7 ± 7.7 F g^{-1} at 1 A g^{-1} and rate performance of 60% capacitance retention (235.9 ± 7.5 F g^{-1} at 500 A g^{-1}).[70]

Sanchez-Sanchez et al. have utilized gallic acid as precursor to fabricate ordered mesoporous carbons (OMCs) through a hard-templating strategy, which indicate the highest surface area of ~1,000 m^2g^{-1}, and the high specific capacitance of 277 F g^{-1} at 1 A g^{-1}, a high rate capability of 60.6% capacitance retention at 12 A g^{-1}, and good cycling stability of 80.6% over 5,000 cycles.[71]

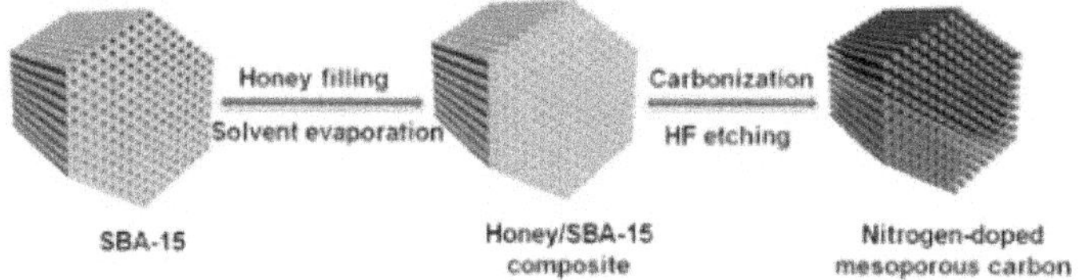

FIGURE 10.5 Schematic diagram of the synthesis of honey-derived OMC.[68]

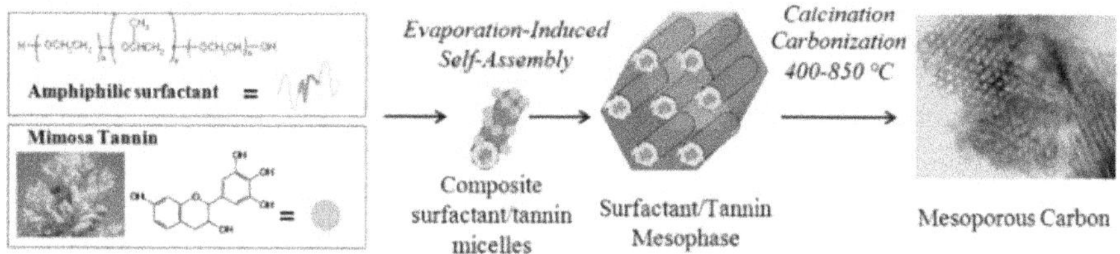

FIGURE 10.6 Schematic diagram of the synthesis of ordered mesoporous polymers and carbons derived from tannins.[72]

However, the drawbacks for the hard template method involve its time-consuming nature, high cost, and the use of highly toxic chemicals (such as HF and NaOH) owing to the purification of the product, which have limited the product's application for large-scale production.

2.3.4.2 Soft template method

The soft templating (ST) strategy involves the cooperative self-assembly of amphiphilic molecules and carbon precursors, which is broadly applied to the fabrication of hollow-structured nanomaterials (Figure 10.6).[72]

Wen et al. have utilized organic modified montmorillonite (OMMT) as template and waste poly (ethylene terephthalate) (PET) beverage bottles as precursor to prepare porous carbon nanosheets (PCNS). The PCNS reveals the specific capacitance of 121 F g^{-1} in 1M TEATFB/PC as well as the superior stability of 90.6% capacitance retention over 5,000 cycles.[73]

Chen et al. have utilized ribose as precursor and oleic acid nano-emulsions as soft template to fabricate flask-like carbonaceous materials, which indicate the specific surface area of 2,335 m^2 g^{-1} and the specific capacitance of 222 F g^{-1} at 1 A g^{-1} and no decrease in capacitance over 10,000 cycles.[74]

Soft template method is a relatively eco-friendly strategy from the view point of green chemistry owing to the removal of the template accomplished in non-toxic or non-corrosive process. Moreover, soft template presents more flexible morphological control and inferior stability than HT.

In brief, although the template method can control the pore size distribution and micro-morphology relatively accurately, it also leads to an increased product cost and complex synthetic procedure, which constraints the widespread application.

Alternatively, the self-template can facilitate the blending of the template and the carbon source, which contributes to overcoming the heterogeneous shortcomings of solid mixing and the complicated operation of liquid mixing. Additionally, template-free approach is a direct strategy to prepare carbon materials without the aid of any structure-guiding agent or metal-based growth catalyst (such as Fe/Ni/Co/Cr) in the synthetic process, which is eco-friendly in the absence of toxic reagents. However, the template-free method is somewhat not skilled in adjusting the morphology and porosity of carbon materials.[75]

2.3.5 Doping of heteroatoms

Heteroatom-doping strategies refer to the introduction of heteroatoms (B, N, O, P, and S) into carbon skeletons usually through pyrolysis and hydrothermal synthesis, which could manipulate the electronic structure and surface chemistry of materials. Heteroatom-doping are generally divided into two strategies: in situ doping and post doping. In situ doping is conducted by the direct pyrolysis of biomass precursors or carbonization/activation of pretreated biomass sources. While for post doping, it is implemented by the post-treatment of electrospun carbon materials in the presence of targeted heteroatom sources.[76]

G. A. Ferrero et al. have utilized soybean waste residue as precursor to develop N-doped microporous carbon of 1.6 at%, which delivers the specific surface area of 2,130 $m^2 g^{-1}$, and the specific capacitance of 258 F g^{-1} at 0.2 A g^{-1} as well as good rate capability of 170 F g^{-1} at 40 A g^{-1}.[77]

Han et al. have reported S-doping carbon materials with bamboo shoot shells as carbon resource and 3,4-Ethylenedioxythiophene (EDOT) as sulfur source, which indicates a high S content of 1.66 at% , SSA of 1,032 $m^2 g^{-1}$ and the capacitance of 302.5 F g^{-1} at 0.5 A g^{-1}.[78]

Xiandong Yue et al. have utilized cornstalk to prepare porous carbon co-doped with N (1.3 at%), B (2.97 at%), and P (0.38 at%), which indicates the specific surface area of 3,123.5 $m^2 g^{-1}$, a high specific capacitance of 342.5 F g^{-1} at 0.5 A g^{-1}, and energy density of up to 26.18 W h kg^{-1}.[79]

3 SUPERCAPACITOR

The decreasing supply of non-renewable energy sources such as fossil fuels and the intensification of the environmental pollution have prompted the search for low cost and renewable energy sources to replace fossil energy. Supercapacitors, as a type of dominant energy conversion/storage devices, have the advantages of fast charge/discharge rate, ultra-long cycling life, high-power density, and strong environmental adaptability, which bridge the energy/power gap between conventional dielectric capacitors and high energy storage batteries. Generally, supercapacitors can satisfy the diverse energy storage needs, such as hybrid vehicles, urban public transportation, military applications, and portable electronic devices.[80]

3.1 Brief history of supercapacitor

As for supercapacitor research/industry to today's commercially viable product over this decade, the first original idea of capacitor is proposed by Lyden Jar in the 18th century. The first nonaqueous electrolyte-based electrochemical capacitor is invented by Robert Rightmire at the Standard OilCo. of Ohio (SOHIO). Raghavendra et al. has reported that the first supercapacitor (EDLC supercapacitor) has been fabricated by General Electric in 1957. In 1971, a new class of electrochemical capacitor (pseudocapacitor) is produced on the basis of RuO_2, followed by extensive research on RuO_2 as a pseudocapacitor material in 1975–1980. In 1989, the U.S. Department of Energy (DOE) starts to support a long-term supercapacitor research on high energy density supercapacitors for application in the Electric and Hybrid Vehicle Program. Afterward, a world-leading supercapacitor manufacturing company, Maxwell Technologies Inc., cooperates with the DOE to develop high-performance supercapacitors, which involve EDLCs, pseudocapacitors, and asymmetric supercapacitors. Supercapacitor companies from all around the world such as Nesscap (Korea), ELTON (Russia), Nippon Chemicon (Japan), and CAP-XX (Australia) have been developing and delivering different types of supercapacitors for various applications. Since 2000, the amount of researches on supercapacitors has continuously and significantly scaled up with respect to the emerging requirement for high-power, high-reliability, and safe energy storage devices. In 2006, the first electric double layer supercapacitor-based buses have been launched in Shanghai (China). In 2016, the researchers at VTT Technical Research center at Finland have developed a hybrid nanoelectrode consisting of porous silicon coated with titanium nitride. Global supercapacitor market has been valuated to be 887 million dollars in the year 2020.[81]

3.2 Mechanisms and Types of Supercapacitor

SC (supercapacitor) devices with different materials and configurations have been extensively studied. On account of energy storage mechanism, SCs could be classified as electric double layer capacitors (EDLCs), pseudocapacitor, and hybrid supercapacitors (Figure 10.7).[81, 82]

EDLC mechanism of supercapacitors: the electrical double layer capacitance is generated physically by the pure electrostatic attraction between ions and the

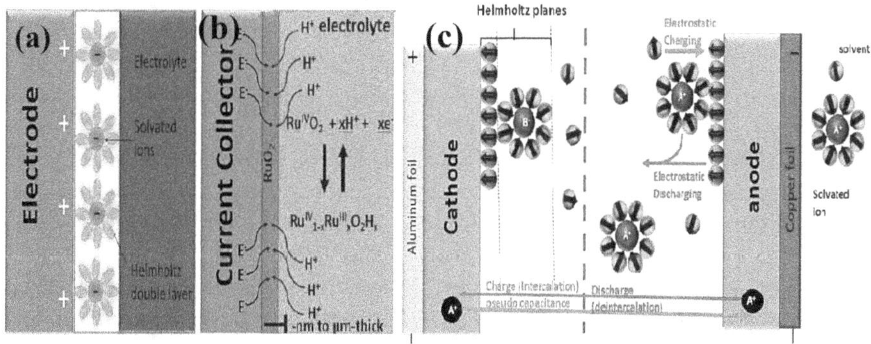

FIGURE 10.7 Charge Storage Mechanism in (a) EDLCs, (b) pseudocapacitor, (c) asymmetric hybrid.[81]

charged electrode surface (e.g., carbon materials). The specific capacitance of EDLC is strongly dependent on the accessible surface area and the surface properties of the electrode materials. The process of charge/discharge of EDLCs only involves charge rearrangement, without any Faradaic reactions. As a result, the electric double layer capacitor can respond immediately to potential changes. Experimentally, EDLCs not only feature high-power density and superb charging-discharging stability but also usually suffer from low-energy density.[83]

According to pseudocapacitor, the pseudocapacitance involves a rapid reversible oxidation/reduction (redox) or Faradaic charge transfer reaction, which is produced by the electroactive substances on the surface of the electrode such as metal oxides (like NiO, Co_3O_4, and RuO_2, conducting polymers (like PPy, PANi, and PTh) and some metal hydroxides ($Co(OH)_2$, $Ni(OH)_2$). Accordingly, fast surface or near-surface reactions with high reversibility contribute to large power density. There are generally two mechanisms for the charge storage taking place in pseudocapacitance. One is redox pseudocapacitance that occurs on the surface of metal oxides and the other is intercalation pseudocapacitance. The redox process is surface limited in pseudocapacitors and the intercalation is diffusion limited in battery type electrodes.[84]

Hybrid supercapacitor combines capacitive type nature with Faradic type to achieve the higher energy density along with its power density. The hybrid capacitors can be further subdivided into three categories. The first category is the asymmetric capacitor, which consists of one EDLC electrode (negative) and another pseudocapacitive electrode (positive). The second category is battery type capacitor, which is composed of one Faradic electrodes and another capacitive electrode. The third category is a composite, which possess the combination of EDLC materials with pseudocapacitor materials.[85]

4 POROUS CARBON MATERIALS FROM BIOMASS IN SUPERCAPACITORS

4.1 Activated Carbon

Activated carbon materials are the most utilized low-cost electrode materials for EDLC electrodes, which indicate a hierarchical permeable structure involving micropores, mesopores, and macropores.

Taniya Purkait et al. have prepared graphene-like nanosheet derived from peanut shells through the activation of KOH and the mechanical exfoliation, which indicates high specific surface area of 2,070 $m^2 g^{-1}$ and 186 F g^{-1} at 0.5 A g^{-1} as well as the highest energy 58.13 Wh kg^{-1} at 375 W kg^{-1}.[86]

Shijiao Song et al. have reported 3D hierarchical porous carbon (3D HPCs) derived from corn husks. The symmetric supercapacitor based on CHHPCs demonstrates a high energy density of 21 W h kg^{-1} at 875 W kg^{-1} and retains as high as 11 W h kg^{-1} at 5,600 W kg^{-1}.[87]

Dong Zhou et al. have reported the interconnected open-channel carbon nanosheets derived from reeds. The assembled supercapacitor indicates the specific capacitance of 147 F g^{-1} at 1 A g^{-1} and energy density of 45.5 W h kg^{-1}at ~750 kW kg^{-1} and the outstanding cycle stability of 90% initial capacitance retention over 10,000 cycles.[88]

4.2 Carbon Fibers

Cellulose as biomass fiber on Earth has been provided with many advantages such as high carbon content, abundant surface chemical properties, superior mechanical properties, high specific surface areas, thermal stability, and ease of processing.

Jie Cai et al. have fabricated carbon nanofibers derived from cellulose acetate nanofibers for supercapacitor, which indicates the specific capacitance of ~241.4 F g^{-1} at 1 A g^{-1} and excellent cycling stability of negligible 0.1% capacitance reduction over 10,000 cycles and maximum power density of ~84.1 kW kg^{-1}.[89]

Cheng-Meng Chen et al. have reported the supercapacitor based on nanocarbon fibers (HCMT-650) prepared with kapok as biomass precursor, which indicates the specific capacitance of 140 F g^{-1} at 1.0 A g^{-1} and high energy density of 48.3 Wh kg^{-1} at 450 W kg^{-1}.[90]

4.3 Carbon Aerogels

Carbon aerogels could provide open channels for continuous unimpeded transportation of electrolyte ions, which could contribute to the performance of carbon aerogels, such as high rate capability and enhanced charge transportation.

Pin Hao et al. have assembled the solid symmetric supercapacitor based on hierarchical porous carbon aerogels derived from bagasse, which delivers the specific capacitance of 142.1 F g^{-1} at 0.5 A g^{-1}, and energy density of ~19.74 Wh kg^{-1} at 0.5 kW kg^{-1} as well as excellent cyclability of 93.9% capacitance retention after 5,000 cycles.[91]

Zhibin Lei et al. have reported the supercapacitor based on aerogels derived from natural cotton, which indicates the high specific capacitance of ~251 F g^{-1} at 0.1 A g^{-1} and superior rate performance of 77% initial capacitance retention as well as a cycling stability of ~3% capacitance decay over 20,000 cycles.[92]

4.4 Carbon Quantum Dot

CDs-derived carbon materials are conductive to the electrochemical performance of symmetrical supercapacitors, owing to the large specific surface area of CDs that provides more surface active sites and edges accessible for the ion adsorption-desorption process.

Zhang et al. have prepared supercapacitor electrode (PWC/MnO_2/GQDs) based on porous wood carbon (PWC) decorated with MnO_2 and GQDs, which indicates a specific capacitance of 188.4 F g^{-1} at 1 A g^{-1} and good rate capability (74.7 F g^{-1} at 20 A g^{-1}) and the cycling stability with 95.3% retention after 2,000 cycles.[93]

Kaner et al. have developed a simple route to prepare 3D turbostratic graphene (3D-ts-graphene) via CDs derived from biomolecule (citric acid and urea), which delivers the specific surface area of 1,405 m^2g^{-1} and the volumetric capacitance of 27.5 mF L^{-1} at the current density of 560 A L^{-1} as well as 94.6% capacitance retention over 20,000 cycles. The supercapacitor indicates an energy density of 17.7 Wh kg^{-1} at a power density of 2.55 kW kg^{-1} and fast charge-discharge cycling rate with a time constant of 3.44 ms.[94]

4.5 Other biomass-based composite carbon

Nanocomposite-based carbon materials refers to the carbon-based materials incorporated along with a conducting polymer (CP) or a metal oxide (MO) or both, which could be provided with high specific surface area and superior capacitance composed of pseudocapacitance and EDL capacitance.

Zhang et al. have fabricated NiCo LDH/IPC composite through NiCo-LDHs sheets grown on interconnected PC (IPC) derived from poplar catkins, which exhibits the specific capacitance of 209.7 mAh g^{-1} at 1 Ag^{-1}. The hybrid supercapacitor based on NiCo LDH/IPC//IPC delivers 29.6 Wh kg^{-1}at 744 W kg^{-1} and good cyclic life with 88% capacitance retention after 4,000 cycles.[95]

Wang et al. have produced graphene/silk fibroin-based carbon nanocomposites (GCN-S), which indicates the specific capacitance of 256 F g^{-1} at 0.5 A g^{-1} and superior rate performance of 188 F g^{-1} at 50 A g^{-1} as well as high stability with 96.3% capacitance retention over 10,000 cycles. The supercapacitor based on GCN-S has achieved high energy density of 22.8 Wh kg^{-1} at 200 W kg^{-1} and still reaches 14.4 Wh kg^{-1} at a high-power density of 40,000 W kg^{-1}.[96]

Peixin Zhang et al. have reported hybrid symmetric supercapacitor based on C/MnO with litchi shell as the carbon source and $KMnO_4$ as Mn source, which achieves high specific capacitance of 162.7 F g^{-1} at 0.5 A g^{-1}, energy density of 57.7 Wh kg^{-1} at 400 W kg^{-1} and excellent reusability of 93.5% capacitance retention after 5,000 cycles.[97]

5 SUMMARY AND PROSPECT

In summary, we have brought together the current advances of the synthetic approaches and mechanisms of porous carbons (PCs) derived from biomass over the years, and demonstrate the effects of different synthesis methods on the nanostructure and properties of the resulting carbon materials as well as the electrochemical performance of the supercapacitors. Although regarded as one of the most potential candidates for energy storage devices in the coming era of pure electric vehicles, the weakness of the supercapacitor such as low-energy density still restricts their wide application in the energy storage market.

The global market size of cellulose-based materials is USD 219.53 billion, which was recorded in 2018, and it is expected to increase up to USD 305.08 billion by 2026. Lignin's global market size exceeds USD730 million in 2018 and the industry anticipates expenses to achieve beyond 1.7 million tons by 2025. The global market size of nanocellulose approached to USD 146.7 million in 2019 and looked forward to grow at a compound annual growth rate (CAGR) of 21.4% from 2020 to 2026.

Advances in PCs based on biomass for high-performance supercapacitors have been made in recent years, but there are still some issues that need to be further developed.

1. Exploiting the relationship between biomass types and synthetic methods. The diversity of biomass composition and structures greatly affect the reproducibility of the PCs. We should rationally deal with the biomass on

account of essential properties such as different composition, nanostructure, and surface properties in the processes of precursor selection, carbonization, and activation.

2. Optimizing the preparation method to achieve cost-effective biomass-derived PCs with controllable structure. Currently, the development of PCs derived from biomass lacks a scalable production method for accurately and effectively controlling the microstructure of PCs with desirable structures (such as size, crystallinity, layer, shape, and location of defects). In addition, the self-template method and the self-activation technique provide new ideas for the synthesis of biomass carbon materials.
3. Exploring PCs derived from biomass with the focus on utilizing and optimizing their natural structure. Natural biomass with a precise structure acts as not only a brilliant carbon resource but also a bio-template for the synthesis of valuable carbon materials. And then, the valuable guidance for the fabrication of PCs should concentrate on the conversion mechanism of the complicated processes in the future work.
4. Optimizing the loading and interfacial interactions to make full use of the synergistic effects between the individual components. The gravimetric and volumetric performances of the biomass-based carbon electrodes are limited to some extent due to their intrinsically low specific capacitance and energy density.
5. Developing proper theoretical model and establishing record files through machine learning or artificial intelligence for calculating physical and structural parameters of biomass. The diversity of biomass resources as a sophisticated criterion is beneficial to select appropriate method to achieve PCs and to predict the capacitive properties of biomass precursor and PCs.
6. Exploring methodical engineering of the PCs from their lab-scale to the pilot-scale with retaining their properties is a necessary task. Extrapolating the specific characteristics of PCs while retaining the electrochemical properties with appropriate preparatory methods would be a formidable task for the future research.
7. Biomass raw materials rarely contain heavy metals or other toxic, harmful substances that lead to environmental hazards, but organic or inorganic pollutants generated in the synthetic process of PCs and their damage to the environmental cannot be neglected.
8. Although PCs derived from natural abundant biomass for supercapacitor is green, cost-saving and sustainable, there are also some obstacles hindering their application, such as (1) Lower efficiency of biomass energy than fossil fuels; (2) Not completely clean; (3) Ecological Devastation; (4) Broad space for biomass plants. Abundant interdisciplinary researches over various science streams need concentrating on the detailed reaction mechanisms to explore new novel sustainable materials to deliver better applications in the domain of sustainable energy and environment. Further studies are vital to minimize the prevailing knowledge gaps.

REFERENCES

1. S. Nanda, R. Azargohar, A.K. Dalai and J.A. Kozinski, An assessment on the sustainability of lignocellulosic biomass for biorefining, *Renew. Sustain. Energy Rev.,* 2015, **50**, 925–941.
2. N. Sarkar, S.K. Ghosh, S. Bannerjee and K. Aikat, Bioethanol production from agricultural wastes: An overview, *Renew. Energy*, 2012, **37** (1), 19–27.
3. Y. Sun, X. L. Shi, Y.L. Yang, G. Suo, L. Zhang, S. Lu and Z.G. Chen, Biomass-derived carbon for high-performance batteries: From structure to properties, *Adv. Funct. Mater.*, 2022, **32**, 2201584.
4. P. Nakhanivej, Q. Dou, P. Xiong and H.S. Park, Two-dimensional pseudocapacitive nanomaterials for high-energy- and high-power-oriented applications of supercapacitors, *Acc. Mater. Res.*, 2021, **2**, 86–96.
5. Y. Gao, J. Remón and A.S. Matharu, Microwave-assisted hydrothermal treatments for biomass valorisation: A critical review, *Green Chem.*, 2021, **23**, 3502–3525.
6. S. Shukla and S. Vyas, Study of biomass briquettes, factors affecting its performance and technologies based on briquettes. *IOSR J. Environ. Sci., Toxicol. Food Technol.*, 2015, **9**, 37–44.
7. W.J. Liu and H.Q. Yu, Thermochemical conversion of lignocellulosic biomass into mass-producible fuels: Emerging technology progress and environmental sustainability evaluation, *ACS Environ. Au,* 2022, **2**, 98–114.
8. H. Zhang, Y. Zhang, L. Bai, Y. Zhang and L. Sun, Effect of physiochemical properties in biomass-derived materials caused by different synthesis methods and their electrochemical properties in supercapacitors, *J. Mater. Chem. A*, 2021, **9**, 12521–12552.
9. Z. Huang, A. Chen, F. Mo, G. Liang, X. Li, Q. Yang, Y. Guo, Z. Chen, Q. Li, B. Dong and C. Zhi, Phosphorene as cathode material for high-voltage, anti-self-discharge zinc ion hybrid capacitors, *Adv. Energy Mater.*, 2020, **10**, 2001024.
10. P. Manasa, S. Sambasivam and F. Ran, Recent progress on biomass waste derived activated carbon electrode materials for supercapacitors applications—A review, *J. Energy Storage*, 2022, **54**, 105290.
11. S. Yu, D. Liu, S. Zhao, B. Bao, C. Jin, W. Huang, H. Chen and Z. Shen, Synthesis of wood derived nitrogen-doped porous carbon-polyaniline composites for supercapacitor electrode materials, *RSC Adv.*, 2015, **5**, 30943–30949.
12. S. Jha, J.A. Okolie, S. Nanda and A.K. Dalai, A review of biomass resources and thermochemical conversion technologies, *Chem. Eng. Technol.,* 2022, **45** (5), 791–799.
13. T.R. Sarker, R. Azargohar, A.K. Dalai and V. Meda, Characteristics of torrefied fuel pellets obtained from co-pelletization of agriculture residues with pyrolysis oil, *Biomass Bioenergy*, 2021, **150**, 106139.
14. Y. Zhao, K.R. Adair and X.L. Sun, Recent developments and insights into the understanding of Na metal anodes for Na-metal batteries, *Energy Environ. Sci.*, 2018, **11**, 2673–2695.
15. R. Azargohar, S. Nanda, A.K. Dalai and J.A. Kozinski, Physico-chemistry of biochars produced through steam gasification and hydro-thermal gasification of canola hull and canola meal pellets, *Biomass Bioenergy*, 2019, **120**, 458–470.
16. M. Biswal, A. Banerjee, M. Deo and S. Ogale, From dead leaves to high energy density supercapacitors, *Energy Environ. Sci.*, 2013, **6**, 1249–1259.
17. M. Fidelis, C.D. Moura, T.K. Junior, N. Pap, P. Mattila, S. Makinen, P. Putnik, D.B. Kovacevic, Y. Tian, B. Yang and D. Granato, Fruit seeds as sources of bioactive compounds: Sustainable production of high value-added ingredients from by-products within circular economy, *Molecules*, 2019, **24**, 3854.
18. J. Wang, L. Shen, Y. Xu, H. Dou and X. Zhang, Lamellar-structured biomass-derived phosphorus-and nitrogen-co-doped porous carbon for high-performance supercapacitors, *New J. Chem.*, 2015, **39**, 9497–9503.

19. A. Gopalakrishnan and S. Badhulika, Effect of self-doped heteroatoms on the performance of biomass-derived carbon for supercapacitor applications, *J. Power Sources*, 2020, **480**, 228830.
20. P. Cheng, S. Gao, P. Zang, X. Yang, Y. Bai, X. Hua, Z. Liu and Z. Lei, Hierarchically porous carbon by activation of shiitake mushroom for capacitive energy storage, *Carbon*, 2015, **93**, 315–324.
21. B. Zhu, B. Liu, C. Qu, H. Zhang, W. Guo, Z. Liang, F. Chen and R. Zou, Tailoring biomass-derived carbon for high-performance supercapacitors from controllably cultivated algae microspheres, *J. Mater. Chem. A*, 2018, **6**, 1523–1530.
22. Y. Li, L.W. Zhou and R.Z. Wang, Urban biomass and methods of estimating municipal biomass resources, *Renew. Sustain. Energy Rev.*, 2017, **80**, 1017–1130.
23. F. Razmjooei, K. Singh, T.H. Kang, N. Chaudhari, J. Yuan and J.S. Yu, Urine to highly porous heteroatom-doped carbons for supercapacitor: A value added journey for human waste, *Sci. Rep.*, 2017, **7**, 10910.
24. X. Mu, Y. Li, X. Liu, C. Ma, H. Jiang, J. Zhu, X. Chen, T. Tang and E. Mijowska, Controllable carbonization of plastic waste into three-dimensional porous carbon nanosheets by combined catalyst for high performance capacitor, *Nanomaterials*, 2020, **10**, 1097.
25. M.A. Yahya, Z.A. Qodah and C.W.Z. Ngah, Agricultural bio-waste materials as potential sustainable precursors used for activated carbon production: A review, *Renew. Sustain. Energy Rev.*, 2015, **46**, 218–235.
26. Y. Zhang, L. Liu, P. Zhang, J. Wang, M. Xu and Q. Deng, Ultra-high surface area and nitrogen-rich porous carbons prepared by a low-temperature activation method with superior gas selective adsorption and outstanding supercapacitance performance, *Chem. Eng. J.*, 2019, **355**, 309–319.
27. M. Karnan, K. Subramani, N. Sudhan, N. Ilayaraja and M. Sathish, Aloe vera derived activated high-surface-area carbon for flexible and high-energy supercapacitors, *ACS Appl. Mater. Interfaces*, 2016, **8** (51), 35191–35202.
28. L. Sun, C. Tian, M. Li, X. Meng, L. Wang, R. Wang, J. Yin and H. Fu, From coconut shell to porous graphene-like nanosheets for high-power supercapacitors, *J. Mater. Chem.*, 2013, **1** (21), 6462–6470.
29. B. Du, H. Zhu, L. Chai, J. Cheng, X. Wang, X. Chen, J. Zhou and R.C. Sun, Effect of lignin structure in different biomass resources on the performance of lignin-based carbon nanofibers as supercapacitor electrode, *Ind. Crop. Prod.*, 2021, **170**, 113745.
30. X. Zhang and L. Wang, Research progress of carbon nanofiber-based precious-metal-free oxygen reaction catalysts synthesized by electrospinning for Zn-Air batteries, *J. Power Sources*, 2021, **507**, 230280.
31. L. Tao, Y. Huang, Y. Zheng, X. Yang, C. Liu, M. Di, S. Larpkiattaworn, M. Nimlos, Z. Zheng and J. Taiwan, Porous carbon nanofiber derived from a waste biomass as anode material in lithium-ion batteries, *Inst. Chem. Eng.*, 2019, **95**, 217–226.
32. S. Zhu, Y. Song, X. Zhao, J. Shao, J. Zhang and B. Yang, The photoluminescence mechanism in carbon dots (graphene quantum dots, carbon nanodots, and polymer dots): Current state and future perspective, *Nano Res.*, 2015, **8**, 355–381.
33. C. Hu, M. Li, J. Qiu and Y.P. Sun, Design and fabrication of carbon dots for energy conversion and storage, *Chem. Soc. Rev.*, 2019, **48**, 2315–2337.
34. S. Bhandari, D. Mondal, S.K. Nataraj and R.G. Balakrishna, Biomolecule-derived quantum dots for sustainable optoelectronics, *Nanoscale Adv.*, 2019, **1**, 913–936.
35. F. Lai, Y. Miao, L. Zuo, H. Lu, Y. Huang and T. Liu, Biomass-derived nitrogen doped carbon nanofiber network: A facile template for decoration of ultrathin nickel-cobalt layered double hydroxide nanosheets as high-performance asymmetric supercapacitor electrode, *Small*, 2016, **12**, 3235–3244.
36. Y. Li, Y.A. Samad, K. Polychronopoulou, S.M. Alhassan and K. Liao, From biomass to high performance solar–thermal and electric–thermal energy conversion and storage materials, *J. Mater. Chem. A*, 2014, **2**, 7759–7765.

37. J. Anjali, V.K. Jose and J.M. Lee, Carbon-based hydrogels: Synthesis and their recent energy applications, *J. Mater. Chem. A*, 2019, **7**, 15491–15518.
38. B.M. Matsagar, R.X. Yang, S. Dutta, Y.S. Ok and K.C.W. Wu, Recent progress in the development of biomass-derived nitrogen-doped porous carbon, *J. Mater. Chem. A*, 2021, **9**, 3703–3728.
39. X.L. Wu, T. Wen, H.L. Guo, S. Yang, X. Wang and A.W. Xu, Biomass-derived sponge-like carbonaceous hydrogels and aerogels for supercapacitors, *ACS Nano*, 2013, **7**, 3589–3597.
40. P. Zhang, L. Wang, S. Yang, J.A. Schott, X. Liu, S.M. Mahurin, C. Huang, Y. Zhang, P. F. Fulvio, M.F. Chisholm and S. Dai, Solid-state synthesis of ordered mesoporous carbon catalysts via a mechanochemical assembly through coordination cross-linking, *Nat. Commun.*, 2017, **8**, 15020.
41. T.A. Khana, A.S. Saudb, S.S. Jamaric, M.H.A. Rahimb, J.W. Parka and H.J. Kim, Hydrothermal carbonization of lignocellulosic biomass for carbon rich material preparation: A review, *Biomass Bioenergy*, 2019, **130**, 1–6.
42. G. Singh, K.S. Lakhi, S. Sil, S.V Bhosale, I. Kim, K. Albahily and A. Vinu, Biomass derived porous carbon for CO_2 capture, *Carbon*, 2019, **148**, 164–186.
43. R.P. Panmand, P. Patil, Y. Sethi, S.R. Kadam, M.V. Kulkarni, S.W. Gosavi, N.R. Munirathnam and B.B. Kale, Unique perforated graphene derived from Bougainvillea flowers for high-power supercapacitors: A green approach, *Nanoscale*, 2017, **9**, 4801–4809.
44. B. Liu, Y. Liu, H. Chen, M. Yang and H. Li, Oxygen and nitrogen co-doped porous carbon nanosheets derived from Perilla frutescens for high volumetric performance supercapacitors, *J. Power Sources*, 2017, **341**, 309–317.
45. M.M. Titirici, A. Thomas, S.H. Yu, J.O. Müller and M. Antonietti, A direct synthesis of mesoporous carbons with bicontinuous pore morphology from crude plant material by hydrothermal carbonization, *Chem. Mater.*, 2007, **19**, 4205–4212.
46. J. Deng, M.M. Li and Y. Wang, Biomass-derived carbon: Synthesis and applications in energy storage and conversion, *Green Chem.*, 2016, **18**, 4824–4854.
47. Y. Zhang, W. Hou, H. Guo, S. Shi and J. Dai, Preparation and characterization of carbon microspheres from waste cotton textiles by hydrothermal carbonization, *J. Renew. Mater.*, 2019, **7**, 1309–1319.
48. N.T. Nguyen, P.A. Le and V.B.T. Phung, Biomass-derived carbon hooks on Ni foam with free binder for high performance supercapacitor electrode, *Chem. Eng. Sci.*, 2021, **229**, 1–12.
49. S.A. Nicolae, H. Au, P. Modugno, H. Luo, A.E. Szego, M. Qiao, L. Li, W. Yin, H. J. Heeres, N. Berge and M.-M. Titirici, Recent advances in hydrothermal carbonisation: From tailored carbon materials and biochemicals to applications and bioenergy, *Green Chem.*, 2020, **22** (15), 4747–4800.
50. Y.Q. Zhao, L. Min, P.Y. Tao, Y.J. Zhang, X.T. Gong, Z. Yang, G.Q. Zhang and H.L. Li, Hierarchically porous and heteroatom doped carbon derived from tobacco rods for supercapacitors, *J. Power Sources*, 2016, **307**, 391–400.
51. W. Du, X. Wang, J. Zhan, X. Sun, L. Kang, F. Jiang, X. Zhang, Q. Shao and Z. Guo, Biological cell template synthesis of nitrogen-doped porous hollow carbon spheres/ MnO_2 composites for high-performance asymmetric supercapacitors, *Electrochim. Acta*, 2019, **296**, 907–915.
52. S. Wang, W. Sun, D.-S. Yang and F. Yang, Conversion of soybean waste to sub-micron porous-hollow carbon spheres for supercapacitor via a reagent and template-free route, *Mater. Today Energy*, 2019, **13**, 50–55.
53. M. Yu, Y. Han, J. Li and L. Wang, CO_2-activated porous carbon derived from cattail biomass for removal of malachite green dye and application as supercapacitors, *Chem. Eng. J.*, 2017, **317**, 493–502.

54. Z. Ling, Z. Wang, M. Zhang, C. Yu, G. Wang, Y. Dong, S. Liu, Y. Wang and J. Qiu, Sustainable synthesis and assembly of biomass-derived B/N co-doped carbon nanosheets with ultrahigh aspect ratio for high-performance supercapacitors, *Adv. Funct. Mater.*, 2016, **26**, 111–119.
55. Y. Liu, Z. Shi, Y. Gao, W. An, Z. Cao and J. Liu, Biomass-swelling assisted synthesis of hierarchical porous carbon fibers for supercapacitor electrodes, *ACS Appl. Mater. Interfaces*, 2016, **8**, 28283–28290.
56. C. Jin, J. Nai, O. Sheng, H. Yuan, W. Zhang, X. Tao and X.W. Lou, Biomass-based materials for green lithium secondary batteries, *Energy Environ. Sci.*, 2021, **14**, 1326–1379.
57. Q.H. Liang, L. Ye, Z.H. Huang, Q. Xu, Y. Bai, F.Y. Kang and Q.H. Yang, A honeycomb-like porous carbon derived from pomelo peel for use in high-performance supercapacitors, *Nanoscale*, 2014, **6**, 13831–13837.
58. Z. Yin, Y. Xu, J. Wu and J. Huang, Effect of pomelo seed-derived carbon on the performance of supercapacitors, *Nanosc. Adv.*, 2021, **3**, 2007–2016.
59. Y. Zhang, X. Song, Y. Xu, H. Shen, X. Kong and H. Xu, Utilization of wheat bran for producing activated carbon with high specific surface area via NaOH activation using industrial furnace, *J. Cleaner Prod.*, 2019, **210**, 366–375.
60. J. Xia, N. Zhang, S. Chong, D. Li, Y. Chen and C. Sun, Three-dimensional porous graphene-like sheets synthesized from biocarbon via low-temperature graphitization for a supercapacitor, *Green Chem.*, 2018, **20**, 694–700.
61. M. Danish and T. Ahmad, A review on utilization of wood biomass as a sustainable precursor for activated carbon production and application, *Renew. Sustain. Energy Rev.*, 2018, **87**, 1.
62. M. Zhang, Z. Song, H. Liu and T. Ma, Biomass-derived highly porous nitrogen-doped graphene orderly supported $NiMn_2O_4$ nanocrystals as efficient electrode materials for asymmetric supercapacitors, *Appl. Surf. Sci.*, 2020, **507**, 145065.
63. D. Wang, J. Nai; L. Xu and T. Sun, A potassium formate activation strategy for the synthesis of ultrathin graphene-like porous carbon nanosheets for advanced supercapacitor applications, *ACS Sustain. Chem. Eng.*, 2019, **7** (23), 18901–18911.
64. S.C. Hu, J. Cheng, W.P. Wang, G.T. Sun, L.L. Hu, M.Q. Zhu and X.H. Huang, Structural changes and electrochemical properties of lacquer wood activated carbon prepared by phosphoric acid-chemical activation for supercapacitor applications, *Renew. Energy*, 2021, **177**, 82–94.
65. M. Chen, X. Kang, T. Wumaier, J. Dou, B. Gao, Y. Han, G. Xu, Z. Liu and L. Zhang, Preparation of activated carbon from cotton stalk and its application in supercapacitor, *J. Solid State Electrochem.*, 2013, **17**, 1005–1012.
66. J. Huang, J. Chen, Z. Yin and J. Wu, A hierarchical porous P-doped carbon electrode through hydrothermal carbonization of pomelo valves for high-performance supercapacitors, *Nanosc. Adv.*, 2020, **2**, 3284–3291.
67. M. Sevilla, G.A. Ferrero and A.B. Fuertes, Beyond KOH activation for the synthesis of superactivated carbons from hydrochar, *Carbon*, 2017, **114**, 50–58.
68. Y. Zhang, L. Chen, Y. Meng, J. Xie, Y. Guo and D. Xiao, Lithium and sodium storage in highly ordered mesoporous nitrogen-doped carbons derived from honey, *J. Power Sources*, 2016, **335**, 20–30.
69. B. Huang, Y. Liu, Q. Guo, Y. Fang, M.M. Titirici, X. Wang and Z. Xie, Porous carbon nanosheets from biological nucleobase precursor as efficient pH independent oxygen reduction electrocatalyst, *Carbon*, 2020, **156**, 179–186.
70. F. Zhang, T.Y. Liu, M.Y. Li, M.H. Yu, Y. Luo, Y.X. Tong, Y. Li, Multiscale pore network boosts capacitance of carbon electrodes for ultrafast charging, *Nano Lett.*, 2017, **17**, 3097–3104.

71. A. Sanchez-Sanchez, M.T. Izquierdo, J. Ghanbaja, G. Medjahdi, S. Mathieu, A. Celzard and V. Fierro, Excellent electrochemical performances of nanocast ordered mesoporous carbons based on tannin-related polyphenols as supercapacitor electrodes, *J. Power Sources*, 2017, **344**, 15–24.
72. S. Schlienger, A.L. Graff, A. Celzard and J. Parmentier, Direct synthesis of ordered mesoporous polymer and carbon materials by a biosourced precursor, *Green Chem.*, 2012, **14**, 313–316.
73. Y. Wen, K. Kierzek, J. Min, X. Chen, J. Gong, R. Niu, X. Wen, J. Azadmanjiri, E. Mijowska and T. Tang, Porous carbon nanosheet with high surface area derived from waste poly (ethylene terephthalate) for supercapacitor applications, *J. Appl. Polym. Sci.*, 2020, **137**, 48338.
74. C. Chen, H. Wang, C. Han, J. Deng, J. Wang, M. Li, M. Tang, H. Jin and Y. Wang, Asymmetric flask like hollow carbonaceous nanoparticles fabricated by the synergistic interaction between soft template and biomass, *J. Am. Chem. Soc.*, 2017, **139**, 2657–2663.
75. Y. Zhu, W. Sun, W. Chen, T. Cao, Y. Xiong, J. Luo, J. Dong, L. Zheng, J. Zhang, X. Wang, C. Chen, Q. Peng, D. Wang and Y. Li, Scale-up biomass pathway to cobalt single-site catalysts anchored on N-doped porous carbon nanobelt with ultrahigh surface area, *Adv. Funct. Mater.*, 2018, **28**, 1802167.
76. C.G. Hu and L.M. Dai, Doping of carbon materials for metal-free electrocatalysis, *Adv. Mater*, 2019, **31**, 1804672.
77. G.A. Ferrero, A.B. Fuertes and M. Sevilla, From soybean residue to advanced supercapacitors, *Sci. Rep.*, 2015, **5**, 16618.
78. J. Han, Q. Li, C. Peng, N. Shu, F. Pan, J. Wang and Y. Zhu, Increasing S dopant and specific surface area of N/S-codoped porous carbon by in-situ polymerization of PEDOT into biomass precursor for high performance supercapacitor, *Appl. Surf. Sci.*, 2020, **502**, 144191.
79. X. Yue, H. Yang, P. An, Z. Gao, H. Lia and F. Ye, Multi-element co-doped biomass porous carbon with uniform cellular pores as a supercapacitor electrode material to realise high value-added utilisation of agricultural waste, *Dalton Trans.*, 2022, **51**, 12125–12136.
80. Y. Zhou, H. Qi, J. Yang, Z. Bo, F. Huang, M.S. Islam, X. Lu, L. Dai, R. Amal, C.H. Wang and Z. Han, Two-birds-one-stone: Multifunctional supercapacitors beyond traditional energy storage, *Energy Environ. Sci.*, 2021, **14**, 1854–1896.
81. K.K. Patel, T. Singhal, V. Pandey, T.P. Sumangala and M.S. Sreekanth, Evolution and recent developments of high performance electrode material for supercapacitors: A review, *J. Energy Storage*, 2021, **44**, 103366.
82. S. Repp, E. Harputlu, S. Gurgen, M. Castellano, N. Kremer, N. Pompe, J. Wörner, A. Hoffmann, R. Thomann, F.M. Emen, S. Weber, K. Ocakoglu and E. Erdem, Synergetic effects of Fe^{3+} doped spinel $Li_4Ti_5O_{12}$ nanoparticles on reduced graphene oxide for high surface electrode hybrid supercapacitors, *Nanoscale,* 2018, **10**, 1877–1884.
83. L.L. Zhang and X.S. Zhao, Carbon-based materials as supercapacitor electrodes, *Chem. Soc. Rev.,* 2009, **38** (9), 2520.
84. J. Xie, P. Yang, Y. Wang, T. Qi, Y. Lei and C.M. Li, Puzzles and confusions in supercapacitor and battery: Theory and solutions, *J. Power Sources,* 2018, **401**, 213–223.
85. H. Li, Y. Zhu, S. Dong, L. Shen, Z. Chen, X. Zhang and G. Yu, Self-assembled Nb2O5 nanosheets for high energy–high power sodium ion capacitors, *Chem. Mater.,* 2016, **28**, 5753–5760.
86. T. Purkait, G. Singh, M. Singh, D. Kumar and R.S. Dey, Large area few-layer graphene with scalable preparation from waste biomass for high-performance supercapacitor, *Sci. Rep.*, 2017, **7**, 15239.
87. S. Song, F. Ma, G. Wu, D. Ma, W. Geng and J. Wan, Facile self-templating large scale preparation of biomass-derived 3D hierarchical porous carbon for advanced supercapacitors, *J. Mater. Chem. A*, 2015, **3**, 18154–18162.

88. D. Zhou, H.L. Wang, N. Mao, Y.R. Chen, Y. Zhou, T.P. Yin, H. Xie, W. Liu, S.G. Chen and X. Wang, High energy supercapacitors based on interconnected porous carbon nanosheets with ionic liquid electrolyte, *Micropor. Mesopor. Mater.*, 2017, **241**, 202–209.
89. J. Cai, H.T. Niu, H.X. Wang, H. Shao, J. Fang, J.R. He, H.G. Xiong, C.J. Ma and T. Lin, High-performance supercapacitor electrode from cellulose-derived, inter-bonded carbon nanofibers, *J. Power Sources*, 2016, **324**, 302–308.
90. Y.F. Cao, L.J. Xie, G.H. Sun, F.Y. Su, Q.Q. Kong, F. Li, W.P. Ma, J. Shi, D. Jiang, C.X. Lu and C.M. Chen, Hollow carbon microtubes from kapok fiber: Structural evolution and energy storage performance, *Sustain. Energy Fuels*, 2018, **2**, 455–465.
91. P. Hao, Z.H. Zhao, J. Tian, H.D. Li, Y.H. Sang, G.W. Yu, H.Q. Cai, H. Liu, C.P. Wong and A. Umar, Hierarchical porous carbon aerogel derived from bagasse for high performance supercapacitor electrode, *Nanoscale*, 2014, **6**, 12120–12129.
92. P. Cheng, T. Li, H. Yu, L. Zhi, Z. Liu and Z. Lei, Biomass-derived carbon fiber aerogel as a binder-free electrode for high-rate supercapacitors, *J. Phys. Chem. C*, 2016, **120**, 2079–2086.
93. W. Zhang, Y. Yang, R. Xia, Y. Li, J. Zhao, L. Lin, J. Cao, Q. Wang, Y. Liu and H. Guo, Graphene-quantum-dots-induced MnO_2 with needle-like nanostructure grown on carbonized wood as advanced electrode for supercapacitors, *Carbon*, 2020, **162**, 114–123.
94. V. Strauss, K. Marsh, M.D. Kowal, M. El-Kady and R.B. Kaner, A simple route to porous graphene from carbon nanodots for supercapacitor applications, *Adv. Mater.*, 2018, **30**, 1704449.
95. X. Zhang, Q. Lu, E. Guo, J. Feng, M. Wei and J. Ma, NiCo layer double hydroxide/biomass-derived interconnected porous carbon for hybrid supercapacitors, *J. Energy Storage*, 2021, **38**, 102514.
96. Y. Wang, Y. Song, Y. Wang, X. Chen, Y. Xia and Z. Shao, Graphene/silk fibroin based carbon nanocomposites for high performance supercapacitors, *J. Mater. Chem. A*, 2015, **3**, 773–781.
97. N. Zhao, L. Deng, D. Luo and P. Zhang, One-step fabrication of biomass-derived hierarchically porous carbon/MnO nanosheets composites for symmetric hybrid supercapacitor, *Appl. Surf. Sci.*, 2020, **526**, 146696.

11 Porous Carbonaceous Materials for Supercapacitors

Yueming Li

1 INTRODUCTION

1.1 Brief history of supercapacitor

It is generally thought that the capacitor was first invented by Pieter van Musschenbroek in University of Leyden in 1746.[1] Later, Faraday constructed the variable capacitors and introduced the concept of dielectric constant. The unit of capacitance was named Faraday in order to commemorate him. Supercapacitor was proposed by Nippon Electric Company (NEC) of Japan in 1978. Later, there are continued progresses achieved by companies and scientist in this field. The detailed development can be referred in the related reviews.[1–3] Supercapacitor, also named ultracapacitor, or electrochemical capacitor, have an energy density greater than conventional capacitors. In general, supercapacitors are composed of two electrodes, electrolyte, and a separator. Supercapacitors have several advantages such as high-power density and long cycling lives. However, the energy density of supercapacitor (generally less than 10 Wh kg^{-1}) should be enhanced. In recent years, hybrid supercapacitor has been proposed including Li-ion hybrid supercapacitor proposed by Amatucci in 2001,[4] Na-ion hybrid supercapacitor,[5–7] K-ion hybrid supercapacitor,[8, 9] Mg-ion hybrid super capacitor,[10] and Zn-ion hybrid supercapacitor.[11, 12]

1.2 Mechanism of supercapacitor

The energy storage mechanism of supercapacitor has not been fully clarified up to now. In general, the storage mechanism can be divided into two categories. One is the storage mechanism by electric double layer (EDL) capacitance, the other is the pseudocapacitance storage mechanism. There are roughly three bilayer models. They are the Helmholtz model (1853–1879),[13] the Gouy-Chapman model (1910–1913),[14, 15] and the Gouy-Chapman-Stern model (1924)[16] which were refined step by step as time progressed (Figure 11.1).

For EDL storage mechanism, the oriented dipoles are formed to store charge electrostatically at the interface between electrolyte and electrode upon polarization. Notably, no charge transfer takes place between electrolyte and electrode in theory. Thus, EDL-based supercapacitors have a high theoretical cycling stability.

DOI: 10.1201/9781003387831-11

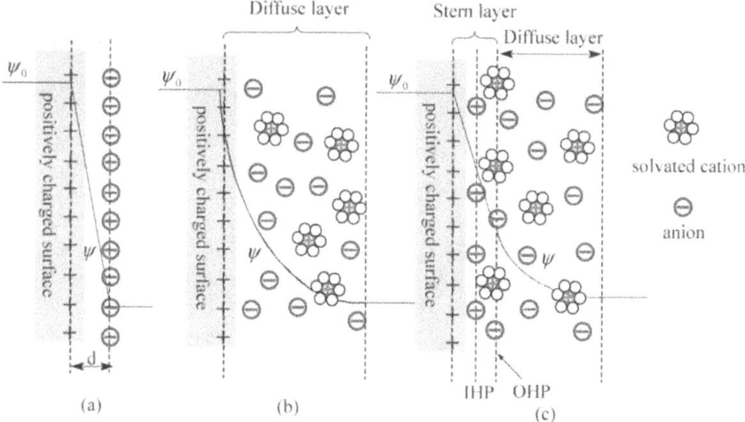

FIGURE 11.1 Schematics of the electric double layer structure showing the arrangement of solvated anions and cations near the electrode/electrolyte interface in the Stern layer and the diffuse layer. (a) Helmholtz model, (b) Gouy-Chapman model, and (c) Gouy-Chapman-Stern model. (Reproduced by permission, copyright of RSC).[17]

The capacitance for an EDL-type parallel-plate supercapacitor can be evaluated based on the following equation (Eqn 11.1):

$$C = \frac{\varepsilon_r \varepsilon_0}{d} A \tag{11.1}$$

The capacitance of electrical double layer capacitor (EDLC) is much higher than that of a conventional dielectric capacitor because the very high specific surface areas (SSA) of porous carbon materials as well as the short distance (serval Ångstroms) of double layer.

Pseudocapacitors are generally considered as a kind of energy storage mechanism between battery type and capacitor type. For pseudocapacitance mechanism, this process involves the fast charge transfer and reversible redox reactions of electroactive species on the electrode surface, which is Faradic.[18–20] In general; the pseudocapacitance is much higher than EDL capacitance. The typical electrode material with pseudocapacitance is ruthenium dioxide. The continuous change in pseudocapacitance electrode materials may lead to a poor cycling stability for supercapacitor. The conducting polymers such as polypyrrole and polyaniline can also display high pseudocapacitance, while the cycling stability should be further improved. The electrochemical redox reaction at the electrode materials promote a Faradic charge transfer similar to that occurring in a battery. Hence, the capacitance in a pseudocapacitor is much higher than that in an EDLC. On the other hand, the redox reactions compromise the power performance of the pseudocapacitor.

As shown in Figure 11.2, pseudocapacitors can be subdivided into three main categories based on electrochemical reaction, such as under-potential deposition,[21, 18] redox pseudocapacitors,[18] and intercalation pseudocapacitors.[22]

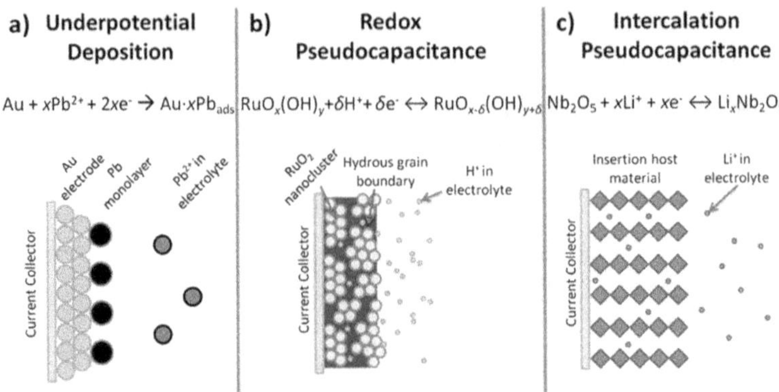

FIGURE 11.2 Different types of reversible redox mechanisms that give rise to pseudocapacitance: (a) underpotential deposition, (b) redox pseudocapacitance, and (c) intercalation pseudocapacitance.[23] (Reproduced by permission, copyright of RSC).

Unlike battery systems, the maximum energy of a capacitor is proportional to the square of the capacitance multiplied by the operating voltage. The voltage of supercapacitor is mainly dependent on the voltage of electrolyte decomposition.[24, 25] In general; aqueous electrolyte and organic liquid electrolytes have been used in commercial supercapacitor. Ionic liquids have a larger potential window, while their ionic conductivity is lower than those of aqueous and organic liquid electrolytes.[26, 27] The maximum voltage window of symmetrical supercapacitor is generally less than the 1.23 V (the voltage of water decomposition). The commercial supercapacitor using organic electrolyte can provide a voltage of 2.7 V.[28, 29]

1.3 The Different Types of Supercapacitor

Based on the mechanisms, supercapacitor can be divided into EDLC, pseudocapacitance supercapacitors, and hybrid supercapacitors. To combine both advantages of EDL and pseudocapacitance capacitors, hybrid supercapacitor has been proposed to store energy using both EDL and Faradaic capacitance. The hybrid supercapacitor has gained more attention due to the higher energy density and excellent rate performance in recent years. Supercapacitor can also be divided into symmetrical supercapacitor and asymmetric supercapacitor based on whether the two electrodes are the same. Due to the low energy density of the conventional supercapacitor, the novel supercapacitors with high energy density such as Li-ion capacitor,[6] Na-ion capacitor,[30] K-ion capacitor,[31] and Zn-hybrid supercapacitors[32] has gained more attention. Furthermore, the introduction of redox electrolyte to supercapacitor has shown certain advantages

2 POROUS CARBON IN SUPERCAPACITOR

The porous carbonaceous materials have been extensively used as electrode materials in supercapacitors due to the large SSA, good chemical stability, fair electrical conductivity, and abundance of carbon sources. The electrochemical performance of carbon materials has a close relationship with SSA, pore size distribution, surface functional groups, doping of heteroatoms and morphology. Thus, carbon materials with high accessible SSA to electrolyte, high electronic conductivity, suitable pore structure, high density, and low cost are needed. It is difficult that one electrode material has all the merits required by supercapacitor.

2.1 Activated Carbon

Activated carbons as amorphous carbon are most frequently used electrode materials in supercapacitor, especially in EDLC. Some carbon can be transformed into highly ordered graphite through high-temperature, which was called as soft carbon. In contrast, some carbon will keep a rigid amorphous structure containing randomly-oriented graphene layer, and these carbons are referred as hard carbon. The graphitizing or non-graphitizing of carbon was largely dependent on carbon precursors. For example, the use of biomass and thermosetting polymer will produce non-graphitizable carbons, which contains porous structure with no need of activation.[33]

Activated carbons have advantages including high SSA (up to 3,000 m^2g^{-1}), abundant micropores (pore size less than 2 nm), and low cost. However, the common activated carbons have a low electronic conductivity. There are a lot of methods which can be used to prepare activated carbon. Among these methods, the preparation of activated carbon by activating biomass-derived porous carbon is an economical, simple, and efficient way. Biomass precursors can be obtained from a wide range of sources including agricultural and forestry residues,[34] industrial biomass residues,[35] and algal biomass.[36] The carbonization of biomass is commonly achieved by pyrolysis,[37] hydrothermal carbonization,[38, 39] and microwave-assisted pyrolysis.[40–42] To achieve a suitable pore structure and a large SSA, the activation toward the pyrolyzed carbon is often necessary.[43]

Activation can be divided into physical activation and chemical activation. In general, physical activation is implanted by heating carbon precursor under oxidizing atmospheres including air, CO_2, and H_2O stream. For comparison, chemical activation is carried out by heating carbon precursor with certain reagents such as alkaline hydroxide, phosphoric acid, and zinc chloride. Compared with physical activation, chemical activation usually has the characteristics of a lower activation temperature, a shorter activation time, relatively high yields, higher tap density, wider pore sizes, and higher SSA.[44] In addition; chemical activation may increase additional steps to remove activation reagent and handling costs. From the view of industrial productions, physical activation is more favorable while chemical activation is the extensively used in lab-scale research.[45]

SSA and pore structure play a vital role in affecting the electrochemical performance of activated carbon however, while there is no liner relationship between SSA and capacitance. Some activated carbon materials have very high SSA, while their capacitance is not as high as expected because not all the pores are available to electrolyte. It is generally accepted that the presence of macropores and large mesopores facilitates ion diffusion, while small mesopores and micropores increase the ion-accessible SSA and thus contribute significantly to capacitance. In recent years, the concept of hierarchical porous carbon materials, which have multiscale pores interconnected and assembled into component layers, were proposed and prepared.[46–48] Template method has been frequently used to prepare templated porous carbons with hierarchical structure.[49–51]

The porous structure of activated carbon can provide a large interface to store charges, however, the well-developed porous structure significantly decreases the electrical conductivity of the activated carbons due to the destroying of the conductive networks.[52] Thus, materials with high electrical conductivity are needed to combine with activated carbon to enhance the power density. Graphene nanosheets, carbon nanotube, and graphene quantum dots, etc. have been proposed as conductive materials to enhance the electron transfer on activated carbon.[53–56] For example, Li et al. reported a self-supporting activated carbon/carbon nanotube/graphene nanosheets as electrode materials in supercapacitor, in which CNT and graphene were used to overcome the low conductivity of activated carbon. Thus, the composites showed an enhanced electrochemical performance.[57]

Doping is an effective strategy to modify the electrochemical performance of activated carbon. The doping to activated carbon can tailor the properties of porous carbon including wettability to electrolyte, functional groups, and interlayer spacing, resulting in modified electrochemical performance. The doping by different elements may play a different effect, and N-doping is most popular among the heteroatoms.[58] In recent years, many elements such as N, P, B, S, O, Si as doping heteroatoms to activated carbon have been studied. As P is in the same group with N, phosphorus-doping (P-doping) is found can enhance the wettability of carbon material as well as widen the working voltage of carbon electrode due to its increased antioxidant capacity of the phosphorus group.[59, 60] For example, the P doped porous carbon prepared by Wang et al. has shown that there is improvement in the areal and volumetric capacitance.[59]

The morphology of activated carbon plays an important role in their packing density on the surface of current collector. There are multiple morphologies for activated carbon, either in non-regular or regular ones. For the common activated carbons, they have a non-regular morphology. In contrast, for activated carbon with regular morphology, the activated carbon with spherical morphology is common. It is proposed that activated carbon with the spherical particle is favorable to achieve a high packing density on the current collector compared with activated carbon with irregular morphology because spheres tend to form close packing.[43] Thus, spherical activated carbon can be thought as one ideal electrode material in supercapacitor. Wickramaratne et al. prepared N-doped porous carbon spheres using resorcinol/formaldehyde as carbon precursors and ethylenediamine (EDA) as doping reagent. The electrochemical measurements showed that high capacitance as well high rate capability can be achieved.[61]

Besides solid carbon microsphere, hollow carbon sphere (HCS) consisting of a carbon shell with adjustable shell thickness and a unique internal cavity, can also be used as electrode materials in supercapacitor because of their special properties, such as regular structure, low density, large specific SSA, and stable physicochemical stability.[62, 63] HCS can be prepared via template method, as shown in Figure 11.3.[64]

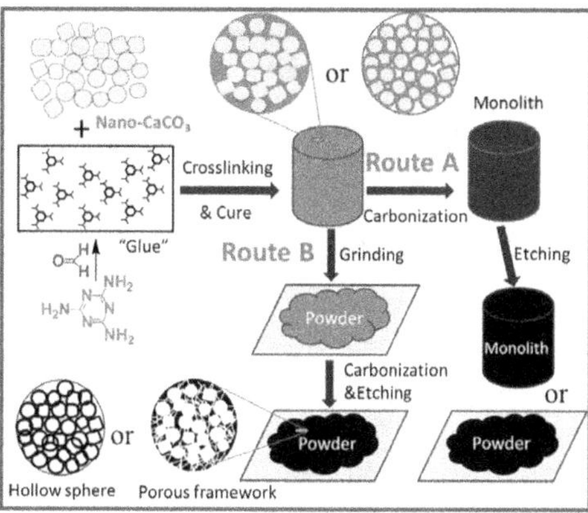

FIGURE 11.3 The preparation route of one hollow carbon sphere[64] (Reproduced by permission, Copyright of Elsevier).

As is known, the power density of supercapacitor is inversely proportional to internal resistance of supercapacitor. The electrical conductivity of activated carbon can play a dominant role on the internal resistance of supercapacitor. The electrical conductivity of activated carbon is related to many factors (such as texture, surface chemistry, and graphitization degree of carbonaceous materials).[65] The total electrical conductivity of powdered carbonaceous materials is the result of comprehensive factors including the intrinsic conductivity of the single carbon particles (intraparticle conductivity), the contact resistance between particles, as well as the volume density of the latter.[66] In general; carbonaceous materials with more perfect structure have higher intraparticle conductivity.[66] In other words, a larger size of graphite like crystallites (or increasing proportion of conjugated carbon in the sp2), less structural defects as well as less porosity for carbon particles will lead to a high electronic conductivity.[67] The larger the total pore volume in activated carbon, the lower its intrinsic electrical conductivity.[66] In most cases, the presence of oxygen heteroatoms in carbonaceous materials is detrimental to carbonaceous materials, while N-doping can enhance the electronic conductivity.[68–71] Notably, the presence of oxygen-containing functional group does not always result to a decrease of electrical conductivity for carbonaceous materials,[72] and N-doping does not always lead to an increase because the conductivity was not solely dependent on functional groups.[73] On the other side, the morphology, particle size, heteroatoms doping as well as external pressure will affect the contact resistance.[66]

2.2 Carbon Nanofiber

One-dimensional (1D) carbon fibers have been taken as one potential electrode materials in supercapacitor especially in wearable devices because of their high mechanical stability and high electrical conductivity.[74] However, there are still some disadvantages for carbon nanofiber as electrode materials in supercapacitor. Compared to activated carbon, the SSA of carbon nano fiber is much lower and the porosity is not as rich as that in activated carbon.[75] Thus, modifications of carbon fiber such as chemical

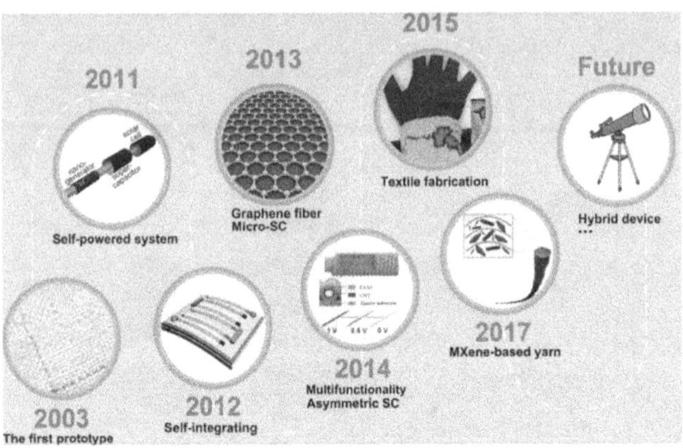

FIGURE 11.4 A brief history of the development of one-dimensional fibrous supercapacitors.[74] (Reproduced with permission, Copyright of Wiley)

activation, physical activation, and heteroatoms doping are frequently applied to improve their electrochemical performance. The detailed development of 1D supercapacitors has been summarized in Figure 11.4.

Carbon nanofiber can be prepared via multiple methods. Howbeit, the SSA of carbon fiber is low compared with activated carbon, and carbon nanofiber has a fair electronic conductivity. Thus, necessary modifications to carbon nanofiber are frequently needed to achieve a suitable pore structure and high electronic conductivity. Heteroatom doping and activation including chemical activation and physical activation are very efficient method to tailor the properties of carbon nanofiber. Electrospinning is one feasible method to prepare carbon nanofiber. Nie et al. have reviewed the recent progress in modifying strategies for carbon nanofiber and their application in supercapacitor.[76] As is known, the power density of supercapacitor is proportional to electrical conductivity of carbon materials in supercapacitor. The graphitization of carbon materials plays a critical role in affecting their electronic conductivity. To achieve a high graphitization, a high treating temperature is needed, resulting in large energy consumption. It is found that some metal species can play a catalytic role during the graphitization process for carbon materials. Some transition metals (Fe, Co, Ni, and Cu) can improve the graphitization at a relatively low temperature. These metals can form solid solution with amorphous carbon, and catalytic graphitization reactions happen during the dissolution–precipitation process. The other graphitization mechanism for high-valence metals is related to metal carbide conversion, in which metals will form metal carbides and metal carbides decompose into graphitic carbon and metal vapor.[76, 77] To increase the electric conductivity of non-graphitic carbon, partly graphitic domains can be introduced through metal-induced graphitization. For example, Chen et al. prepared the N-doped nanoporous graphic carbon fiber by Ni-induced graphitization. First, the electrospun carbon/Ni nanofibers were prepared through transforming the amorphous carbon in to a graphitic carbon with the help of Ni catalysis.[78] The excessive graphitization will decrease the pores and heteroatoms in carbon materials, resulting to a low specific capacitance. Thus, a balance between graphitization degree and pore structure should be achieved to obtain a good electrochemical performance.

The pore structure and SSA play a key role in determining the specific capacitance of carbon materials. For carbon nanofiber, the SSA of conventional carbon nanofiber is considerably lower than those of activated carbon. It is very important to introduce more pores and enlarge the SSA of carbon nanofiber. To create pores in carbon nanofiber, template method is frequently used. The templates can be metal oxide, metal hydroxide such as $Mg(OH)_2$,[79] non-metal oxide such as SiO_2,[46] organic and polymer materials,[80] metal organic frameworks (MOFs),[81] NaCl,[82] and ZnO.[83–85]

Polyacrylonitrile (PAN) is one of the most frequently used polymer to prepare carbon nanofiber because the carbon yield using PAN is high compared most polymers. To produce special pore structure, the second polymer can be introduced into electrospinning precursors. The pore structure can form because the different polymers have different thermal stability, molecular weight, and cross-linking degree. Many polymers such as Poly(vinylpyrrolidone) (PVP), PS, polyvinylidene fluoride, polyethylene oxide, Nafion, polysulfone, pitch, poly(methyl methacrylate) (PMMA), poly(m-aminophenol), pluronic F127 can be combined with PAN to form electrospinning solvents.[76]

2.3 Graphene Nanosheets

Graphene nanosheets, as one-layer or few-layer graphite with two dimensional (2D) structures, have become a hot spot because of their special properties such as high conductivity, 2D structure, and high theoretical SSA (2,630 $m^2 g^{-1}$). Notably, the actual SSA of graphene nanosheets is often much lower than that of theoretical value due to the restack of graphene nanosheets, and thus the electrochemical performance of graphene as electrode materials in supercapacitor is not ideal as expected. Thus, activation or doping are often used to modify the microstructure of graphene nanosheets to enhance the electrochemical performance of graphene nanosheets.[86] In addition, another drawback of graphene is that graphene nanosheets prepared via conventional method such as the reduction of graphene oxide show a low packing density, which will decrease the volumetric density of supercapacitor. Thus, it is very important to develop graphene-based electrode with high packing density because volumetric energy density is more important from the point view of practical application.[87]

Although graphene nanosheets have a large SSA in theory, their pore structures are not as rich as those in activated carbon. To increase the in-plane pores, H_2O_2 was used to further oxidize graphene oxide to obtain defective sites and vacancies. Using this strategy, holey graphene was obtained proposed by Duan and his collegaue.[88] The nanopores or nanoholes on graphene flakes are helpful for the vertical transportation of the electrolyte ions between the graphene layers, which can shorten the ion pathway as well as to enlarge the ion adsorption surface. The prepared holey graphene nanosheets have displayed better electrochemical performance because of increased contact area between electrolyte and graphene compared with graphene prepared by routine method.

When used as electrode materials, graphene can be fabricated in multiple structures: including graphene nanosheets powder (0 dimensional), one-dimensional fiber or graphene yarn, graphene nanosheets film, and three dimensional (3D) freestanding graphene.[89, 90] And the different structures show their own advantages in supercapacitor. For instance, 1D graphene yarns and 2D graphene nanosheets have shown great potential to be applied in wearable devices considering their freestanding merits.[91, 92] Various methods can be used to assemble 1D graphene-based fibers, and wet-spinning method is most frequently used method. Kou et al. reported a coaxial polyelectrolyte coated graphene-CNT core-sheath fiber via wet-spinning method. The prepared composites showed a high capacitance in length (5.3 mF cm^{-1}), area (177 mF cm^{-2}) as well as volume (158 F cm^{-3}).[93] Notably, there are also disadvantages for wet-spinning method, such as post treatment to reduce GO to graphene and the possible introduction of impurity.[94] Another strategy to obtain graphene nanofiber is hydrothermal treatment. Graphene oxide suspension was put into a glass pipeline under hydrothermal treatment at 230°C for 2 h, yielding a flexible graphene fiber with strong mechanical strength and high electrical conductivity.[95] Yu et al. reported a carbon fiber consisted of aligned single-walled carbon nanotubes and N-doped graphene nanosheets with silica capillary column functioning as a hydrothermal microreactor.[96] The prepared carbon fibers display a medium SSA, and high electrical conductivity. As electrode materials in micro-supercapacitor, the prepared carbon fibers exhibited a high volumetric capacity in both H_2SO_4 and polyvinyl alcohol (PVA)/H_3PO_4 electrolytes. Furthermore, the assembled supercapacitor showed high-power density as well as long cycle lives.[96]

The 2D graphene film has shown great potential in flexible thin film supercapacitor because their large SSA, high conductivity, strong mechanical stability, and good chemical stability.[97] Thus, quite a number of strategies such as layer-by-layer doping,[92] soft interfacial self-assembly,[98] vacuum filtration,[99, 100] and electrophoretic deposition method[101] have been developed to fabricate 2 D graphene structure including graphene films and graphene membrane. This 2D graphene film can be used in flexible solid-state supercapacitor. For example, Wu et al. prepared a freestanding graphene film via vacuum filtration of graphene oxide suspension. Electrochemical results disclose that a flexible supercapacitor using such the fabricated films as electrodes shows a high volumetric energy density (14.8 Wh L^{-1} at 53.6W L^{-1}), and excellent capacitance retention.[102]

To prevent the restack of graphene nanosheets, constructing graphene with three-dimensional (3D) structure was proposed to achieve a high SSA and suitable pore structure. As electrode materials in supercapacitor, the 3D structure helps more inner layer of graphene nanosheets available to the electrolyte ions, resulting to an enhanced electrochemical performance. Hard templates such as polymer microspheres or soft template such as surfactants can be used to construct 3D graphene nanosheets.[103] Graphene foam (GF) belongs to 3D graphene architecture assembled by graphene nanosheets building block. GF can be obtained by chemical vapor deposition using metal form as template.[104] Graphene foams have shown some special properties such as ultralow density, high mechanical strength and compressibility, high electronic conductivity, and a large SSA and abundant pore structure.[105] Several groups have summarized systemically the progress in the synthesis, characterization, properties, and applications of 3D graphene nanosheets in the past few years.[105–109] There have been a lot of reports on the preparation of 3D graphene. For example: (i)

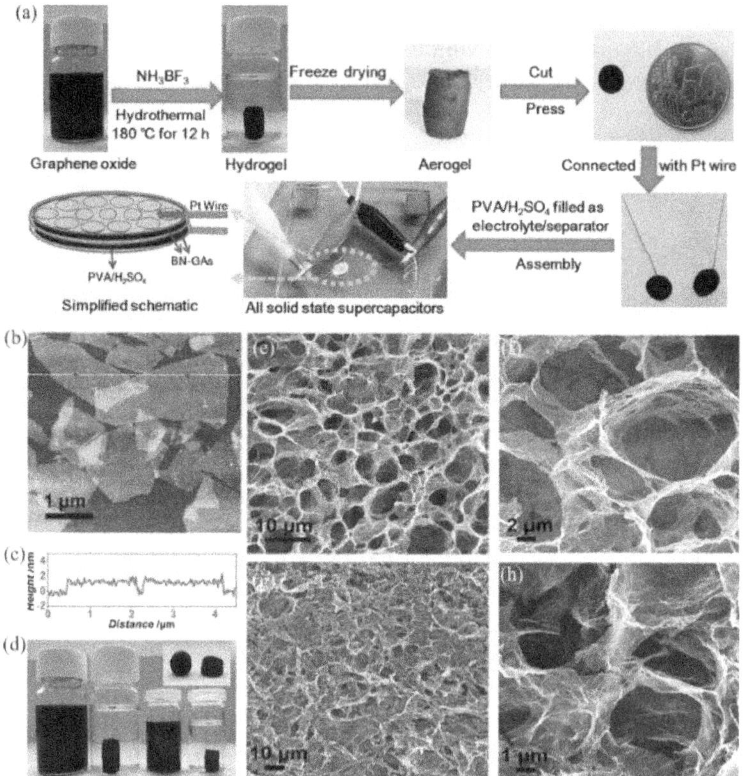

FIGURE 11.5 The preparation of B,N-doped 3D graphene and the corresponding characterization (Reproduced with permission, Copyright of Wiley).[111]

template-assisted method (ii) self-assembly method, (iii) 3D printing method,[110] and hydrothermal method,[111] shown in Figure 11.5.

Porous metal substrates are often used as hard template as well as catalyst via chemical vapor deposition.[104] The metal templates can be easily removed by acid etching. Self-assembly without templates is another method to prepare 3D graphene nanosheets. The self-assembly of graphene nanosheets can be carried out via hydrothermal or solvothermal treatment of graphene oxide suspension with or without template.[112] The templates using this method can be silica microsphere,[113] Poly(styrene) (PS) spheres, ice crystal,[114] and PMMA spheres.[115, 116] 3D printing can also be used as one efficient method to produce 3D graphene materials in a large scale. For example, 3D aerogel with macroscopic structure has been fabricated using 3D printing method.[117]

3D graphene nanosheets have been used as electrode materials in supercapacitor. For example, a 120 µm thick 3D graphene hydrogel thin film can be obtained via hydrothermal method, which shows superior performance including high gravimetric and areal capacitance, high cycling stability, and good mechanical flexibility.[118]

Although 3D graphene nanosheets have shown unique advantages as electrode material in supercapacitor, their low density will lower the volumetric density. Thus, the devolving graphene nanosheet with high density is very important. Li et al. prepared reduced graphene oxide with a high density with the help of inorganic zinc compounds, thus the volumetric capacitance of the as-prepared graphene oxide is high.[87]

2.4 Other Carbon Forms

MOFs with high SSA and large pore volume have been investigated for many years, and recently, great efforts have been made to use MOFs for electrochemical applications.[119] However, their low electrical conductivity and poor stability still pose significant challenges for their practical application in supercapacitors.[120–122] MOFs can be used as precursors for the preparation of high porosity nanoporous carbons (NPCs). Their parent MOFs can be prepared by simple coordination chemistry with an infinite combination of organic and inorganic components, and their porous structure, pore volume, SSA, etc. can be controlled.[123] These unique properties of MOF-derived NPCs make them useful for many technical applications. Compared to carbonaceous materials prepared using conventional precursors, MOF-derived carbon offers significant advantages in terms of simple synthesis and inherent diversity, allowing precise control of porous structure, pore volume, and SSA.[123] Wang et al.[124] prepared ZIF-8/PAN fibers by electrostatic spinning of ZIF-8/PAN/N,N-Dimethylformamide (DMF) hybrid solution, and then carbonized in nitrogen atmosphere to obtain nanoporous carbon fiber (NPCF) materials. On the one hand, this one-dimensional (1D) nitrogen-doped carbon material has high electrical conductivity, which facilitates the transfer of electrons and charges during charging and discharging; on the other hand, the porous carbon from ZIF-8 is uniformly distributed in the nitrogen-doped carbon from PAN, which makes full use of the large SSA of the MOFs-derived carbon material. The specific capacitance of the prepared NPC reached 332 F g^{-1}, which is higher than that of PAN-derived carbon (PAN-C) and ZIF-8-derived nanoporous carbon (ZIF-8-NPC). The retention of specific capacitance of porous carbon was 98.9% after 5,000 cycles at a current density of 1 A g^{-1}.

Although there have been lot of porous carbonaceous materials, activated carbons are still used as the main active material in supercapacitor considering the electrochemical performance, cost, and source availability. Notably, MOFs-derived carbon might be difficult to be applied in practical situation because this process is slightly complicated and time consuming. In addition, the use of MOFs will lead to an increased cost. For comparison, graphene nanosheets might be used in future if their price decreases to an affordable level. For carbon nanofiber, they might have great potential in wearable devices or flexible devices considering their good mechanical strength.

3 SUMMARY AND FUTURE PROSPECT

Supercapacitor has become one necessary energy storage devices because of their high-power density and long cycling lives. However, their energy density should be further improved. The porous carbonaceous materials are indispensable to develop supercapacitor with high energy density. The relationship between microstructure of carbonaceous materials including pore structure, surface chemistry, and electrochemical performance in supercapacitor should be further clarified. Nano carbonaceous material can provide a higher gravimetric specific capacitance, while the volumetric specific capacitance should be improved. The hybrid supercapacitor including Zn-ion hybrid supercapacitor, K-ion hybrid supercapacitor will gain more attention due to the high energy density, wide potential window, and fair cycling stability.

REFERENCES

1. J. Ho, T.R. Jow and S. Boggs, Historical introduction to capacitor technology, *IEEE Electr. Insul. M*, 2010, **26**, 20–25.
2. N.S. Shaikh, S.B. Ubale, V.J. Mane, J.S. Shaikh, V.C. Lokhande, S. Praserthdam, C.D. Lokhande and P. Kanjanaboos, Novel electrodes for supercapacitor: Conducting polymers, metal oxides, chalcogenides, carbides, nitrides, MXenes, and their composites with graphene, *J. Alloys. Compd*, 2022, **893**, 161998.
3. Y. Shao, M.F. El-Kady, J. Sun, Y. Li, Q. Zhang, M. Zhu, H. Wang, B. Dunn and R.B. Kaner, Design and mechanisms of asymmetric supercapacitors, *Chem. Rev.*, 2018, **118**, 9233–9280.
4. G.G. Amatucci, F. Badway, A. Du Pasquier and T. Zheng, An asymmetric hybrid nonaqueous energy storage cell, *J. Electrochem. Soc.*, 2001, **148**, A930.
5. K. Kuratani, M. Yao, H. Senoh, N. Takeichi, T. Sakai and T. Kiyobayashi, Na-ion capacitor using sodium pre-doped hard carbon and activated carbon, *Electrochim. Acta*, 2012, **76**, 320–325.
6. H. Wang, C. Zhu, D. Chao, Q. Yan and H.J. Fan, Nonaqueous hybrid lithium-ion and sodium-ion capacitors, *Adv. Mater.*, 2017, **29** (46), 1702093.
7. D. Chao, C. Zhu, P. Yang, X. Xia, J. Liu, J. Wang, X. Fan, S.V. Savilov, J. Lin, H.J. Fan and Z.X. Shen, Array of nanosheets render ultrafast and high-capacity Na-ion storage by tunable pseudocapacitance, *Nat. Commun.*, 2016, **7**, 12122.
8. L. Fan, K. Lin, J. Wang, R. Ma and B. Lu, A nonaqueous potassium-based battery–supercapacitor hybrid device, *Adv. Mater.*, 2018, **30**, 1800804.
9. A. Le Comte, Y. Reynier, C. Vincens, C. Leys and P. Azaïs, First prototypes of hybrid potassium-ion capacitor (KIC): An innovative, cost-effective energy storage technology for transportation applications, *J. Power Sources*, 2017, **363**, 34–43.
10. H.D. Yoo, I. Shterenberg, Y. Gofer, R.E. Doe, C.C. Fischer, G. Ceder and D. Aurbach, A magnesium-activated carbon hybrid capacitor, *J. Electrochem. Soc.*, 2014, **161**, A410–A415.
11. L. Dong, X. Ma, Y. Li, L. Zhao, W. Liu, J. Cheng, C. Xu, B. Li, Q.-H. Yang and F. Kang, Extremely safe, high-rate and ultralong-life zinc-ion hybrid supercapacitors, *Energy Storage Mater.*, 2018, **13**, 96–102.
12. H.D. Yoo, S.-D. Han, R.D. Bayliss, A.A. Gewirth, B. Genorio, N.N. Rajput, K.A. Persson, A.K. Burrell and J. Cabana, "Rocking-chair"-type metal hybrid supercapacitors, *ACS Appl. Mater. Interf.*, 2016, **8**, 30853–30862.
13. H. Helmholtz, Studien über electrische Grenzschichten, *Annalen der Physik und Chemie*, 1879, **243**, 337–382.
14. D.L. Chapman, LI. A contribution to the theory of electrocapillarity, *Lond., Edinb., Dublin Philos. Mag J. Sci.*, 2010, **25**, 475–481.
15. M. Gouy, Sur la constitution de la charge électrique à la surface d'un électrolyte, *Journal de Physique Théorique et Appliquée*, 1910, **9**, 457–468.
16. O. Stern, The theory of the electrolytic double layer, *Zeitschrift fuer Elektrochemie und Angewandte Physikalische Chemie*, 1924, **30**, 508–516.

17. L.L. Zhang and X.S. Zhao, Carbon-based materials as supercapacitor electrodes, *Chem. Soc. Rev.*, 2009, **38**, 2520–2531.
18. V. Augustyn, P. Simon and B. Dunn, Pseudocapacitive oxide materials for high-rate electrochemical energy storage, *Energy Environ. Sci.*, 2014, **7**, 1597–1614.
19. T. Brousse, D. Bélanger and J.W. Long, To be or not to be pseudocapacitive?, *J. Electrochem. Soc.*, 2015, **162**, A5185–A5189.
20. C. Costentin and J.M. Saveant, Energy storage: Pseudocapacitance in prospect, *Chem. Sci.*, 2019, **10**, 5656–5666.
21. Y. Jiang and J. Liu, Definitions of pseudocapacitive materials: A brief review, *Energy Environ. Mater.*, 2019, **2**, 30–37.
22. V. Augustyn, J. Come, M.A. Lowe, J.W. Kim, P.L. Taberna, S.H. Tolbert, H.D. Abruna, P. Simon and B. Dunn, High-rate electrochemical energy storage through Li^+ intercalation pseudocapacitance, *Nat. Mater.*, 2013, **12**, 518–522.
23. V. Augustyn, P. Simon and B. Dunn, Pseudocapacitive oxide materials for high-rate electrochemical energy storage, *Energy Environ. Sci.*, 2014, **7**, 1597–1614.
24. V. Augustyn, P. Simon and B. Dunn, High-voltage supercapacitors based on aqueous electrolytes, *ChemElectroChem*, 2018, **6**, 976–988.
25. Z. Dai, C. Peng, J.H. Chae, K.C. Ng and G.Z. Chen, Cell voltage versus electrode potential range in aqueous supercapacitors, *Sci. Rep.*, 2015, **5**, 9854.
26. Y. Wu and C. Cao, The way to improve the energy density of supercapacitors: Progress and perspective, *Sci. China Mater.*, 2018, **61**, 1517–1526.
27. K.L. Van Aken, M. Beidaghi and Y. Gogotsi, Formulation of ionic-liquid electrolyte to expand the voltage window of supercapacitors, *Angew. Chem. Int. Ed.*, 2015, **54**, 4806–4809.
28. P. Simon and Y. Gogotsi, Perspectives for electrochemical capacitors and related devices, *Nat. Mater.*, 2020, **19**, 1151–1163.
29. F. Zhang, Y. Lu, X. Yang, L. Zhang, T. Zhang, K. Leng, Y. Wu, Y. Huang, Y. Ma and Y. Chen, A flexible and high-voltage internal tandem supercapacitor based on graphene-based porous materials with ultrahigh energy density, *Small*, 2014, **10**, 2285–2292.
30. B. Wang, X. Gao, L. Xu, K. Zou, P. Cai, X. Deng, L. Yang, H. Hou, G. Zou and X. Ji, Advanced carbon materials for sodium-ion capacitors, *Batteries Supercaps*, 2021, **4**, 538–553.
31. M. Liu, L. Chang, Z. Le, J. Jiang, J. Li, H. Wang, C. Zhao, T. Xu, P. Nie and L. Wang, Emerging potassium-ion hybrid capacitors, *ChemSusChem*, 2020, **13**, 5837–5862.
32. Q. Liu, H. Zhang, J. Xie, X. Liu and X. Lu, Recent progress and challenges of carbon materials for Zn-ion hybrid supercapacitors, *Carbon Energy*, 2020, **2**, 521–539.
33. A.G. Pandolfo and A.F. Hollenkamp, Carbon properties and their role in supercapacitors, *J. Power Sources*, 2006, **157**, 11–27.
34. M.I. Abdurrahman, S. Chaki and G. Saini, Stubble burning: Effects on health & environment, regulations and management practices, *Env. Adv.*, 2020, **2**, 100011.
35. S. Rawat, A. Kumar, A. Narani and T. Bhaskar, Isolation of lignin from spruce and pine wood: Role of structural difference for potential value addition, *Biomass Biofuels Biochem.*, 2021, 173–191.
36. H. Chowdhury, B. Loganathan, I. Mustary, F. Alam and S.M.A. Mobin, Chapter 12, Algae for biofuels: The third generation of feedstock, in *Second and Third Generation of Feedstocks*, Edited by A. Basile and F. Dalena, Elsevier, 2019, 323–344.
37. J. Wei, C. Tu, G. Yuan, Y. Liu, D. Bi, L. Xiao, J. Lu, B.K.G. Theng, H. Wang, L. Zhang and X. Zhang, Assessing the effect of pyrolysis temperature on the molecular properties and copper sorption capacity of a halophyte biochar, *Environ. Pollut.*, 2019, **251**, 56–65.
38. M.-M. Fu, C.-H. Mo, H. Li, Y.-N. Zhang, W.-X. Huang and M.H. Wong, Comparison of physicochemical properties of biochars and hydrochars produced from food wastes, *J. Clean Prod.*, 2019, **236**, 117637.

39. A. Kumar, K. Saini and T. Bhaskar, Hydochar and biochar: Production, physicochemical properties and techno-economic analysis, *Bioresour. Technol.*, 2020, **310**, 123442.
40. S. Ethaib, R. Omar, S.M.M. Kamal, D.R. Awang Biak and S.L. Zubaidi, Microwave-assisted pyrolysis of biomass waste: A mini review, *Processes*, 2020, **8** (9), 1190.
41. L. Zhu, H. Lei, L. Wang, G. Yadavalli, X. Zhang, Y. Wei, Y. Liu, D. Yan, S. Chen and B. Ahring, Biochar of corn stover: Microwave-assisted pyrolysis condition induced changes in surface functional groups and characteristics, *J. Anal. Appl. Pyrolysis*, 2015, **115**, 149–156.
42. R.K. Liew, W.L. Nam, M.Y. Chong, X.Y. Phang, M.H. Su, P.N. Y. Yek, N.L. Ma, C.K. Cheng, C.T. Chong and S.S. Lam, Oil palm waste: An abundant and promising feedstock for microwave pyrolysis conversion into good quality biochar with potential multi-applications, *Process Saf. Environ.*, 2018, **115**, 57–69.
43. Y. Li, Z. Pu, Q. Sun and N. Pan, A review on novel activation strategy on carbonaceous materials with special morphology/texture for electrochemical storage, *J. Energy Chem.*, 2021, **60**, 572–590.
44. D.-W. Kim, H.-S. Kil, K. Nakabayashi, S.-H. Yoon and J. Miyawaki, Structural elucidation of physical and chemical activation mechanisms based on the microdomain structure model, *Carbon*, 2017, **114**, 98–105.
45. J. Yin, W. Zhang, N. A. Alhebshi, N. Salah and H. N. Alshareef, Synthesis strategies of porous carbon for supercapacitor applications, *Small Methods*, 2020, **4**, 1900853.
46. T. Liu, F. Zhang, Y. Song and Y. Li, Revitalizing carbon supercapacitor electrodes with hierarchical porous structures, *J. Mater. Chem. A*, 2017, **5**, 17705–17733.
47. G. Wang, H. Wang, X. Lu, Y. Ling, M. Yu, T. Zhai, Y. Tong and Y. Li, Solid-state supercapacitor based on activated carbon cloths exhibits excellent rate capability, *Adv. Mater.*, 2014, **26**, 2676–2682.
48. K.A. Cychosz, R. Guillet-Nicolas, J. Garcia-Martinez and M. Thommes, Recent advances in the textural characterization of hierarchically structured nanoporous materials, *Chem. Soc. Rev.*, 2017, **46**, 389–414.
49. K. Naoi and P. Simon, New materials and new configurations for advanced electrochemical capacitors, *Electrochem. Soc. Interf.*, 2008, **17**, 34.
50. C. Vix-Guterl, E. Frackowiak, K. Jurewicz, M. Friebe, J. Parmentier and F. Béguin, Electrochemical energy storage in ordered porous carbon materials, *Carbon*, 2005, **43**, 1293–1302.
51. S. Liu, Y. Zhao, B. Zhang, H. Xia, J. Zhou, W. Xie and H. Li, Nano-micro carbon spheres anchored on porous carbon derived from dual-biomass as high rate performance supercapacitor electrodes, *J. Power Sources*, 2018, **381**, 116–126.
52. Y. Qing, Y. Jiang, H. Lin, L. Wang, A. Liu, Y. Cao, R. Sheng, Y. Guo, C. Fan, S. Zhang, D. Jia and Z. Fan, Boosting the supercapacitor performance of activated carbon by constructing overall conductive networks using graphene quantum dots, *J. Mater. Chem. A*, 2019, **7**, 6021–6027.
53. W.-L. Song, X. Li and L.-Z. Fan, Biomass derivative/graphene aerogels for binder-free supercapacitors, *Energy Storage Mater.*, 2016, **3**, 113–122.
54. X. Zhang, Y. Jiao, L. Sun, L. Wang, A. Wu, H. Yan, M. Meng, C. Tian, B. Jiang and H. Fu, GO-induced assembly of gelatin toward stacked layer-like porous carbon for advanced supercapacitors, *Nanoscale*, 2016, **8**, 2418–2427.
55. M. Enterría, F. Martín-Jimeno, F. Suárez-García, J. Paredes, M. Pereira, J. Martins, A. Martínez-Alonso, J. Tascón and J. Figueiredo, Effect of nanostructure on the supercapacitor performance of activated carbon xerogels obtained from hydrothermally carbonized glucose-graphene oxide hybrids, *Carbon*, 2016, **105**, 474–483.
56. Y. Song, J. Yang, K. Wang, S. Haller, Y. Wang, C. Wang and Y. Xia, In-situ synthesis of graphene/nitrogen-doped ordered mesoporous carbon nanosheet for supercapacitor application, *Carbon*, 2016, **96**, 955–964.

57. X. Li, Y. Tang, J. Song, W. Yang, M. Wang, C. Zhu, W. Zhao, J. Zheng and Y. Lin, Self-supporting activated carbon/carbon nanotube/reduced graphene oxide flexible electrode for high performance supercapacitor, *Carbon*, 2018, **129**, 236–244.
58. T. Lin, I.-W. Chen, F. Liu, C. Yang, H. Bi, F. Xu and F. Huang, Nitrogen-doped mesoporous carbon of extraordinary capacitance for electrochemical energy storage, *Science*, 2015, **350**, 1508–1513.
59. F. Wang, J.Y. Cheong, Q. He, G. Duan, S. He, L. Zhang, Y. Zhao, I.-D. Kim and S. Jiang, Phosphorus-doped thick carbon electrode for high-energy density and long-life supercapacitors, *Chem. Eng. J.*, 2021, **414**, 128767.
60. Z. Wu, L. Li, J.M. Yan and X.B. Zhang, Materials design and system construction for conventional and new-concept supercapacitors, *Adv. Sci.*, 2017, **4**, 1600382.
61. N.P. Wickramaratne, J. Xu, M. Wang, L. Zhu, L. Dai and M. Jaroniec, Nitrogen enriched porous carbon spheres: Attractive materials for supercapacitor electrodes and CO_2 adsorption, *Chem. Mater.*, 2014, **26**, 2820–2828.
62. R. Atchudan, S. Perumal, T.N.J.I. Edison and Y.R. Lee, Facile synthesis of monodisperse hollow carbon nanospheres using sucrose by carbonization route, *Mater. Lett.*, 2016, **166**, 145–149.
63. G. Wang, X. Pan, J.N. Kumar and Y. Liu, One-step synthesis of hollow carbon nanospheres in non-coordinating solvent, *Carbon*, 2015, **83**, 180–182.
64. G. Yang, H. Han, T. Li and C. Du, Synthesis of nitrogen-doped porous graphitic carbons using nano-$CaCO_3$ as template, graphitization catalyst, and activating agent, *Carbon*, 2012, **50**, 3753–3765.
65. A. Barroso Bogeat, Understanding and tuning the electrical conductivity of activated carbon: A state-of-the-art review, *Crit. Rev. Solid State*, 2021, **46**, 1–37.
66 F. Maillard, P.A. Simonov and E.R. Savinova, *Carbon Materials as Supports for Fuel Cell Electrocatalysts*, Edited by P. Serp, J. L. Figueiredo John Wiley, 2009.
67. L.R. Radovic, C. Moreno-Castilla and J. Rivera-Utrilla, Carbon materials as adsorbents in aqueous solutions, *Chem. Phys. Carbon*, 2001, 227–406.
68. M.E. Ramos, J.D. González, P.R. Bonelli and A.L. Cukierman, Effect of process conditions on physicochemical and electrical characteristics of denim-based activated carbon cloths, *Ind. Eng. Chem. Res.*, 2007, **46**, 1167–1173.
69. M.J. Mostazo-López, R. Ruiz-Rosas, E. Morallón and D. Cazorla-Amorós, Generation of nitrogen functionalities on activated carbons by amidation reactions and Hofmann rearrangement: Chemical and electrochemical characterization, *Carbon*, 2015, **91**, 252–265.
70. M. Inagaki, M. Toyoda, Y. Soneda and T. Morishita, Nitrogen-doped carbon materials, *Carbon*, 2018, **132**, 104–140.
71. D.P. Kim, C. Lin, T. Mihalisin, P. Heiney and M. Labes, Electronic properties of nitrogen-doped graphite flakes, *Chem. Mater.*, 1991, **3**, 686–692.
72 C. Zhang, D. Long, B. Xing, W. Qiao, R. Zhang, L. Zhan, X. Liang and L. Ling, The superior electrochemical performance of oxygen-rich activated carbons prepared from bituminous coal, *Electrochem. Commun*, 2008, **10**, 1809–1811.
73. J.D. Wiggins-Camacho and K.J. Stevenson, Effect of nitrogen concentration on capacitance, density of states, electronic conductivity, and morphology of N-doped carbon nanotube electrodes, *J. Phys. Chem. C*, 2009, **113**, 19082–19090.
74. S. Zhai, H.E. Karahan, C. Wang, Z. Pei, L. Wei and Y. Chen, 1D supercapacitors for emerging electronics: Current status and future directions, *Adv. Mater.*, 2020, **32**, e1902387.
75. S.C. Alan B. Dalton, E. Muñoz, J.M. Razal, V.H. Ebron, J.P. Ferraris, J.N. Coleman, B.G. Kim and R.H. Baughman, Super-tough carbon-nanotube fibres-these extraordinary composite fibres can be woven into electronic textiles, *Nature*, 2003, **423**, 703.

76. G. Nie, X. Zhao, Y. Luan, J. Jiang, Z. Kou and J. Wang, Key issues facing electrospun carbon nanofibers in energy applications: On-going approaches and challenges, *Nanoscale*, 2020, **12**, 13225–13248.
77. A. Ōya and S. Ōtani, Catalytic graphitization of carbons by various metals, *Carbon*, 1979, **17**, 131–137.
78. T.Q. Chen, Y. Liu, L.K. Pan, T. Lu, Y.F. Yao, Z. Sun, D.H.C. Chua and Q. Chen, Electrospun carbon nanofibers as anode materials for sodium ion batteries with excellent cycle performance, *J. Mater. Chem. A*, 2014, **2**, 4117–4121.
79. J. Tan, Y. Han, L. He, Y. Dong, X. Xu, D. Liu, H. Yan, Q. Yu, C. Huang and L. Mai, In situ nitrogen-doped mesoporous carbon nanofibers as flexible freestanding electrodes for high-performance supercapacitors, *J. Mater. Chem. A*, 2017, **5**, 23620–23627.
80. P. Ramakrishnan, S.-G. Park and S. Shanmugam, Three-dimensional hierarchical nitrogen-doped arch and hollow nanocarbons: Morphological influences on supercapacitor applications, *J. Mater. Chem. A*, 2015, **3**, 16242–16250.
81. C. Wang, Y.V. Kaneti, Y. Bando, J. Lin, C. Liu, J. Li and Y. Yamauchi, Metal–organic framework-derived one-dimensional porous or hollow carbon-based nanofibers for energy storage and conversion, *Mater. Horizons*, 2018, **5**, 394–407.
82. R. Singhal and V. Kalra, Using common salt to impart pseudocapacitive functionalities to carbon nanofibers, *J. Mater. Chem. A*, 2015, **3**, 377–385.
83. L. Wang, G. Zhang, X. Zhang, H. Shi, W. Zeng, H. Zhang, Q. Liu, C. Li, Q. Liu and H. Duan, Porous ultrathin carbon nanobubbles formed carbon nanofiber webs for high-performance flexible supercapacitors, *J. Mater. Chem. A*, 2017, **5**, 14801–14810.
84. S. Wang, Z. Cui, J. Qin and M. Cao, Thermally removable in-situ formed ZnO template for synthesis of hierarchically porous N-doped carbon nanofibers for enhanced electrocatalysis, *Nano Res.*, 2016, **9**, 2270–2283.
85. W. Yang, J. Zhou, S. Wang, W. Zhang, Z. Wang, F. Lv, K. Wang, Q. Sun and S. Guo, Freestanding film made by necklace-like N-doped hollow carbon with hierarchical pores for high-performance potassium-ion storage, *Energy Environ. Sci.*, 2019, **12**, 1605–1612.
86. Y. Zhu, S. Murali, M.D. Stoller, K.J. Ganesh, W. Cai, P.J. Ferreira, A. Pirkle, R.M. Wallace, K.A. Cychosz, M. Thommes, D. Su, E.A. Stach and R.S. Ruoff, Carbon-based supercapacitors produced by activation of graphene, *Science*, 2011, **332**, 1537–1541.
87. Y. Li and D. Zhao, Preparation of reduced graphite oxide with high volumetric capacitance in supercapacitors, *Chem. Commun.*, 2015, **51**, 5598–5601.
88. Y. Xu, Z. Lin, X. Zhong, X. Huang, N.O. Weiss, Y. Huang and X. Duan, Holey graphene frameworks for highly efficient capacitive energy storage, *Nat. Commun.*, 2014, **5**, 4554.
89. Q. Ke and J. Wang, Graphene-based materials for supercapacitor electrodes–A review, *J. Materiomics*, 2016, **2**, 37–54.
90. R. Dubey and V. Guruviah, Review of carbon-based electrode materials for supercapacitor energy storage, *Ionics*, 2019, **25**, 1419–1445.
91. Y. Meng, Y. Zhao, C. Hu, H. Cheng, Y. Hu, Z. Zhang, G. Shi and L. Qu, All-graphene core-sheath microfibers for all-solid-state, stretchable fibriform supercapacitors and wearable electronic textiles, *Adv. Mater.*, 2013, **25**, 2326–2331.
92. F. Gunes, H.-J. Shin, C. Biswas, G.H. Han, E.S. Kim, S.J. Chae, J.-Y. Choi and Y.H. Lee, Layer-by-layer doping of few-layer graphene film, *ACS Nano*, 2010, **4**, 4595–4600.
93. L. Kou, T. Huang, B. Zheng, Y. Han, X. Zhao, K. Gopalsamy, H. Sun and C. Gao, Coaxial wet-spun yarn supercapacitors for high-energy density and safe wearable electronics, *Nat. Commun.*, 2014, **5**, 1–10.
94. L. Chen, Y. Liu, Y. Zhao, N. Chen and L. Qu, Graphene-based fibers for supercapacitor applications, *Nanotechnology*, 2016, **27**, 032001.
95. Z. Dong, C. Jiang, H. Cheng, Y. Zhao, G. Shi, L. Jiang and L. Qu, Facile fabrication of light, flexible and multifunctional graphene fibers, *Adv. Mater.*, 2012, **24**, 1856–1861.

96. D. Yu, K. Goh, H. Wang, L. Wei, W. Jiang, Q. Zhang, L. Dai and Y. Chen, Scalable synthesis of hierarchically structured carbon nanotube–graphene fibres for capacitive energy storage, *Nat. Nanotech.*, 2014, **9**, 555–562.
97. Y. Shao, M.F. El-Kady, L.J. Wang, Q. Zhang, Y. Li, H. Wang, M. F. Mousavi and R.B. Kaner, Graphene-based materials for flexible supercapacitors, *Chem. Soc. Rev.*, 2015, **44**, 3639–3665.
98. S. Gan, L. Zhong, T. Wu, D. Han, J. Zhang, J. Ulstrup, Q. Chi and L. Niu, Spontaneous and fast growth of large-area graphene nanofilms facilitated by oil/water interfaces, *Adv. Mater.*, 2012, **24**, 3958–3964.
99. S. Zhang, Y. Li and N. Pan, Graphene based supercapacitor fabricated by vacuum filtration deposition, *J. Power Sources*, 2012, **206**, 476–482.
100. X. Yang, C. Cheng, Y. Wang, L. Qiu and D. Li, Liquid-mediated dense integration of graphene materials for compact capacitive energy storage, *Science*, 2013, **341**, 534–537.
101. M. Wang, J. Oh, T. Ghosh, S. Hong, G. Nam, T. Hwang and J.-D. Nam, An interleaved porous laminate composed of reduced graphene oxide sheets and carbon black spacers by in situ electrophoretic deposition, *RSC Adv.*, 2014, **4**, 3284–3292.
102. D.-Y. Wu, W.-H. Zhou, L.-Y. He, H.-Y. Tang, X.-H. Xu, Q.-S. Ouyang and J.-J. Shao, Micro-corrugated graphene sheet enabled high-performance all-solid-state film supercapacitor, *Carbon*, 2020, **160**, 156–163.
103. C. Li and G. Shi, Three-dimensional graphene architectures, *Nanoscale*, 2012, **4**, 5549–5563.
104. Z. Chen, W. Ren, L. Gao, B. Liu, S. Pei and H.-M. Cheng, Three-dimensional flexible and conductive interconnected graphene networks grown by chemical vapour deposition, *Nat. Mater.*, 2011, **10**, 424–428.
105. L. Jiang and Z. Fan, Design of advanced porous graphene materials: from graphene nanomesh to 3D architectures, *Nanoscale*, 2014, **6**, 1922–1945.
106. Y. Ma and Y. Chen, Three-dimensional graphene networks: Synthesis, properties and applications, *Natl Sci. Rev.*, 2015, **2**, 40–53.
107. S. Mao, G. Lu and J. Chen, Three-dimensional graphene-based composites for energy applications, *Nanoscale*, 2015, **7**, 6924–6943.
108. X. Cao, Z. Yin and H. Zhang, Three-dimensional graphene materials: preparation, structures and application in supercapacitors, *Energy Environ. Sci.*, 2014, **7**, 1850–1865.
109. Z. Chen, L. Jin, W. Hao, W. Ren and H.-M. Cheng, Synthesis and applications of three-dimensional graphene network structures, *Mater. Today Nano*, 2019, **5**, 100027.
110. Y. Wu, J. Zhu and L. Huang, A review of three-dimensional graphene-based materials: Synthesis and applications to energy conversion/storage and environment, *Carbon*, 2019, **143**, 610–640.
111. Z.-S. Wu, A. Winter, L. Chen, Y. Sun, A. Turchanin, X. Feng and K. Müllen, Three-dimensional nitrogen and boron co-doped graphene for high-performance all-solid-state supercapacitors, *Adv. Mater.*, 2012, **24**, 5130–5135.
112. Y. Xu, K. Sheng, C. Li and G. Shi, Self-assembled graphene hydrogel via a one-step hydrothermal process, *ACS Nano*, 2010, **4**, 4324–4330.
113. D. Huang, X. Li, S. Wang, G. He, W. Jiang, J. Hu, Y. Wang, N. Hu, Y. Zhang and Z. Yang, Three-dimensional chemically reduced graphene oxide templated by silica spheres for ammonia sensing, *Sens. Actuat. B: Chem.*, 2017, **252**, 956–964.
114. L. Qiu, J.Z. Liu, S.L. Chang, Y. Wu and D. Li, Biomimetic superelastic graphene-based cellular monoliths, *Nat. Commun.*, 2012, **3**, 1–7.
115. D. Su, Macroporous 'bubble'graphene film via template-directed ordered-assembly for high rate supercapacitors, *Chem. Commun.*, 2012, **48**, 7149–7151.
116. S. Han, D. Wu, S. Li, F. Zhang and X. Feng, Porous graphene materials for advanced electrochemical energy storage and conversion devices, *Adv. Mater.*, 2014, **26**, 849–864.

117. C. Zhu, T. Han, E.B. Duoss, A.M. Golobic, J.D. Kuntz, C.M. Spadaccini and M.A. Worsley, Highly compressible 3D periodic graphene aerogel microlattices, *Nat. Commun.*, 2015, **6**, 1–8.
118. Y. Xu, Z. Lin, X. Huang, Y. Liu, Y. Huang and X. Duan, Flexible solid-state supercapacitors based on three-dimensional graphene hydrogel films, *ACS Nano*, 2013, **7**, 4042–4049.
119. D. Xu, Y. Pan, L. Zhu, Y. Yusran, D. Zhang, Q. Fang, M. Xue and S. Qiu, Simple coordination complex-derived Ni NP anchored N-doped porous carbons with high performance for reduction of nitroarenes, *CrystEngComm*, 2017, **19**, 6612–6619.
120. M. Eddaoudi, D.F. Sava, J.F. Eubank, K. Adil and V. Guillerm, Zeolite-like metal-organic frameworks (ZMOFs): Design, synthesis, and properties, *Chem. Soc. Rev.*, 2015, **44**, 228–249.
121. H.C. Zhou and S. Kitagawa, Metal-organic frameworks (MOFs), *Chem. Soc. Rev.*, 2014, **43**, 5415–5418.
122. R.R. Salunkhe, J. Tang, Y. Kamachi, T. Nakato, J.H. Kim and Y. Yamauchi, Asymmetric supercapacitors using 3D nanoporous carbon and cobalt oxide electrodes synthesized from a single metal-organic framework, *ACS Nano*, 2015, **9**, 6288–6296.
123. R.R. Salunkhe, Y.V. Kaneti, J. Kim, J.H. Kim and Y. Yamauchi, Nanoarchitectures for metal-organic framework-derived nanoporous carbons toward supercapacitor applications, *Acc. Chem. Res.*, 2016, **49**, 2796–2806.
124. C. H. Wang, C. Liu, J. S. Li, X.Y. Sun, J. Y. Shen, W. Q. Han, L. J. Wang, Electrospun metal–organic framework derived hierarchical carbon nanofibers with high performance for supercapacitors, *Chem. Commun.* 2017, **53**, 1751–1754.

Index

Note: *Italic* page numbers refer to figures.

For Product Safety Concerns and Information please contact our EU
representative GPSR@taylorandfrancis.com
Taylor & Francis Verlag GmbH, Kaufingerstraße 24, 80331 München, Germany

www.ingramcontent.com/pod-product-compliance
Lightning Source LLC
LaVergne TN
LVHW010548110826
845149LV00003B/603

* 9 7 8 1 0 3 2 4 8 1 8 9 0 *